AF352946

Innovation Ecosystems

Innovation Ecosystems: A Sustainability Perspective

Editor

António Abreu

MDPI • Basel • Beijing • Wuhan • Barcelona • Belgrade • Manchester • Tokyo • Cluj • Tianjin

Editor
António Abreu
Polytechnic Institute of Lisbon
Portugal

Editorial Office
MDPI
St. Alban-Anlage 66
4052 Basel, Switzerland

This is a reprint of articles from the Special Issue published online in the open access journal *Sustainability* (ISSN 2071-1050) (available at: https://www.mdpi.com/journal/sustainability/special_issues/innovation_ecosystems).

For citation purposes, cite each article independently as indicated on the article page online and as indicated below:

LastName, A.A.; LastName, B.B.; LastName, C.C. Article Title. *Journal Name* **Year**, *Volume Number*, Page Range.

ISBN 978-3-0365-0834-4 (Hbk)
ISBN 978-3-0365-0835-1 (PDF)

Contents

About the Editor

António Abreu (Ph.D.), before joining the academic world, had an industrial career in manufacturing industries with management positions. He is currently a professor of Industrial Engineering at the Polytechnic Institute of Lisbon (ISEL), where he is now an associate professor with a habilitation position. He is a member of several national and international associations, e.g., he is the co-founder of SOCOLNET (Society of Collaborative Networks), and a member of IFAC—International Federation of Automatic Control at TC5.3 Enterprise Integration and Networking. As a researcher, he has been involved in several European research projects, and several national projects. He has been involved in the organization, and program committees of several national and international conferences. He has been invited by editors of several international journals to review manuscripts. He has more than 100 publications in books, journals, and conference proceedings in the area of collaborative networked organizations, logistics, project management, quality management, open innovation, and lean management.

 *sustainability*

Editorial

Innovation Ecosystems: A Sustainability Perspective

António Abreu [1,2]

1   Department of Mechanical Engineering, Polytechnic Institute of Lisbon, 1959-007 Lisboa, Portugal;
     ajfa@dem.isel.ipl.pt
2   CTS Uninova, 2829-516 Caparica, Portugal

Academic Editor: Marc A. Rosen

Received: 6 January 2021; Accepted: 4 February 2021; Published: 6 February 2021

In the last decade, the increasing globalization of markets and revolution 4.0, has caused profound changes in the best way to manage the innovation process. The innovation methods of the past are not well adapted to the turbulence of the modern world.

In order to be competitive, companies must develop capabilities that will allow them to react rapidly to market demands.

The development of new complex products/services requires access to a distinct set of resources and skills that companies do not normally have. Thus, in order to ensure their level of competitiveness, companies are confronted with the following dilemma: to develop the skills and resources needed from their own assets, they sometimes need to make significant investments, or, alternatively, use the skills and resources that can be made available by other companies in the context of an innovation ecosystem.

However, despite the fact that collaboration among companies in an innovation ecosystem had been considered unusual and indeed suspicious by many Small and Medium Enterprise (SME) managers until a few years ago, nowadays it is commonly assumed that many companies will participate in an innovation ecosystem. Literature in the field has pointed out that participating in an innovation ecosystem brings benefits to the involved entities. Underlying these expectations are, amongst others, the following factors: the sharing of risks and resources, the joining of complementary skills and capacities, and access to new/wider markets and new knowledge.

In fact, there is an intuitive assumption that, when a company is a member of a long-term networked structure, it will operate more effectively in pursuit of its goals.

However, it has been difficult to support this assumption due to the lack of models that support mechanisms explaining innovation processes in an innovation ecosystem environment.

This e-book comprises an edition of the Special Issue entitled "Innovation Ecosystems: A Sustainability Perspective", published by the journal *Sustainability*, and includes a collection of thirteen papers that discuss theoretical approaches, case studies, and surveys focused on issues related to open innovation and its mechanisms in order to support the promotion and sustainability of innovation ecosystems.

Concerning theoretical contributions, Jütting, in a systematic literature review, explores and conceptualizes the idea of mission-oriented innovation ecosystems and presents a typology [1]

Dias et al. propose a functional holistic model which integrates the strategic, organizational, and operational levels, as well as a set of factors to take into account supporting innovative processes [2]. Based on the Panarchy model that describes the evolutionary nature of complex adaptive systems, Boyer proposes an evolutionary and sustainability perspective of the innovation ecosystem [3].

Santos et al. introduce a framework to evaluate the risk level of system development in open innovation environment based on a fuzzy logic approach [4]. Nunes and Abreu propose an open innovation risk management model based on concepts from social network analysis to estimate the outcome likelihood (success or failure) of ongoing open innovation projects [5].

Munodawafa and Johl, based on results achieved from a systematic literature review of eco-innovation, suggest that organizational stakeholders, resources, and capabilities are critical factors in the definition of an innovation strategy. Furthermore, the authors conclude that resource-based and stakeholder theories are frequently utilized to explain eco-innovation processes [6].

Sarri et al. introduce a new methodological proposal to help the development of smart management as a means to support the progressive development of technological innovations and their adoption in wine farms [7].

Zandebasiri et al. discuss the advantages of using Multi-Criteria Decision Making (MCDM) methodologies supported by concepts from game theory in decision-making processes to ensure sustainable forest management [8].

Regarding studies and surveys, Costa and Matias discuss how open innovation can improve sustainable innovation ecosystems and drive the digital transition [9].

Wurster et al. discuss the results of a survey conducted in Germany with the purpose of creating an empirical foundation for the specification of software for sustainable automotive products, particularly sustainable tyres [10].

Yang et al. analyze the formation and evolution of BIM in China from the perspective of an innovation ecosystem [11]. Chaminade and Randelli, in their work. analyze the role of innovation ecosystems as a driver of wine industry transformation in the Panzano region [12].

Yordanova et al., in their work, analyze the role of a university in the development of technopreneurial intentions among Bulgarian STEM (STEM refers to any subjects that fall under the disciplines of science, technology, engineering, or mathematics) students [13].

Last but not least, as the guest editor of this e-book, I would like to express my profound gratitude for the opportunity to publish with MDPI. This acknowledgment extends to the *Sustainability* Editorial Office and especially to Mrs. Debbie Li, who has supported me constantly throughout this process.

It was a great pleasure to work in such conditions. I look forward to collaborating with the *Sustainability* journal in the future.

Funding: This research received no external funding.

Conflicts of Interest: The author declares no conflict of interest.

References

1. Jütting, M. Exploring Mission-Oriented Innovation Ecosystems for Sustainability: Towards a Literature-Based Typology. *Sustainability* **2020**, *12*, 6677. [CrossRef]
2. Dias, A.S.M.E.; Abreu, A.; Navas, H.V.G.; Santos, R. Proposal of a Holistic Framework to Support Sustainability of New Product Innovation Processes. *Sustainability* **2020**, *12*, 3450. [CrossRef]
3. Boyer, J. Toward an Evolutionary and Sustainability Perspective of the Innovation Ecosystem: Revisiting the Panarchy Model. *Sustainability* **2020**, *12*, 3232. [CrossRef]
4. Santos, R.; Abreu, A.; Dias, A.; Calado, J.M.F.; Anes, V.; Soares, J. A Framework for Risk Assessment in Collaborative Networks to Promote Sustainable Systems in Innovation Ecosystems. *Sustainability* **2020**, *12*, 6218. [CrossRef]
5. Nunes, M.; Abreu, A. Managing Open Innovation Project Risks Based on a Social Network Analysis Perspective. *Sustainability* **2020**, *12*, 3132. [CrossRef]
6. Munodawafa, R.T.; Johl, S.K. A Systematic Review of Eco-Innovation and Performance from the Resource-Based and Stakeholder Perspectives. *Sustainability* **2019**, *11*, 6067. [CrossRef]

7. Sarri, D.; Lombardo, S.; Pagliai, A.; Perna, C.; Lisci, R.; De Pascale, V.; Rimediotti, M.; Cencini, G.; Vieri, M. Smart Farming Introduction in Wine Farms: A Systematic Review and a New Proposal. *Sustainability* **2020**, *12*, 7191. [CrossRef]

8. Zandebasiri, M.; Filipe, J.A.; Soosani, J.; Pourhashemi, M.; Salvati, L.; Mata, M.N.; Mata, P.N. An Incomplete Information Static Game Evaluating Community-Based Forest Management in Zagros, Iran. *Sustainability* **2020**, *12*, 1750. [CrossRef]

9. Costa, J.; Matias, J.C.O. Open Innovation 4.0 as an Enhancer of Sustainable Innovation Ecosystems. *Sustainability* **2020**, *12*, 8112. [CrossRef]

10. Wurster, S.; Heß, P.; Nauruschat, M.; Jütting, M. Sustainable Circular Mobility: User-Integrated Innovation and Specifics of Electric Vehicle Owners. *Sustainability* **2020**, *12*, 7900. [CrossRef]

11. Yang, Y.; Zhang, Y.; Xie, H. Exploring Cultivation Path of Building Information Modelling in China: An Analysis from the Perspective of an Innovation Ecosystem. *Sustainability* **2020**, *12*, 6902. [CrossRef]

12. Chaminade, C.; Randelli, F. The Role of Territorially Embedded Innovation Ecosystems Accelerating Sustainability Transformations: A Case Study of the Transformation to Organic Wine Production in Tuscany (Italy). *Sustainability* **2020**, *12*, 4621. [CrossRef]

13. Yordanova, D.; Filipe, J.A.; Pacheco Coelho, M. Technopreneurial Intentions among Bulgarian STEM Students: The Role of University. *Sustainability* **2020**, *12*, 6455. [CrossRef]

 *sustainability*

Article

A Systematic Review of Eco-Innovation and Performance from the Resource-Based and Stakeholder Perspectives

Russell Tatenda Munodawafa and Satirenjit Kaur Johl *

Department of Management and Humanities, Universiti Teknologi PETRONAS, 32610 Seri Iskandar, Perak, Malaysia; russell_17003483@utp.edu.my
* Correspondence: satire@utp.edu.my; Tel.: +60-5-368-7756

Received: 7 August 2019; Accepted: 17 September 2019; Published: 1 November 2019

Abstract: The growing concerns surrounding the precarious state of the biosphere have triggered organizations to develop and implement innovations that curb environmental degradation (eco-innovation). However, eco-innovation is a risky proposition for organizations and their stakeholders, due to uncertainty of outcome. Despite the high investment risk of eco-innovation, the literature that assesses eco-innovation outcomes from an organizational performance perspective is scant. Thus, this paper uses a systematic approach to review eco-innovation and performance literature. The eco-innovation and performance literature reviewed in this paper is sourced from the Scopus and Web of Science (WoS) scientific databases. Results from this systematic review suggest that the capital market stakeholder group—an essential stakeholder group—has received little attention in the eco-innovation and performance literature. This is alarming, as this stakeholder group is expected to act in the best interests of the organization—as well as the other stakeholders—especially during strategy formulation and implementation. This paper also finds that the resource-based view and stakeholder theory are frequently utilized in explaining eco-innovation. However, the natural resource-based view is least utilized, despite growing environmental pressures. A multi-theoretical perspective can help to overcome the limitations of one theory, as well as help to unearth additional organizational factors which could potentially catalyze the eco-innovation and performance relationship.

Keywords: eco-innovation; cleaner production; strategy; performance; natural resource-based view; stakeholder theory

---

1. Introduction

Growing concerns about the state of the natural environment are mainly centered upon waste management, greenhouse gas (GHG) emissions, pollution and contamination management, natural resource management, as well as food, water and energy security [1]. The rising prominence of these natural environmental challenges has trigged global and local policy adjustments, as well as industry initiatives to stymie these challenges. Examples of such initiatives includes the establishment of 17 Sustainable Development Goals (SDGs) by the United Nations (UN) in 2015. These 17 SDG goals are a global attempt, led by the United Nations, to direct the world towards sustainable development [2]. The SDGs, which incorporate approximately 169 targets under the "Transforming our world: The 2030 Agenda for Sustainable Development" manifesto, are largely based upon the recommendations of the Brundtland Commission of 1987 [3]. In this report, experts and scientists acknowledged the negative effects that anthropogenic activities were theoretically having on the biosphere [4].

According to the report, corrective action needs to be taken if humanity is to avert an imminent natural environmental crisis. Remedial and preventive action will help ensure the continuous, uninhibited operation of the biosphere, which serves the dual functions of resource provider and anthropogenic emissions sink [5]. Thus, development needs to be sustainable; i.e., it needs to focus on satisfying present needs without negatively impacting future generations' ability to satisfy their own needs [6]. As a result, environmental sustainability, one of the three pillars of sustainable development, has become the nexus for international agencies, governments, academics and industry practitioners worldwide [7]. Further driving attention towards sustainable development is the growth of the consumer consciousness of environmental sustainability issues [8], as well as a stricter enforcement of environmental regulations [9]. Hence, to meet the targets of sustainable development, industrial activities should be undertaken within the threshold of the natural environment, vis-à-vis sustainable development.

For industry to usher in sustainable development, organizations need to shift from focusing solely upon profit. Instead, organizations must continue their pursuit of profit without neglecting the interests of planet and people, as the protection of the biosphere will influence current and future competitiveness for organizations [10]. Environmental challenges are thus a potential boon for organizations to increase their competitiveness and performance. This is because the increasing prominence of environmental issues enables organizations to integrate environmental initiatives (eco-initiatives) into their strategy. These eco-initiatives require the engagement of all stakeholder groups to reap the benefits [11]. In turn, the organization's problem-solving capability vis-à-vis environmental innovativeness is improved [12]. In fact, organizations that channel their resources and capabilities towards addressing environmental challenges will realize greater competitive advantages [13]. For instance, firms that embed waste and GHG emission minimization into their ethos by practicing continuous improvements of their products can develop the strategic capability of pollution prevention. This capability could potentially foster lower production costs [13]. Therefore, for organizations to improve on their environmental innovativeness, they must direct their resources towards eco-innovation [12].

Eco-innovation is defined as innovation which encompasses the development of new ideas, technologies, behaviors, products, or processes that result in the reduction of environmental burdens whilst simultaneously improving economic performance [14]. Also included in this definition of eco-innovation are organizational changes —where stakeholders (groups or individuals, whose actions affect the organization and vice-versa [15]) are the key elements [16].

In addition, improving environmental sustainability through reducing the usage of non-renewable resources, and minimizing hazardous waste generation, could possibly lead to a permanently regenerative economy [17]. Consequently, eco-innovation's emphasis on the environment also makes it an ideal catalyst in the transition from a linear economy to a closed loop, circular systems of production and consumption—i.e., the "circular economy" (CE). This capability makes eco-innovation more distinct from other types of innovation [18]. For example, applications of eco-innovations, such as micro fuel cells for combined heat and power production, enables buildings to lower their CO_2 and other GHG emissions [19]. Thus, capabilities exhibited by eco-innovations enhance sustainable development mainly through: Reducing non-renewable resource dependence, materials and energy consumption, as well as waste and hazardous materials production [20,21].

Eco-innovation is not just the domain of one industry. It is applicable across multiple industries within the wide expanse of the consumption and production system [22]. Hence, a multidisciplinary approach is paramount towards addressing issues related to the environment, economy and society at large. Because of its technological nature [23], eco-innovation has been addressed in a number of engineering-oriented studies [24]. However, eco-innovation needs to be receiving attention from a strategic perspective as well, due to its potential to renew business models and inspire value creation activities [25]. Eco-innovation can also be viewed from non-technological angles, such as institutional and organizational contexts, i.e., its governance and management [26]. Anzola-Román et al. [27] posit that the non-technological activities of an organization can also be a meaningful source of innovation.

When looked at as a process, the output (eco-innovation) can be attributed to its strategy, routines and resources [27]. These aspects are all a function of the various stakeholders of the organization [28]. Hence, having a firmer understanding of eco-innovation and its role in the transition to sustainable development will help influencers in the eco-innovation process—such as the stakeholders of the organization—to better adjust and calibrate their organization's resources and capabilities towards better performance [18].

Eco-innovation is also a risky proposition for organizations. This is due to the uncertain outcome and impact of any type of innovation, including eco-innovation [29]. Consequently, the linkage between an organization and its stakeholders is critical in minimizing the risks associated with the uncertainty of eco-innovation [30]. Organizations should make eco-innovation the main focal point of their strategic planning, so as to possibly minimize their risk, and also to enhance their strategic and economic performance [31].

Despite these assertions, studies that analyze eco-innovation from a strategic perspective are scant. Specifically, the mapping of organizational strategic factors to eco-innovation (such as resources, capabilities and stakeholder groups, e.g., top management), remains obscure [32]. This is alarming, given that strategic organizational systems can benefit from eco-innovation. Also concurring with this point is Tyl et al. [33], who stated that eco-innovation literature ought to capture how different stakeholders understand and interpret the challenges to the natural environment. Literature that captures the stakeholders' insight into the environmental concept would help to effectively gauge the environmental value proposition for an organization's stakeholders [33].

In addition, eco-innovation literature that calls into focus organization-level determinants and drivers is currently limited. The 'bottom line' or non-technical contextual outcomes of eco-innovation literature, such as performance, have received limited coverage [34]. Also, organizational structural factors play a fundamental role in shaping the eco-innovation capabilities of organizations. However, studies discussing these eco-innovations seem to overlook the role of key stakeholder groups, such as the top management and shareholders. Yet, these stakeholders have a direct bearing on the internal eco-innovative capabilities of the organization, as they influence the culture and structural aspects of such organizations [35]. In fact, a critical gap was pointed out by He, Miao, Wong and Lee [26], who found that most eco-innovation literature that assesses its theory development is mostly skewed towards technical outcomes. As a result, the existing literature fails to explicitly account for internal and external factors, such as capabilities and stakeholders, that account for the competitive performance outcomes of eco-innovation [26]. In addition, theory usage in eco-innovation literature has predominantly focused on the resource-side of the value creation process. As a result, demand-side perspectives—where other stakeholders are present and can potentially influence eco-innovation value creation eco-system—has been largely neglected. Hence, paying attention to contextual factors such as market dynamics, together with the stakeholder groups behind them, can help in further understanding eco-innovation and performance [26]. Knowing the performance outcomes of eco-innovations is essential, given the high financial risk involved in eco-innovations [36]. Thus, this paper aims to address this current gap in literature by looking at the role of the organizational and contextual factors towards eco-innovation, and subsequently, performance. By addressing this gap, potential, resource, capability and stakeholder factors that can be mapped to the eco-innovation and performance relationship, can be unearthed, providing the opportunity to suggest future eco-innovation and performance research directions.

Therefore, this paper seeks to answer the following research questions:

1. What are the current organizational and contextual factors that influence the eco-innovation and performance literature?
2. What are the key organizational and contextual factors that can potentially be mapped to the eco-innovation and performance relationship?
3. What are the possible future research avenues for eco-innovation and performance literature?

The remainder of this paper is structured as follows: Section 2 presents the background and theoretical development of eco-innovation. Section 3 describes the methods used in this paper. Section 4 presents the findings, whilst Section 5 discusses the findings. Lastly, Section 6 concludes the paper.

2. Background

Innovation as a concept was first brought to prominence in the 1930s by Joseph Aloïs Schumpeter, who saw innovation as the industrial or commercial application of new business models, production methods, processes, products, or even supply source to strategically encourage economic development [37]. The initial interpretation of innovation during this era related it to changes of varying magnitudes (i.e. small or large scale) that significantly amplify holistic changes in entire industries or market segments. This indicates that, by definition, innovation is connected to change. Innovation, however, should not only be restricted to change abstractly. Newness or novelty must also be present. Thus, the element of novelty (newness) in applications (industrial or commercial) demarcates innovation from invention, even if the entity behind both the innovation and invention are the same [37].

Innovation is, hence, a function of value creation, and subsequently competitive advantage [38]. This viewpoint on innovation is also supported by Kotsemir, Abroskin and Meissner [29], who stated that innovation encapsulates changes, whose aim is to create value. This value addition should positively impact the performance of the organization on numerous fronts, such as operational efficiency, improved working practices, or better flexibility in an ever-changing business environment. Hence, innovation is a source of competitive advantage, through it being a capability [39–41]. As a capability, innovation is arrived at through utilizing organizational resources, which conjures sustainable competitive advantage [42].

Innovation also encompasses other facets which include, but not solely, technical development [43]. Other aspects about innovation include marketing, financing, strategic planning and stakeholder engagements, such as governments and supply chains [44,45]. This fact highlights the cross-cutting, multifaceted nature of innovation.

Innovation definitions, however, have traditionally viewed it from a product perspective—i.e., making innovation synonymous with new products. There are, however, other dimensions to innovation such as strategy, process or value-adding service innovations—which represent a new, generally overlooked frontier [29,46,47]. These innovation dimensions—which are a function of an organization's resources—represent an organization's innovativeness or innovation capability when they are aggregated [48–50].

Integrating environmental initiatives into an organization's strategy has convincingly demonstrated its ability to buttress an organization's innovativeness and competitiveness [51]. Hence, environmental innovativeness is another important frontier, as environmental issues become increasingly prominent [52,53]. As organizations become more pro-active in addressing environmental concerns, environmental innovativeness becomes an increasingly important capability [54,55]. This is because environmental innovativeness presents organizations state-of-the-art problem solving vis-à-vis the implementation of eco-initiatives [12]. These eco-initiatives—which require the engagement of all stakeholders at all levels—can assist in creating a competitive advantage through several means, such as cost reduction [11,56] or environmental compliance [57].

Furthermore, when environmental initiatives are integrated into novel commercial or industrial applications, economic development can thus be stimulated without causing irreversible harm to the natural environment [58]. Green, McMeekin and Irwin [58] were amongst the early scholars to point out the increasing need of incorporating the natural environment into novel applications. As a result, the term eco-innovation was formulated and introduced by Rennings [14]. Eco-innovation was thus defined as encompassing the development of new ideas, technologies, behaviors, products, or processes that result in a reduction of environmental burdens, whilst concurrently improving economic performance [14].

Although Rennings [14] was one of the first to coin the term eco-innovation, several attempts to define this innovation have been made by various scholars and government bodies. For instance, the European Commission's definition of eco-innovation placed more emphasis on the environmental performance (positive or negative) of innovations, instead of simply focusing on the aim of the eco-innovation [59]. Meanwhile the Organization for Economic Co-operation and Development (OECD) defined eco-innovation as: "The production, assimilation or exploitation of a product, production process, service or management or business method that is novel to the organization (developing or adopting it) and which results, throughout its life cycle, in a reduction of environmental risk, pollution and other negative impacts of resources use (including energy use) compared to relevant alternatives." [60].

Managing resources, economies, and the environment, in a responsible manner, without harming the prospects of future generations, is critical. Corporations, consumers, policy makers and researchers play a leading role in reducing the degradation to the natural environment and ensuring sustainable development [61]. Because these stakeholders are the engines for economic development and growth, this paper adopts the OECD's definition of eco-innovation. Carrillo-Hermosilla et al. [62] also support this definition of eco-innovation, since they argue that economic stimulation through the reduction of any negative environmental impact should be the distinguishing hallmark of eco-innovation from other innovation. Hence, eco-innovation seeks to positively influence both environmental and economic performance simultaneously through means such as cost reduction and improved efficiency.

Eco-innovation and its impact on performance has been studied by various scholars who encompass a wide spectrum of theoretical backgrounds. As a result, several theories on eco-innovation have been postulated by various scholars within areas such as Finance, Economics, Law & Strategic Management [63], Engineering [64], Urban Planning [65] and Environmental Science [66]. However, Pham, Paillé and Halilem [12] argued that environmental innovativeness is the blueprint to eco-innovation. Therefore, from a theoretical perspective, eco-innovation is a competency that can be arrived at from utilizing resources. This helps to explain why research on eco-innovation and performance has largely been studied from the resource-based view [67].

The definition of eco-innovation, as posited earlier, is inclusive of organizational changes, and according to Carrillo-Hermosilla, del González and Könnölä [16], the stakeholders are key in organizational factors. In fact, there have been increasing calls for the integration of stakeholder viewpoints into eco-innovation [24,33]. For instance, Bag and Gupta [68] indicate that further research should attempt to integrate the resource-based view with other theories, especially in observing eco-innovation. Therefore, the theories applied in this study include the resource-based view and the stakeholder theory.

One of the basic building blocks of an organization are its resources and/or capabilities [69]. An organizational resource is loosely defined as being anything that could be classified as a strength of the organization, and are key to its success [70]. Creating the conditions for the optimum usage of an organization's resources could possibly help to generate higher returns of time [70]. Whereas, capabilities factor in the ability of an organization to effectively and efficiently utilize its resources [71].

From the perspective of an organization, products/services and resources are closely related, as most products/services require the utilization of several resources, and most resources can be inputs to several products/services. In order for an organization to reconfigure and realign its competencies and resources towards accomplishing eco-innovation—and in turn—better performance, its various stakeholders play a fundamental role in this process [72]. Therefore, looking at the resource profile of an organization can be useful in helping it fine tune its product-market activities.

The resource-based view postulates that a competitive advantage can be built and sustained from the strategic harnessing of an organization's resources [73,74]. An organization is able to coordinate and cross-functionally integrate the exploitation of its resources to develop a competency.

The degree to which an organization attains a competitive advantage, and therefore higher performance, is determined by the value (V), rareness (R), imitability (I) and Organization (O) of its competencies [74–76]. In addition, a resource-based view of strategy—a function of stakeholders such as top management who formulate strategy in conjunction with the organization's ownership—helps to further optimize the resource utilization of organizations. Efficient and effective exploitation of an organization's valuable, rare resources and capabilities can help establish a competitive advantage. This in turn can lead to higher performance, benefiting the organization's ownership [77,78].

3. Materials and Methods

Building an exceptional literature review requires a methodical and systematic approach. Hence, the use of a systematic approach towards the literature review, i.e., the systematic literature review—is highly recommended by scholars due to its rigorous, formal and scientific nature [79]. Utilizing a systematic literature review procedure is beneficial to research from the management and organizational domain, as it enables researchers to deduce context-sensitive research logically and reliably, improving the overall output [80]. Therefore, this paper conducted a systematic review of the available literature, using the 3-stage principles and procedures formulated by Tranfield et al. [81] to answer the research questions. These three main stages consisted of planning, execution and reporting. This procedure, which commenced on the 4 February, and finalized on the 3 May 2019, was selected as it has been geared towards research in the management domain.

As part of the planning stage, a review protocol was established. The protocol would guide the literature review process through defining the review parameters. The first condition for the studies to be included is that they must be investigating eco-innovation and performance as the main concept. This inclusion condition was put in place for several reasons. Firstly, to ensure that the literature remains within the focused scope of this paper. Secondly, to ensure that the paper does not deviate away from the main domain of the focal research. Thirdly, eco-innovation as a concept is transdisciplinary, covering a wide expanse of technologies, practices and services. Hence, numerous synonyms have been used in conjunction with eco-innovation [82]. Consequently, semi-independent and autonomous sub-fields will emerge, which may result in the researchers deviating away from the scope of the research [81]. Lastly, focusing on studies that investigate eco-innovation and performance as the main concept will help to answer the research questions. Hence, focusing exclusively on studies that investigate eco-innovation ensures the paper remains within the bounds of the research scope. Only empirical research papers were included, as they featured empirical results of investigations on eco-innovation. Therefore, review articles, books and book chapters where excluded.

Eco-innovation is a research area that has yet to mature—with scholarly research publications beginning around the year 2000 [83]. The second inclusion criteria, therefore, specifically focuses on research that falls between the years 2000 and 2018. In addition, this paper also excluded records that were not available in English.

Following the definition of review parameters, the steps to be followed during execution and reporting were established. These steps which assisted the paper in compiling the publications for further analysis are described in the framework illustrated by Figure 1.

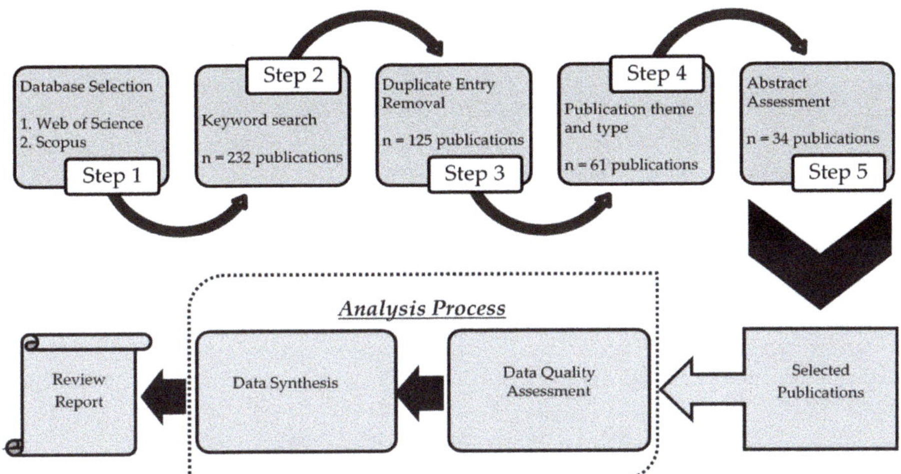

Figure 1. Illustration of the Review Process (Authors' own elaboration).

3.1. Step 1: Database Selection

This paper undertook its information search from two major scientific databases of knowledge, Scopus and the Web of Science (WoS). This paper utilized these two databases, as they are accessible to the researchers via the institution's information center subscription. As a result, the researchers can access top tier publications from the likes of Science Direct, Emerald, Wiley and Springer. Furthermore, these two scholarly databases are the current leading research publication institutions where current, up to date journal publications can be sourced [84]. Hence, utilizing these databases affords this paper access to abundant sources of knowledge within the scope of this paper, as these databases feature top tier publications [85].

3.2. Step 2: Keyword Search

The keywords to be utilized in this paper were in line with answering the research questions. Hence, the strategy for searching the databases for articles needed the usage of the AND operator. The AND operation would be utilized in searching for research articles that follow the concepts laid out in the inclusion and exclusion criteria. The performance word was utilized so as to include other precursor synonyms to refer to an organization such as firm or corporate.

Following the search location and strategy, Table 1 summarizes the keywords that were utilized in the search databases:

Table 1. Keyword Search.

Theory	Search String	Scopus	WoS	Total
Resource-based	Eco-innovation AND Performance AND resource-based	40	62	102
Stakeholder	Eco-innovation AND Performance AND stakeholder	85	45	130

3.3. Step 3: Duplicate Entry Removal

After combining the search results from the two databases, the results returned a total of 232 records as of 31 December 2018. The publication results were then compiled to create a full list of the publication records. Thereafter, the search string results were validated by removing duplicate publication records from the Scopus and WoS Databases. This was a necessary step, given the preceding step of combining publication records from both databases into the one list. Hence, some duplicate publication records were expected to be seen—as some publication records are available on both Scopus and WoS. EndNote

Version X8 was used to facilitate this process. After removing duplicate publication records from the search string results, 125 publication records remained.

3.4. Step 4: Publication Theme and Type

Next, the inclusion and exclusion criteria were then applied to the remaining 125 search results. Articles need to be empirically analyzing eco-innovation and performance for final inclusion in this paper. After the application of inclusion and exclusion criteria, a total of 61 papers where then selected to undergo an examination of the title and abstract, to make sure that the records are in the field of eco-innovation and performance.

3.5. Step 5: Abstract Assessment

At this stage, an abstract analysis was conducted on the remaining 61 papers. Only papers that fulfilled the inclusion criteria as specified in the review protocol where selected. Studies whose abstract did not, or was not, in the context of eco-innovation and performance, were not selected. Therefore, in total, 34 papers from the 61 where then selected for final in-depth qualitative synthesis. This resulted in 34 publication records being selected to undergo further qualitative synthesis, helping the paper remain within its scope.

3.6. Step 6: Analysis Process

An assessment of the quality of the selected articles was then undertaken prior to data synthesis. This was carried out in order to provide a picture on the relevance and recency of the selected studies and their impact. Journal Impact Factor and CiteScore®quartile ranking, as well as citations, are a proxy indicator of the quality of study, as studies accepted into these publications have undergone rigorous research quality assessment [86].

Next, after quality assessment, the selected articles were synthesized in order to gather the data from the articles. Main ideas were analyzed, and their effect, i.e. eco-innovation and its impact on performance through content analysis. Once data was extracted and synthesized, this paper then discussed the findings. Gaps identified were also discussed, and recommendations for possibly addressing the gaps where then presented.

4. Results

4.1. Data Quality Assessment

Most of the selected studies in this paper came from journal sources that are ranked in the top Quartile according to Scopus's CiteScore®ranking system in their respective fields. Journal of Cleaner Production, Business, Strategy and the Environment, as well as Sustainability—which are also top ranked journals—are some of the source titles for a number of the selected studies. The source title for one of the studies is not covered under Scopus's CiteScore®system. Instead, it is ranked in the Web of Science's Emerging Sources Citation Index (ESCI).

Despite the concept of eco-innovation being established in the year 2000, its connection to performance did not gain much attention in the first 10 years post 2000. There were limited attempts to connect eco-innovation to performance from a strategic and stakeholder perspective, with the notable attempts being made post 2010. Only from the year 2016 did the publication numbers for eco-innovation and performance literature begin to gain traction. Hence, most of the analyzed publications are quite recent; i.e., within the last three years, as eco-innovation and performance research is yet to mature.

4.2. Data Synthesis

Stakeholders play a critical role in an organization's performance. Because of the influence of their actions and the influence of the organization's actions on them, their synergy is pivotal to key areas of the organization such as performance. And one of the key organizational aspects that is inextricably linked to performance is strategy [87]. The strategy, which determines the bearing of an organization, is often the culmination of stakeholders' synergy. The performance of an organization is not just critical for its own success. It is also of interest to the numerous stakeholders. Hence, the bearing of an organization's strategic direction is a function of the various stakeholders, as the organization relies on them for its performance. The stakeholders in return are also reliant on the organization's performance [88].

Stakeholders can be classified according to their various abilities to wield influence; namely primary (e.g., owners, suppliers, employees); secondary (e.g., government, pressure groups); and outside (e.g., market forces, societal trends) [89].

However, the stakeholder groups have recently been reclassified according to major groups; namely: Capital market stakeholders (e.g., shareholders, suppliers of capital); product market stakeholders (e.g., customers, suppliers, host communities); and organizational stakeholders (e.g., employees, managers, non-managers) [90].

The capital market stakeholder group—consisting of shareholders (institutional or individual)—is one of the most important stakeholder group, as they assume the risk and provide to finance the firm, and expect to generate a return on their risk and financial investment. This group also has the ability to influence the firm through a number of means, such as selecting and appointing the executives/top management (principals) to run the organization [90]. In addition, these executives are pivotal in strategy formulation [91]. Meanwhile, product market and organizational stakeholders, such as employees, suppliers and managers buttress strategy implementation [92], highlighting the important functions of stakeholders for organizations. Figure 2 illustrates the distribution of literature focusing on the role of stakeholder groups in the eco-innovation and performance literature. As can be witnessed, a large percentage of eco-innovation and performance literature has been dedicated to the product market and organizational groups, yet only a fraction has focused on the capital market stakeholder group.

Figure 2. Eco-innovation and Performance: Stakeholder groups.

Furthermore, one of the keys to achieving excellent organizational performance through sustainable value creation is to align strategy and eco-innovation [93]. In order to align its strategy and eco-innovation, an organization needs to consider its resources and capabilities (the ability for an organization to efficiently and effectively utilize its resources) [70].

Effective and efficient utilization of valuable, rare and inimitable organizational resources can enable an organization to develop capabilities that optimize its product-market activities. Optimization of product-market activities, in turn, help an organization's performance through a competitive advantage [74]. Resources thus enable organizations to develop unique innovations vis-à-vis their capabilities. Therefore, resources and capabilities are indispensable to the strategy, innovativeness and performance of organizations [94].

Thus, in terms of resources, capabilities and stakeholders, eco-innovation and performance research has mostly focused on identifying various sets of drivers. These drivers themselves emanate from the various stakeholders interacting with the organization, as highlighted in Figure 3.

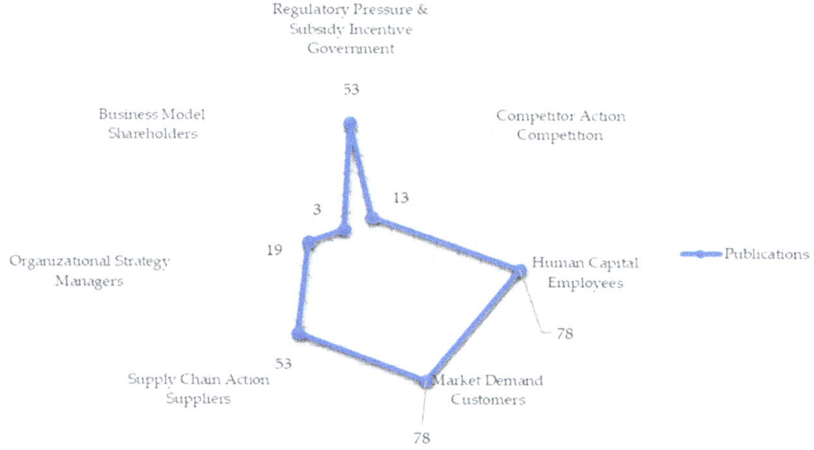

Figure 3. Eco-innovation and Performance: Drivers and Stakeholders.

Figure 3 also presents an illustration of the identified drivers and the stakeholder factor behind the driver. The identified driver behind each factor has been grouped into their respective stakeholder group (see Figure 2). Most literature identified human capital, market demand, supply chain requirements and regulations as key drivers for the eco-innovation and performance relationship. Meanwhile, organizational strategy, competitors and the business model were also pointed out as drivers, but did not receive as much attention as the other drivers. However, the publications highlighting the capital market stakeholder group are few. Only a handful of studies have paid attention to this group's influence towards the eco-innovation and performance relationship.

Furthermore, the selected publications were also assessed from an industry perspective, as highlighted by Figure 4. Even though eco-innovation and performance literature has been growing post 2010, there seems to be a disproportionate focus of literature, with much of it derived from manufacturing.

A high proportion of the selected publication has focused on the manufacturing industry. Less than 10 percent of studies emanated from the services sector, indicating a coverage that is heavily skewed towards the manufacturing sector.

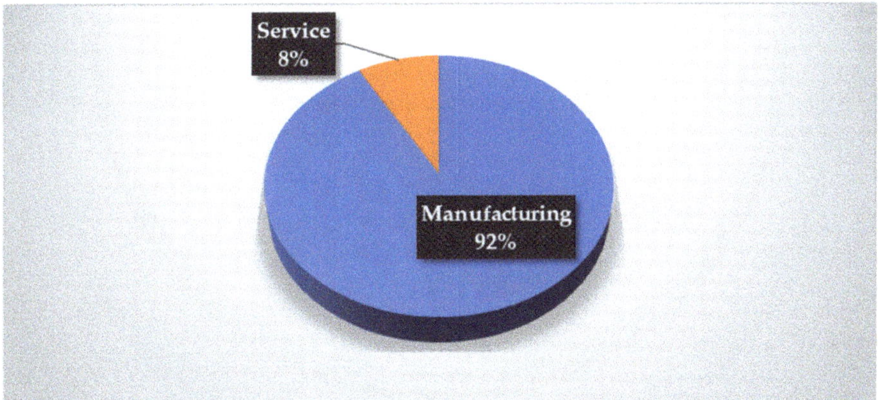

Figure 4. Industry Coverage of Eco-innovation and Performance.

4.2.1. Eco-innovation and Performance: Regulatory Effects

Environment regulations have been a prominent driver in the adoption of eco-innovation by organizations. Environmental policies formulated by governments, especially those that seek to curb waste generation, environmental pollution and GHG emissions, would require organizations to comply in order to operate—regardless of their size [95,96]. Regulation has been found to be a strong antecedent to eco-innovation [97]. In fact, meeting regulatory standards has proven to cause a threefold increase in the likelihood of organizations engaging in eco-innovation [98], underscoring the importance of environmental policy formulation, enforcement and monitoring by the authorities, as also suggested by Fernando, Wah and Shaharudin [95]. Hence, when organizations eco-innovate to make their market offerings comply with environmental regulations, they derive better performance overall as a result of compliance with set environmental regulations. Furthermore, potential revenue streams are created as a result of environmental compliance [99]. Perhaps this may also be unique in cases where regulatory authorities incentivized environmental compliance [100–102]. In addition to the potential revenue streams that are created, compliance with environmental regulation was also found to boost an organization's environmental performance [103].

One of the ways that can help an organization to improve or comply on environmental regulation is by engaging in dialog with stakeholders [104,105]. A possible reason why stakeholder engagement is important emanates from the earlier findings of Blum-Kusterer and Hussain [106], Galliano et al. [107] and Ratten [108], who all agreed that apart from regulation, pressure from the various stakeholders can influence organizations to adopt eco-innovation or increase their eco-innovation practices. An example of this is illustrated by the findings of Bossle et al. [109], who stated that it is not only government that pressures organizations to eco-innovate so as comply with regulation. In fact, it was noted by other researchers that government played a minute role in making organizations eco-innovate [109]. Pressure to eco-innovate came from the other stakeholders (competitors, customers and suppliers). As a result of normative pressures from this group of stakeholders, organizations that do not eco-innovate find themselves not being able to compete in the market, hence the need to eco-innovate. Tsai and Liao [110] concur with this argument, when they also found that it was the lure of revenue and performance, rather than regulations, that enticed organizations to eco-innovate, underlining the importance of the product market stakeholders.

4.2.2. Eco-innovation and Performance: Market Dynamics

The lure of potential revenue and performance is, in fact proving to be the underlying reason for organizations eco-innovating. This is due to the mixed results of regulation as a driver for eco-innovation [109–111]. In fact, the growing consumer awareness of environmental issues is creating a distinct set of consumers who engage in sustainable consumption and are willing to pay premium pricing to products that are eco-friendly [112]. Hence, it is the allure of financial performance through increased turnover from eco-products. An example is illustrated by the growing consumption of eco-products, as well as other value propositions created through the utilization of eco-processes and cleaner production technologies [112–114]. Interestingly however, other studies found that eco-innovation did not contribute to performance [104,114].

4.2.3. Eco-innovation and Performance: Organizational Factors

It is not just the market as well that increases the propensity of organizations eco-innovating. Employees are also increasingly playing a pivotal in organizations' eco-innovative capabilities, thus contributing to performance. For instance, it was argued that employees have a major role to play in the outcomes of eco-innovation and the performance of organizations [115].

This could be due to employees being important sources of knowledge, and therefore being a critical resource and stakeholder [99,114]. In fact, managers are also an important key in the eco-innovation and performance chain [111,116,117]. In fact, it was pointed out that environmental concern by the management of an organization buttressed eco-innovation—subsequently contributing positively towards performance [118]. This viewpoint is further corroborated by the argument that increased managerial concerns for the environment directly increased the performance of organizations which eco-innovate [109].

5. Discussion

This section presents the discussion points and gaps identified from the systematic review of the literature.

5.1. Eco-innovation and Performance: Drivers

The literature identifies and discusses drivers of eco-innovation. These drivers are important, as they play a key role in eco-innovation adoption—with performance then derived from eco-innovation. However, whilst drivers of eco-innovation and their impact on performance are subsequently covered and discussed by the literature, the majority of the factors are derived from the product market and organizational stakeholder group. Hence, most research analyzing the eco-innovation and performance relationship looks at the influence that stakeholders who come from the product market and organizational groups impose on the relationship. However, studies that mentioned top management, shareholders and competitors as stakeholders are scant, as illustrated by Figure 3.

This is alarming, given the role capital market stakeholders play in organizations vis-à-vis strategy formulation. Furthermore, the capital market stakeholder group itself can be the source of the talent, skills, capabilities and knowledge of multiple disciplines, as has been witnessed by charismatic top management and investors who are concerned about and support the natural environment.

The capital market stakeholder group needs to put in place the necessary steps for dynamic interactions amongst stakeholders, to have efficient and effective usage of resources and capabilities. For organizations to improve their capability to eco-innovate, the organization's capital market stakeholder group needs to configure their commitment to environmental issues and reassess their current capabilities. The capability of the organization to eco-innovate can then have an impact on other stakeholders through competitive advantage. Competitive advantage can in turn lead to superior performance. Hence, higher levels of commitment to the environment would be result in the capital market stakeholder group's eco-innovation strategy boosting its positive impact on other stakeholder groups—both internal and external.

5.2. Eco-innovation and Performance: Top Management Commitment

Commitment to the environment by the top management has the potential to impact an organization's ability to eco-innovate. When eco-innovation is approached from the stakeholder theoretical lens, each stakeholder group has a dynamic two-way relationship with the organization. Top management is an internal stakeholder and a core influencer in the system of the organization through strategy formulation. As a key stakeholder, top management's commitment levels to the environment are critical, given their function in strategy formulation. In fact, amongst the stakeholder groups, top management may possess the most influence in driving the organization towards an eco-innovation strategy [109]. Whilst the top management is an integral part of the organization, and therefore a key stakeholder, other stakeholders are also important in other functional areas of the organization.

Thus, other stakeholder groups should continue to pay attention to the top management's commitment levels to the environment in order to ensure that they are able to formulate eco-innovation strategies that positively impact the organization, and consequently other stakeholder groups, such as shareholders, employees, customers and governments through its financial and environmental performance.

5.3. Eco-innovation and Performance: Shareholders' Role

This paper also identified that shareholders are a pivotal stakeholder group, as shareholding was also found to be a significant top management factor. This is due to shareholders being one of the key stakeholders of an organization. However, shareholders—who fall under the capital market stakeholder group—have received little coverage in eco-innovation and performance literature, as illustrated by Figures 3 and 4 earlier.

Hence, there is a shortfall in literature that analyzes the eco-innovation and performance relationship from the capital market stakeholder group (shareholders). Yet this stakeholder group is pivotal in both the decision making and performance of organizations. As providers of finance to the organization, shareholders are exposed to the highest risk—that is, their financial interests in the organization. Although developing new and radical technologies creates a risk for all stakeholders, the shareholders are the most exposed, as their value may be diminished or damaged in cases of unsuccessful innovations. In the case of eco-innovations and cleaner production technologies, they often require resources such as finance, and capabilities committed to them with an uncertain outcome. The uncertainty of eco-innovations may thus be of concern to shareholders and other providers of capital, that is, those financing the organization.

This stakeholder group is also closely related to all other stakeholders, but most importantly to the top management, as the top management is appointed by this group. Hence, top management's role as an agent and key stakeholder is to create value for the shareholders. Hence, it is important for top management to focus on the relationship of the organization to this important stakeholder group. Keeping shareholders satisfied is important for top management, who must then balance between satisfying shareholders and other stakeholder groups. Top management must also balance the risk that the eco-innovation strategy poses to its shareholders, versus the potential benefits and value creation to other key stakeholders of the organization.

Top management's incorporation of the environment in strategy could create additional value—keeping shareholders satisfied—if the top management's eco-innovation strategy is creating value for the capital market group. If the shareholders are also concerned about the environment, then it will be a more manageable task for top management to align the interests of all stakeholders. Galliano, Goncalves and Triboulet [107] support this notion by highlighting the importance of an organization's ownership in its eco-innovation practices. As important as the shareholders are to the organization, at this juncture however, it is worth noting the concerns raised earlier by Gelter [119] and Tyl, Vallet, Bocken and Real [33], who warned that dominance from this group could potentially be harmful for the interests of other stakeholders. For example, the motor vehicle industry was rocked by the emissions scandals [120] to the extent that some of the automakers have had to announce their intent of focusing on vehicles with hybrid or battery electric drive trains [121,122], saving shareholders from having their value further diminished.

5.4. Eco-innovation and Performance: Industry Perspectives

In addition, most of the investigations on the eco-innovation and performance relationship has extensively focused on the manufacturing industry, as earlier illustrated by Figure 4 This is not strange though, as most environmental regulations seem to be skewed towards ensuring the manufacturing industry lowers its negative environmental impact, a viewpoint shared by Pacheco et al. [123], Spedding [124] and Wang et al. [125].

Manufacturing has thus been extensively discussed from an eco-innovation and performance perspective. This is further buttressed by the fact that most organizations adopt or engage in eco-innovation practices towards their production operations.

For instance, Cheng et al. [126] found a strong relationship between eco-product and performance, as well as eco-process and performance, respectively. However, eco-innovation from the perspective of the service sector has received little coverage. This could be problematic, particularly as organizations are no longer solely focusing on providing products, but are also expected to be providing solutions, such as consultancy to customers of their products, so that the customers get the most out of their purchase [103]. Some organizations even package their market offering as an entire system, providing initial consultancy, set up, the product itself, and after sales service support/vendor support [127–129]. Hence, it is important to investigate the eco-innovation practices of companies that are moving towards servitization in the form of product-service system and service provisions.

5.5. Eco-innovation and Performance: Geographical Context Dynamics

Furthermore, most research on eco-innovation and performance has been concentrated on developed markets. This is an important fact, as different geographical and economic contexts have different effects on eco-innovation. This is because of different contexts exhibiting differences in terms of market dynamics, cleaner production technology availability, environmental regulation requirements, as well as different levels of enforcement [130]. For instance, Tsai and Liao [110] found that environmental regulation in their study context was light. Penalties for violation were light, incentives for compliance were low, and sentencing for environmental conviction was lenient. Meanwhile, Jabbour et al.'s [131] study reveals that the lack of regulatory incentives was a hinderance to eco-innovation in that particular context. Consequently, any motivation for compliance may be low in these contexts. However, when compared to the study of Dong et al. [132], the research context found that regulation increased the eco-innovation practices of organizations. Hence, investigating eco-innovation and performance from other geographical contexts would help in understanding the influence of that particular location's environmental regulations—which subsequently influence eco-innovation and performance.

5.6. Eco-innovation and Performance: Theoretical and Managerial Implications

This paper made several contributions to theory. Firstly, by analyzing eco-innovation and performance from the resource-based views and stakeholder theory, this paper contributed towards addressing the current limitations in eco-innovation research by identifying and analyzing eco-innovation factors from these theoretical perspectives. Most research that analyzes eco-innovation tends to utilize one distinct theory, such as a resource-based view. However, using more than one theory could help to overcome the potential weaknesses inherent in that theory. Therefore, eco-innovation and performance can be analyzed from a multi-dimension perspective to widen the scope of factors that can impact eco-innovation strategy in an organization—and subsequently performance.

Secondly, this literature review unearthed a few emerging areas such as top management environmental concern and shareholder environmental concern that have received little attention from a theoretical viewpoint. Thirdly, this paper proves that the resource-based views and stakeholder theory enjoy a dynamic relationship and are complementary to each other. The identified stakeholders that are important to these theories where top management and shareholders/ownership.

This paper showed that top management and shareholders are key stakeholders—as their environmental concern can greatly influence an organization's eco-innovation strategy orientation, which in turn impacts performance—subsequently impacting other stakeholders. External stakeholders can impact and be impacted by the top management, for example environmental pressure groups/lobbyists. However, although external stakeholders do impose pressure, it is top management who can have a positive bearing on eco-innovation and performance. The top management's commitment to eco-innovation can assist organizations in building up their abilities to infuse eco-innovation into their competencies and comply with environmental regulations whilst shareholders provide the necessary financial resources to enable eco-innovation.

5.7. Eco-innovation and Performance: Future Research Avenues

Given that innovations of any kind face a high degree of uncertainty, approaching eco-innovation strategy with more information and knowledge from multiple theoretical backgrounds could help mitigate this risk. One potential way of obtaining more information and knowledge to drive eco-innovation and performance is through the utilization of predictive analytics technology such as Big Data Analytics. Therefore, future reviews could explore eco-innovation and performance from an information and knowledge perspective, given the potential that data driven analytics capabilities have in driving eco-innovation and performance, and also the ability which Big Data Analytics have in reducing the high degree of uncertainty related to eco-innovations. Reducing uncertainties, and therefore risk, is important for stakeholders, particularly the shareholders who are most exposed to financial risk.

Secondly, eco-innovating successfully may also mean altering existing business models, or a completely radical, disruptive business model. For example, there could be eco-innovation in the form of new clean technologies, such as the emergence of new energy vehicle organizations with distinct eco-innovation capabilities. From the resource-based view, organizations must have the skills, capabilities and knowledge of multiple disciplines in order to increase the success of eco-innovations. Having the necessary resources is becoming increasingly important, even for established industry powerhouses, as they are ever more coming under threat from new players with a flair for eco-innovation. New players which have skills, capabilities and knowledge across a wide range of disciplines introduce disruptive cleaner eco-innovative products and technologies. Therefore, future research can possibly empirically investigate if organization characteristics such as age, size, sector, as well as the ownership structure, exhibits any differences in eco-innovation and performance.

Thirdly, it is not just organizational resources and capabilities that can be sources of competitive advantage. Interestingly, top management, as well as owners in of themselves, can potentially be sources of competitive advantage vis-à-vis their resource management and strategy formulation, as well as their knowledge, visions and leadership traits. The skills and capabilities of top management can then drive eco-innovation, ultimately resulting in a competitive advantage under the resource-based theories, and further enhancing value for stakeholders. Hence, when shareholders appoint or assess their agents (top management), having a look at the talent and skill sets of their agents is crucial, especially where the agent's skills and abilities complement their environmental concern. This fit or alignment of interests, skills and abilities can potentially be a great asset for organizations. In fact, such is the importance of top management in the competitive advantage strategy formulation and uncertainty reduction that many companies with innovation flair often "poach" talented top management from successful organizations. Hence, future research could empirically test organizations to see if top management characteristics such as education, age, domain experience, technical expertise and leadership style have any bearing on eco-innovation and performance.

Lastly, organizations stand to benefit by assessing the challenges and opportunities afforded by the state of the natural environment. Therefore, tackling these environmental challenges requires organizations to be familiar with and garner the necessary experience to achieve eco-innovation. Hence, top management and shareholders, through their concern for the environment, could themselves become key drivers of eco-innovation—helping to create value for all stakeholders through sustainable development for a better tomorrow.

6. Conclusions

In this paper, organizational stakeholders, resources and capabilities were identified as important determinants of innovation and strategy. In an era of mounting environmental challenges, organizational resources, together with stakeholder synergies, need to be channeled towards addressing environmental challenges. This ability by organizations to innovatively tackle challenges to the natural environment can be the source of a sustainable competitive advantage. As a result, there is a growing importance of eco-innovations towards organizational competitiveness. Due to this growing importance, literature pertaining to eco-innovation and performance from a resource-based view and stakeholder theory was reviewed using a systematic method.

Firstly, the systematic review process allowed the research questions to be answered. Second, based upon the findings, gaps were identified. Contributions and pathways for future research were then recommended. Human capital, market demand, regulation and supply chain requirements were major contributors towards the eco-innovation and performance relationship. However, not enough attention has been paid to the capital market group of stakeholders. This is alarming, given the important role they play as capital providers to the organization.

The capital market stakeholder group can also influence the firm through several other means—such as selecting and appointing the executives/top management (principals) to formulate organizational strategy. The owners, as principals, wield significant power, especially when their ownership of shares is significantly large. However, it is also important that despite their importance, the views of ownership group should not be too dominant, as the interests of the other stakeholders must also be protected. In addition, the stakeholder theory—although a fundamental theory that explains organizational behavior beyond the interests of the shareholders—often is focused on human elements. However, it is not just human elements that could influence and provide ethical guidelines and heuristics for the organization. The natural environment, for instance, is overlooked by the stakeholder theory, yet the natural environment is a primary but non-human stakeholder whose interests must also be considered by the firm's decision makers.

Therefore, in order to overcome this limitation of the stakeholder theory when investigating the eco-innovation and performance relationship, this paper calls for the unique approach of complementing the stakeholder theory with the natural resource-based view (NRBV) for two main reasons. Firstly, the NRBV, by its nature, was introduced to compensate for the shortfall in the RBV. It compensates for the shortcoming of the RBV by incorporating the natural environment in the framework [13], which is important as markets and organizations are dependent upon the functionality of the natural environment. As natural environment challenges mount, firms that channel their resources and develop capabilities towards addressing environmental challenges will realize greater competitive advantage. For instance, firms that embed waste and GHG emission minimization by practicing continuous improvements of their products are able to develop the strategic capability of pollution prevention. This capability would help create lower production costs. In addition to the pollution prevention and product stewardship capabilities, firms ought to engage in cleaner production. Developing or using cleaner production technologies requires firms to have the capability of eco-innovation. Therefore, the NRBV is essential, as it simultaneously overcomes the limitations in both the stakeholder and RBV. Secondly, the stakeholder theory would add an interesting reinforcement to the eco-innovation and performance relationship, as it takes into account the ability of stakeholders to pressure the organization to remain competitive through eco-innovation. In fact, the stakeholder theory would be instrumental in considering organization specific attributes such as age, size and sector. These attributes are important as organizations that may be resource-disadvantaged due to these organizational attributes. Hence, such organizations may compensate for their resource-disadvantage through better management of stakeholder relationships. Better management of stakeholder relationships might therefore enable these resource-disadvantaged organizations to eco-innovate, and in turn improve performance [133]. Taking this strategic perspective would support organizations that strive to address the mounting environmental challenges while concurrently generating economic returns through a sustained competitive advantage—key aspects of sustainable development.

This paper was not without limitations. Firstly, the literature reviewed in this paper originated from two databases only—the Scopus and Web of Science. Secondly, this paper focused on publications written in English. Third, this paper focused on empirical research and excluded book chapters and conference proceedings. Fourth, other synonyms of eco-innovation were not utilized in this paper.

Thus, future research could overcome these limitations by replicating this study and incorporating other scholarly databases such as Google Scholar. Another strategy to overcome limitations could be assembling a multi-language team of researchers.

Such a multi-language team will allow the limitations of one language to be overcome and will broaden the articles under review to more than one language. Furthermore, to increase the number of articles selected for analysis, book chapters, conference proceedings and other types of publications can be incorporated into future research. Lastly, using synonyms of eco-innovation could give future research an opportunity to understand the evolutionary nature of the subject matter.

Author Contributions: Conceptualization, R.T.M. and S.K.J.; methodology, R.T.M.; data curation, R.T.M.; validation, R.T.M. and S.K.J.; formal analysis, R.T.M.; investigation, R.T.M. and S.K.J.; resources, R.T.M.; writing—original draft preparation, R.T.M.; writing—review and editing, R.T.M. and S.K.J.; visualization, R.T.M.; supervision, S.K.J.; project administration, S.K.J.; funding acquisition, R.T.M. and S.K.J.

Funding: This research was supported by the Yayasan UTP Fundamental Research Grant (YUTP-FRG), grant number 015LC0-016.

Acknowledgments: The authors would like to thank the Department of Management and Humanities, and the Centre for Graduate Studies for their support. The authors would also like to express gratitude to the four anonymous reviewers, whose comments and feedback helped to improve this paper.

Conflicts of Interest: The authors declare no conflict of interest. The funders had no role in the design of the study; in the collection, analyses, or interpretation of data; in the writing of the manuscript, or in the decision to publish the results.

References

1. Al-Ansari, T.; Korre, A.; Nie, Z.; Shah, N. Integration of greenhouse gas control technologies within the energy, water and food nexus to enhance the environmental performance of food production systems. *J. Clean. Prod.* **2017**, *162*, 1592–1606. [CrossRef]
2. Hajer, M.; Nilsson, M.; Raworth, K.; Bakker, P.; Berkhout, F.; De Boer, Y.; Rockström, J.; Ludwig, K.; Kok, M. Beyond Cockpit-ism: Four Insights to Enhance the Transformative Potential of the Sustainable Development Goals. *Sustainability* **2015**, *7*, 1651–1660. [CrossRef]
3. United Nations. *Transforming Our World: The 2030 Agenda for Sustainable Development*; United Nations: New York, NY, USA, 2015; pp. 1–35.
4. Dahle, K. Toward governance for future generations: How do we change course? *Futures* **1998**, *30*, 277–292. [CrossRef]
5. Noble, I.; Scholes, R.J. Sinks and the Kyoto Protocol. *Clim. Policy* **2001**, *1*, 5–25. [CrossRef]
6. Ashby, M.F. Chapter 1—Background: Materials, Energy and Sustainability. In *Materials and Sustainable Development*; Ashby, M.F., Ed.; Butterworth-Heinemann: Boston, MA, USA, 2016; pp. 1–25.
7. Olawumi, T.O.; Chan, D.W.M. A scientometric review of global research on sustainability and sustainable development. *J. Clean. Prod.* **2018**, *183*, 231–250. [CrossRef]

8. Ricci, E.C.; Banterle, A.; Stranieri, S. Trust to Go Green: An Exploration of Consumer Intentions for Eco-friendly Convenience Food. *Ecol. Econ.* **2018**, *148*, 54–65. [CrossRef]

9. Wang, Y.; Shen, N. Environmental regulation and environmental productivity: The case of China. *Renew. Sustain. Energy Rev.* **2016**, *62*, 758–766. [CrossRef]

10. Stock, T.; Obenaus, M.; Kunz, S.; Kohl, H. Industry 4.0 as enabler for a sustainable development: A qualitative assessment of its ecological and social potential. *Process. Saf. Environ. Prot.* **2018**, *118*, 254–267. [CrossRef]

11. Ball, P.; Lunt, P. Lean eco-efficient innovation in operations through the maintenance organisation. *Int. J. Prod. Econ.* **2018**. [CrossRef]

12. Pham, D.D.T.; Paillé, P.; Halilem, N. Systematic review on environmental innovativeness: A knowledge-based resource view. *J. Clean. Prod.* **2019**, *211*, 1088–1099. [CrossRef]

13. Hart, S.L. A Natural-Resource-Based View of the Firm. *Acad. Manag. Rev.* **1995**, *20*, 986–1014. [CrossRef]

14. Rennings, K. Redefining innovation—Eco-innovation research and the contribution from ecological economics. *Ecol. Econ.* **2000**, *32*, 319–332. [CrossRef]

15. Donaldson, T.; Preston, L.E. The Stakeholder Theory of the Corporation: Concepts, Evidence, and Implications. *Acad. Manag. Rev.* **1995**, *20*, 65–91. [CrossRef]

16. Carrillo-Hermosilla, J.; del González, P.R.; Könnölä, T. What is eco-innovation? In *Eco-Innovation: When Sustainability and Competitiveness Shake Hands*; Palgrave Macmillan UK: London, UK, 2009; pp. 6–27.

17. Prieto-Sandoval, V.; Jaca, C.; Ormazabal, M. Towards a consensus on the circular economy. *J. Clean. Prod.* **2018**, *179*, 605–615. [CrossRef]

18. De Jesus, A.; Antunes, P.; Santos, R.; Mendonça, S. Eco-innovation in the transition to a circular economy: An analytical literature review. *J. Clean. Prod.* **2018**, *172*, 2999–3018. [CrossRef]

19. Von Geibler, J.; Bienge, K.; Schüwer, D.; Berthold, O.; Dauensteiner, A.; Grinewitschus, V.; Hoffmann, D.; Renner, W.; Ostermeyer, Y. Identifying business opportunities for green innovations: A quantitative foundation for accelerated micro-fuel cell diffusion in residential buildings. *Energy Rep.* **2018**, *4*, 226–242. [CrossRef]

20. Sjöström, J.; Talanquer, V. Eco-reflexive chemical thinking and action. *Curr. Opin. Green Sustain. Chem.* **2018**, *13*, 16–20. [CrossRef]

21. Gude, V.G. Sustainable chemistry and chemical processes for a sustainable future. *Resour. Effic. Technol.* **2017**, *3*, 249–251. [CrossRef]

22. Dewick, P.; Foster, C. Focal Organisations and Eco–innovation in Consumption and Production Systems. *Ecol. Econ.* **2018**, *143*, 161–169. [CrossRef]

23. World Economic Forum. *The Role of Technology Innovation in Accelerating Food Systems Transformation*; World Economic Forum: Geneva, Switzerland, 2018.

24. Prendeville, S.; O'Connor, F.; Palmer, L. Material selection for eco-innovation: SPICE model. *J. Clean. Prod.* **2014**, *85*, 31–40. [CrossRef]

25. Löbberding, L.; Madlener, R. System Cost Uncertainty of Micro Fuel Cell Cogeneration and Storage. *Energy Procedia* **2017**, *142*, 2824–2830. [CrossRef]

26. He, F.; Miao, X.; Wong, C.W.Y.; Lee, S. Contemporary corporate eco-innovation research: A systematic review. *J. Clean. Prod.* **2018**, *174*, 502–526. [CrossRef]

27. Anzola-Román, P.; Bayona-Sáez, C.; García-Marco, T. Organizational innovation, internal R&D and externally sourced innovation practices: Effects on technological innovation outcomes. *J. Bus. Res.* **2018**, *91*, 233–247.

28. Ruiz-Jiménez, J.M.; del Mar Fuentes-Fuentes, M. Management capabilities, innovation, and gender diversity in the top management team: An empirical analysis in technology-based SMEs. *BRQ Bus. Res. Q.* **2016**, *19*, 107–121. [CrossRef]

29. Kotsemir, M.; Abroskin, A.; Meissner, D. Innovation Concepts and Typology–An Evolutionary Discussion. *High. Sch. Econ.* **2013**. [CrossRef]

30. Del Río, P.; Carrillo-Hermosilla, J.; Könnölä, T.; Bleda, M. Resources, capabilities and competences for eco-innovation. *Technol. Econ. Dev. Econ.* **2016**, *22*, 274–292. [CrossRef]

31. Bossle, M.B.; Dutra de Barcellos, M.; Vieira, L.M.; Sauvée, L. The drivers for adoption of eco-innovation. *J. Clean. Prod.* **2016**, *113*, 861–872. [CrossRef]

32. Xavier, A.F.; Naveiro, R.M.; Aoussat, A.; Reyes, T. Systematic literature review of eco-innovation models: Opportunities and recommendations for future research. *J. Clean. Prod.* **2017**, *149*, 1278–1302. [CrossRef]

33. Tyl, B.; Vallet, F.; Bocken, N.M.P.; Real, M. The integration of a stakeholder perspective into the front end of eco-innovation: A practical approach. *J. Clean. Prod.* **2015**, *108*, 543–557. [CrossRef]

34. Del Río, P.; Peñasco, C.; Romero-Jordán, D. What drives eco-innovators? A critical review of the empirical literature based on econometric methods. *J. Clean. Prod.* **2016**, *112*, 2158–2170. [CrossRef]

35. Salim, N.; Ab Rahman, M.N.; Abd Wahab, D. A systematic literature review of internal capabilities for enhancing eco-innovation performance of manufacturing firms. *J. Clean. Prod.* **2019**, *209*, 1445–1460. [CrossRef]

36. Hazarika, N.; Zhang, X. Evolving theories of eco-innovation: A systematic review. *Sustain. Prod. Consum.* **2019**, *19*, 64–78. [CrossRef]

37. Schumpeter, J.A.; Opie, R.; Elliott, J.E. *The Theory of Economic Development: An Inquiry into Profits, Capital, Credit, Interest, and the Business Cycle*; Routledge: New York, NY, USA, 2017.

38. Von Krogh, G.; Ichijo, K.; Nonaka, I. *Enabling Knowledge Creation: How to Unlock the Mystery of Tacit Knowledge and Release the Power of Innovation*; Oxford University Press: Oxford, UK, 2011; pp. 1–302.

39. Välikangas, L. Manage innovation as a corporate capability. *Chem. Eng. Prog.* **2003**, *99*, 64–69.

40. Vanhaverbeke, W.; Peeters, N. Embracing Innovation as Strategy: Corporate Venturing, Competence Building and Corporate Strategy Making. *Creat. Innov. Manag.* **2005**, *14*, 246–257. [CrossRef]

41. Kodama, M. Developing strategic innovation in large corporations—The dynamic capability view of the firm. *Knowl. Process. Manag.* **2017**, *24*, 221–246. [CrossRef]

42. Breznik, L.; Lahovnik, M. Innovation capability as a source of competitive advantage in Slovenian information technology firms. *Tech. Technol. Educ. Manag.* **2012**, *7*, 1132–1143.

43. Pavitt, K. Sectoral patterns of technical change: Towards a taxonomy and a theory. *Res. Policy* **1984**, *13*, 343–373. [CrossRef]

44. Geels, F.W. From sectoral systems of innovation to socio-technical systems: Insights about dynamics and change from sociology and institutional theory. *Res. Policy* **2004**, *33*, 897–920. [CrossRef]

45. Choi, T.Y.; Krause, D.R. The supply base and its complexity: Implications for transaction costs, risks, responsiveness, and innovation. *J. Oper. Manag.* **2006**, *24*, 637–652. [CrossRef]

46. Eiriz, V.; Faria, A.; Barbosa, N. Firm growth and innovation: Towards a typology of innovation strategy. *Innov. Manage. Policy Pract.* **2013**, *15*, 97–111. [CrossRef]

47. Zhang, L.; Zhou, R.; Xu, J. An exploratory case study of the relationship between strategy and brand value based on innovation choice. In *Proceedings of the 9th International Conference on Management Science and Engineering Management (ICMSEM 2015), Karlsruhe, Germany, 21–23 July 2015*; Machado, V.C., Xu, J., Nickel, S., Hajiyev, A., Eds.; Springer Verlag: Berlin/Heidelberg, Germany, 2015; pp. 1055–1063.

48. Persaud, A. Enhancing synergistic innovative capability in multinational corporations: An empirical investigation. *J. Prod. Innov. Manag.* **2005**, *22*, 412–429. [CrossRef]

49. Wang, C.L.; Ahmed, P.K. The development and validation of the organisational innovativeness construct using confirmatory factor analysis. *Eur. J. Innov. Manag.* **2004**, *7*, 303–313. [CrossRef]

50. Vicente, M.; Abrantes, J.L.; Teixeira, M.S. Measuring innovation capability in exporting firms: The INNOVSCALE. *Int. Mark. Rev.* **2015**, *32*, 29–51. [CrossRef]

51. Lefebvre, E.; Lefebvre, L.A.; Talbot, S. Environmental Initiatives, Innovativeness and Competitiveness: Some Empirical Evidence. In Proceedings of the 2000 IEEE Engineering Management Society. EMS—2000 (Cat. No.00CH37139), Albuquerque, NM, USA, 15 August 2000; pp. 674–679.

52. Singh, P.J.; Mittal, V.K.; Sangwan, K.S. Development and validation of performance measures for environmentally conscious manufacturing. *Int. J. Serv. Oper. Manag.* **2013**, *14*, 197–220. [CrossRef]

53. Mousavi, S.; Bossink, B.; van Vliet, M. Dynamic capabilities and organizational routines for managing innovation towards sustainability. *J. Clean. Prod.* **2018**, *203*, 224–239. [CrossRef]

54. Sakalayen, Q.M.H.; Chen, P.S.L.; Cahoon, S. Investigating the strategies for Australian regional ports' involvement in regional development. *Int. J. Shipp. Transp. Logist.* **2016**, *8*, 153–174. [CrossRef]

55. Vickers, I. Cleaner production: Organizational learning or business as usual? An example from the domestic appliance industry. *Bus. Strategy Environ.* **2000**, *9*, 255–268. [CrossRef]

56. Khanna, M.; Kumar, S. Corporate Environmental Management and Environmental Efficiency. *Environ. Resour. Econ.* **2011**, *50*, 227–242. [CrossRef]

57. Walls, J.L.; Phan, P.H.; Berrone, P. Measuring environmental strategy: Construct development, reliability, and validity. *Bus. Soc.* **2011**, *50*, 71–115. [CrossRef]

58. Green, K.; McMeekin, A.; Irwin, A. Technological trajectories and R&D for environmental innovation in UK firms. *Futures* **1994**, *26*, 1047–1059.

59. Kemp, R.; Pearson, P. *Final Report MEI Project about Measuring Eco-Innovation*; European Commission: Brussels, Belgium, 2007.

60. Organisation for Economic Co-operation and Development (OECD). *Oslo Manual*; European Commission, Eurostat: Luxembourg, 2005.

61. Skeete, J.P. Examining the role of policy design and policy interaction in EU automotive emissions performance gaps. *Energy Policy* **2017**, *104*, 373–381. [CrossRef]

62. Carrillo-Hermosilla, J.; Del Río, P.; Könnölä, T. Diversity of eco-innovations: Reflections from selected case studies. *J. Clean. Prod.* **2010**, *18*, 1073–1083. [CrossRef]

63. Coenen, L.; Díaz López, F.J. Comparing systems approaches to innovation and technological change for sustainable and competitive economies: An explorative study into conceptual commonalities, differences and complementarities. *J. Clean. Prod.* **2010**, *18*, 1149–1160. [CrossRef]

64. González-García, S.; García Lozano, R.; Buyo, P.; Pascual, R.C.; Gabarrell, X.; Rieradevall, I.; Pons, J.; Moreira, M.T.; Feijoo, G. Eco-innovation of a wooden based modular social playground: Application of LCA and DfE methodologies. *J. Clean. Prod.* **2012**, *27*, 21–31. [CrossRef]

65. Huang-Lachmann, J.T.; Lovett, J.C. How cities prepare for climate change: Comparing Hamburg and Rotterdam. *Cities* **2016**, *54*, 36–44. [CrossRef]

66. Nag, S.; Mondal, A.; Bar, N.; Das, S.K. Biosorption of chromium (VI) from aqueous solutions and ANN modelling. *Environ. Sci. Pollut. Res.* **2017**, *24*, 18817–18835. [CrossRef] [PubMed]

67. López, F.J.D.; Montalvo, C. A comprehensive review of the evolving and cumulative nature of eco-innovation in the chemical industry. *J. Clean. Prod.* **2015**, *102*, 30–43. [CrossRef]

68. Bag, S.; Gupta, S. Antecedents of Sustainable Innovation in Supplier Networks: A South African Experience. *Glob. J. Flex. Syst. Manag.* **2017**, *18*, 231–250. [CrossRef]

69. Wheelen, T.L.; Hunger, J.D.; Hoffman, A.N.; Bamford, C.E. *Strategic Management and Business Policy: Globalization, Innovation and Sustainability*; Pearson Education: London, UK, 2014.

70. Wernerfelt, B. A resource-based view of the firm. *Strateg. Manag. J.* **1984**, *5*, 171–180. [CrossRef]

71. Hall, R. A framework linking intangible resources and capabiliites to sustainable competitive advantage. *Strateg. Manag. J.* **1993**, *14*, 607–618. [CrossRef]

72. Kiefer, C.P.; Carrillo-Hermosilla, J.; Del Río, P.; Callealta Barroso, F.J. Diversity of eco-innovations: A quantitative approach. *J. Clean. Prod.* **2017**, *166* (Suppl. C), 1494–1506. [CrossRef]

73. Backman, C.A.; Verbeke, A.; Schulz, R.A. The Drivers of Corporate Climate Change Strategies and Public Policy: A New Resource-Based View Perspective. *Bus. Soc.* **2017**, *56*, 545–575. [CrossRef]

74. Barney, J.B. Is the Resource-Based "View" a Useful Perspective for Strategic Management Research? Yes. *Acad. Manag. Rev.* **2001**, *26*, 41–56.

75. Barney, J.B. Firm Resources and Sustained Competitive Advantage. *J. Manag.* **1991**, *17*, 99–120. [CrossRef]

76. Chatzoglou, P.; Chatzoudes, D.; Sarigiannidis, L.; Theriou, G. The role of firm-specific factors in the strategy-performance relationship: Revisiting the resource-based view of the firm and the VRIO framework. *Manag. Res. Rev.* **2018**, *41*, 46–73. [CrossRef]

77. Bobillo, A.M.; Rodríguez-Sanz, J.A.; Tejerina-Gaite, F. Corporate governance drivers of firm innovation capacity. *Rev. Int. Econ.* **2018**, *26*, 721–741. [CrossRef]

78. Yadav, P.L.; Han, S.H.; Kim, H. Sustaining Competitive Advantage Through Corporate Environmental Performance. *Bus. Strategy Environ.* **2017**, *26*, 345–357. [CrossRef]

79. Kitchenham, B.A. Systematic reviews. In Proceedings of the 10th International Symposium on Software Metrics (METRICS 2004), Chicago, IL, USA, 11–17 September 2004.

80. Fisch, C.; Block, J. Six tips for your (systematic) literature review in business and management research. *Manag. Rev. Q.* **2018**, *68*, 103–106. [CrossRef]

81. Tranfield, D.; Denyer, D.; Smart, P. Towards a Methodology for Developing Evidence-Informed Management Knowledge by Means of Systematic Review. *Br. J. Manag.* **2003**, *14*, 207–222. [CrossRef]

82. Karakaya, E.; Hidalgo, A.; Nuur, C. Diffusion of eco-innovations: A review. *Renew. Sustain. Energy Rev.* **2014**, *33*, 392–399. [CrossRef]

83. González-Moreno, Á.; Sáez-Martínez, F.J. Eco-innovation: Insights from a literature review AU—Díaz-García, Cristina. *Innovation* **2015**, *17*, 6–23.

84. Powell, K.R.; Peterson, S.R. Coverage and quality: A comparison of Web of Science and Scopus databases for reporting faculty nursing publication metrics. *Nurs. Outlook* **2017**, *65*, 572–578. [CrossRef] [PubMed]
85. Morioka, S.N.; de Carvalho, M.M. A systematic literature review towards a conceptual framework for integrating sustainability performance into business. *J. Clean. Prod.* **2016**, *136*, 134–146. [CrossRef]
86. Teixeira da Silva, J.A.; Memon, A.R. CiteScore: A cite for sore eyes, or a valuable, transparent metric? *Scientometrics* **2017**, *111*, 553–556. [CrossRef]
87. Peteraf, M.A. The cornerstones of competitive advantage: A resource-based view. *Strateg. Manag. J.* **1993**, *14*, 179–191. [CrossRef]
88. Freeman, R.E. *Strategic Management: A Stakeholder Approach*; Cambridge University Press: Cambridge, UK, 2015; pp. 1–276.
89. Waddock, S.A.; Bodwell, C.; Graves, S.B. Responsibility: The new business imperative. *Acad. Manag. Exec.* **2002**, *16*, 132–148. [CrossRef]
90. Hitt, M.A.; Ireland, R.D.; Hoskisson, R.E. *Strategic Management: Competitiveness & Globalization: Concepts*, 12th ed.; Cengage Learning: Boston, MA, USA, 2016.
91. Díaz-Fernández, M.C.; González-Rodríguez, M.R.; Simonetti, B. Top management team's intellectual capital and firm performance. *Eur. Manag. J.* **2015**, *33*, 322–331. [CrossRef]
92. Hyväri, I. Roles of Top Management and Organizational Project Management in the Effective Company Strategy Implementation. *Procedia Soc. Behav. Sci.* **2016**, *226*, 108–115. [CrossRef]
93. Dobni, C.B.; Sand, C. Strategy shift: Integrating strategy and the firm's capability to innovate. *Bus. Horiz.* **2018**, *61*, 797–808. [CrossRef]
94. Knight, G.A.; Cavusgil, S.T. Innovation, organizational capabilities, and the born-global firm. *J. Int. Bus. Stud.* **2004**, *35*, 124–141. [CrossRef]
95. Fernando, Y.; Wah, W.X.; Shaharudin, M.S. Does a firm's innovation category matter in practising eco-innovation? Evidence from the lens of Malaysia companies practicing green technology. *J. Manuf. Technol. Manag.* **2016**, *27*, 208–233. [CrossRef]
96. Muscio, A.; Nardone, G.; Stasi, A. How does the search for knowledge drive firms' eco-innovation? Evidence from the wine industry. *Ind. Innov.* **2017**, *24*, 298–320. [CrossRef]
97. Horbach, J.; Rammer, C.; Rennings, K. Determinants of eco-innovations by type of environmental impact—The role of regulatory push/pull, technology push and market pull. *Ecol. Econ.* **2012**, *78*, 112–122. [CrossRef]
98. Sanni, M. Drivers of eco-innovation in the manufacturing sector of Nigeria. *Technol. Forecast. Soc. Chang.* **2018**, *131*, 303–314. [CrossRef]
99. Wu, K.J.; Liao, C.J.; Chen, C.C.; Lin, Y.H.; Tsai, C.F.M. Exploring eco-innovation in dynamic organizational capability under incomplete information in the Taiwanese lighting industry. *Int. J. Prod. Econ.* **2016**, *181*, 419–440. [CrossRef]
100. Ma, D.; Ye, J.M.; Zhang, Y.R. Can Firm Exploit Economic Gainsfrom Eco-Innovation? An Empirical Investigation of Listed Companiesin China. *Transform. Bus. Econ.* **2018**, *17*, 33–53.
101. Portillo-Tarragona, P.; Scarpellini, S.; Moneva, J.M.; Valero-Gil, J.; Aranda-Usón, A. Classification and measurement of the firms' resources and capabilities applied to eco-innovation projects from a resource-based view perspective. *Sustainability* **2018**, *10*, 3161. [CrossRef]
102. Tsai, K.H.; Liao, Y.C. Sustainability Strategy and Eco-Innovation: A Moderation Model. *Bus. Strategy Environ.* **2017**, *26*, 426–437. [CrossRef]
103. Fernando, Y.; Wah, W.X. The impact of eco-innovation drivers on environmental performance: Empirical results from the green technology sector in Malaysia. *Sustain. Prod. Consum.* **2017**, *12*, 27–43. [CrossRef]
104. Ganapathy, S.P.; Natarajan, J.; Gunasekaran, A.; Subramanian, N. Influence of eco-innovation on Indian manufacturing sector sustainable performance. *Int. J. Sustain. Dev. World Ecol.* **2014**, *21*, 198–209. [CrossRef]
105. Park, M.S.; Bleischwitz, R.; Han, K.J.; Jang, E.K.; Joo, J.H. Eco-innovation indices as tools for measuring eco-innovation. *Sustainability* **2017**, *9*, 2206. [CrossRef]
106. Blum-Kusterer, M.; Hussain, S.S. Innovation and corporate sustainability: An investigation into the process of change in the pharmaceuticals industry. *Bus. Strategy Environ.* **2001**, *10*, 300–316. [CrossRef]
107. Galliano, D.; Goncalves, A.; Triboulet, P. Eco-Innovations in Rural Territories: Organizational Dynamics and Resource Mobilization in Low Density Areas. *J. Innov. Econ. Manag.* **2017**, *3*, 35–62. [CrossRef]
108. Ratten, V. Eco-innovation and competitiveness in the Barossa Valley wine region. *Compet. Rev.* **2018**, *28*, 318–331. [CrossRef]

109. Bossle, M.B.; De Barcellos, M.D.; Vieira, L.M. Why food companies go green? The determinant factors to adopt eco-innovations. *Br. Food J.* **2016**, *118*, 1317–1333. [CrossRef]
110. Tsai, K.H.; Liao, Y.C. Innovation Capacity and the Implementation of Eco-innovation: Toward a Contingency Perspective. *Bus. Strategy Environ.* **2017**, *26*, 1000–1013. [CrossRef]
111. Zhang, J.A.; Walton, S. Eco-innovation and business performance: The moderating effects of environmental orientation and resource commitment in green-oriented SMEs. *R D Manag.* **2017**, *47*, E26–E39. [CrossRef]
112. Severo, E.A.; de Guimaraes, J.C.F.; Dorion, E.C.H. Cleaner production, social responsibility and eco-innovation: Generations' perception for a sustainable future. *J. Clean. Prod.* **2018**, *186*, 91–103. [CrossRef]
113. Liao, Y.C.; Tsai, K.H. Innovation intensity, creativity enhancement, and eco-innovation strategy: The roles of customer demand and environmental regulation. *Bus. Strategy Environ.* **2018**, *28*, 316–326. [CrossRef]
114. Ryszko, A. Proactive Environmental Strategy, Technological Eco-Innovation and Firm Performance-Case of Poland. *Sustainability* **2016**, *8*, 156. [CrossRef]
115. Alhyasat, W.M.K.; Sharif, Z.M.; Alhyasat, K.M. The mediating effect of eco-innovation between motivation and organization performance in Jordan Industrial Estates Company in Jordan. *Int. J. Eng. Technol.* **2018**, *7*, 414–423. [CrossRef]
116. Aboelmaged, M. Direct and indirect effects of eco-innovation, environmental orientation and supplier collaboration on hotel performance: An empirical study. *J. Clean. Prod.* **2018**, *184*, 537–549. [CrossRef]
117. Marín-Vinuesa, L.M.; Scarpellini, S.; Portillo-Tarragona, P.; Moneva, J.M. The Impact of Eco-Innovation on Performance Through the Measurement of Financial Resources and Green Patents. *Organ. Environ.* **2018**. [CrossRef]
118. Scarpellini, S.; Marín-Vinuesa, L.M.; Portillo-Tarragona, P.; Moneva, J.M. Defining and measuring different dimensions of financial resources for business eco-innovation and the influence of the firms' capabilities. *J. Clean. Prod.* **2018**, *204*, 258–269. [CrossRef]
119. Gelter, M. The dark side of shareholder influence: Managerial autonomy and stakeholder orientation in comparative corporate governance. *Harv. Int. Law J.* **2009**, *50*, 129–194.
120. Li, L.; McMurray, A.; Xue, J.; Liu, Z.; Sy, M. Industry-wide corporate fraud: The truth behind the Volkswagen scandal. *J. Clean. Prod.* **2018**, *172*, 3167–3175. [CrossRef]
121. Sushandoyo, D.; Magnusson, T. Strategic niche management from a business perspective: Taking cleaner vehicle technologies from prototype to series production. *J. Clean. Prod.* **2014**, *74*, 17–26. [CrossRef]
122. Borthakur, S.; Subramanian, S.C. Optimized Design and Analysis of a Series-Parallel Hybrid Electric Vehicle Powertrain for a Heavy Duty Truck. *IFAC PapersOnLine* **2018**, *51*, 184–189. [CrossRef]
123. De Jesus Pacheco, D.A.; ten Caten, C.S.; Jung, C.F.; Ribeiro, J.L.D.; Navas, H.V.G.; Cruz-Machado, V.A. Eco-innovation determinants in manufacturing SMEs: Systematic review and research directions. *J. Clean. Prod.* **2017**, *142*, 2277–2287. [CrossRef]
124. Spedding, L.S. Chapter 16—Environmental due diligence and risk management: Sustainability and corporate governance. In *Due Diligence Handbook*; Spedding, L.S., Ed.; CIMA Publishing: Oxford, UK, 2009; pp. 535–641.
125. Wang, Y.; Liu, J.; Hansson, L.; Zhang, K.; Wang, R. Implementing stricter environmental regulation to enhance eco-efficiency and sustainability: A case study of Shandong Province's pulp and paper industry, China. *J. Clean. Prod.* **2011**, *19*, 303–310. [CrossRef]
126. Cheng, C.C.J.; Yang, C.L.; Sheu, C. The link between eco-innovation and business performance: A Taiwanese industry context. *J. Clean. Prod.* **2014**, *64*, 81–90. [CrossRef]
127. Laperche, B.; Picard, F. Environmental constraints, Product-Service Systems development and impacts on innovation management: Learning from manufacturing firms in the French context. *J. Clean. Prod.* **2013**, *53*, 118–128. [CrossRef]
128. Wasserbaur, R.; Sakao, T. Analysing interplays between PSS business models and governmental policies towards a circular economy. *Procedia CIRP* **2018**, *73*, 130–136. [CrossRef]
129. Wang, Y.-Y.; Wang, Y.-S.; Lin, T.-C. Developing and validating a technology upgrade model. *Int. J. Inf. Manag.* **2018**, *38*, 7–26. [CrossRef]
130. Lee, K.H.; Min, B. Green R&D for eco-innovation and its impact on carbon emissions and firm performance. *J. Clean. Prod.* **2015**, *108*, 534–542.
131. Jabbour, C.J.C.; Jugend, D.; de Sousa Jabbour, A.B.L.; Govindan, K.; Kannan, D.; Leal Filho, W. "There is no carnival without samba": Revealing barriers hampering biodiversity-based R&D and eco-design in Brazil. *J. Environ. Manag.* **2018**, *206*, 236–245.

132. Dong, Y.; Wang, X.; Jin, J.; Qiao, Y.; Shi, L. Effects of eco-innovation typology on its performance: Empirical evidence from Chinese enterprises. *J. Eng. Technol. Manag.* **2014**, *34*, 78–98. [CrossRef]
133. Andries, P.; Stephan, U. Environmental Innovation and Firm Performance: How Firm Size and Motives Matter. *Sustainability* **2019**, *11*, 3585. [CrossRef]

Case Report

An Incomplete Information Static Game Evaluating Community-Based Forest Management in Zagros, Iran

Mehdi Zandebasiri [1], **José António Filipe** [2,*], **Javad Soosani** [3], **Mehdi Pourhashemi** [4], **Luca Salvati** [5,6], **Mário Nuno Mata** [7,8] **and Pedro Neves Mata** [9,10]

[1] Department of Forestry, Faculty of Natural Resources, Behbahan Khatam Alanbia University of Technology, Behbahan 63616-47189, Iran; mehdi.zandebasiri@yahoo.com

[2] Department of Mathematics, ISTA—School of Technology and Architecture, University Institute of Lisbon (ISCTE-IUL), Information Sciences, Technologies and Architecture Research Center (ISTAR-IUL), Business Research Unit-IUL (BRU-IUL), 1649-026 Lisbon, Portugal

[3] Department of Forestry, Faculty of Agriculture & Natural Resources, Lorestan University, Khorramabad 6814-94414, Iran; soosani.j@lu.ac.ir

[4] Forest Research Division, Research Institute of Forests and Rangelands, Agricultural Research Education and Extension Organization (AREEO), Tehran 14968-13111, Iran; pourhashemi@rifr-ac.ir

[5] Council for Agricultural Research and Economics (CREA), Research Center for Forestry and Wood, Viale S. Margherita 80, I-52100 Arezzo, Italy; luca.salvati@crea.gov.it

[6] Global Change Research Institute of the Czech Academy of Sciences, Lipová 9, CZ-37005 České Budějovice, Czech Republic

[7] Lisbon Accounting and Business School Lisbon Polytechnic Institute, Avenida Miguel Bombarda 20, 1069-035 Lisbon, Portugal; mnmata@iscal.ipl.pt

[8] Polytechnic Institute of Santarém, School of Management and Technology (ESGTS-IPS), 2001-904 Santarém, Portugal

[9] ISTA—University Institute of Lisbon (ISCTE-IUL), 1649-026 Lisbon, Portugal; pedro_mata@iscte-iul.pt

[10] ESCS—Escola Superior de Comunicação Social, Instituto Politécnico de Lisboa, 1549-014 Lisbon, Portugal

* Correspondence: jose.filipe@iscte.pt

Received: 30 January 2020; Accepted: 23 February 2020; Published: 26 February 2020

Abstract: The present study adopts a game theory approach analyzing land-use planning in Zagros forests, Iran. A Static Game of Incomplete Information (SGII) was applied to the evaluation of participatory forest management in the study area. This tool allows a complete assessment of sustainable forest planning producing two modeling scenarios based on (i) high and (ii) low social acceptance. According to the SGII results, the Nash Bayesian Equilibrium (NBE) strategy suggests the importance of landscape protection in forest management. The results of the NBE analytical strategy show that landscape protection with barbed wires is the most used strategy in local forest management. The response to the local community includes cooperation in conditions of high social acceptance and noncooperation in conditions of low social acceptance. Overall, social acceptance is an adaptive goal in forest management plans.

Keywords: decision making; forest management; Nash Bayesian Equilibrium (NBE); Harsanyi's Transformation (HT)

---

1. Introduction

Forestry refers to the use of forest landscapes to achieve specific objectives [1]. Community forestry results from the management of woodlands when the aim is to create a specific benefit for the neighboring communities [2,3]. Forest management issues are crucial for environmental sustainability since they involve the intervention of several different stakeholders (e.g., the state, private companies, local forest users). Coordinating local commitments and engaging civil society to reconcile management asymmetry and build consensus can promote sustainable outcomes [4,5]. In these regards, a modeling approach, providing a set of interactive scenarios in forest management contributes to more effective policies addressing socioeconomic problems of Local Communities (LCs). For the adequate use of forest resources, communities should take part in their management. The idea of a Community Based Forest Management (CBFM) was first developed in the mid-1970s [4]. CBFM is a way of authorizing Local Communities (LCs) of a forest to participate in the decision-making processes [4]. This kind of forest managed by CBFM is called 'community forests', which provide economic and environmental services to more than half a billion LCs worldwide [6]. LCs are a source of a wide range of strengths but also threats for forest policy makers. LCs have goals that usually are diverse from the ones of official forest managers, in terms of Wood Products (WP)-fuel wood, bark or cork, and Non-Wood Forest Products (NWFP), as the case of mushrooms, goat feeding or fruits, for instance.

Identifying and determining LCs' goals is the basis for any attempt to manage forests, especially those in problematic socioeconomic situations. The choice of an adaptive strategy within LCs choices is one of the most influential decisions in CBFM. One option for modeling LC choices when designing future forest strategies is integrating already possible strategies for social values of management systems. It is remarkable to note that recent studies found a negative correlation between the formalization of forest monitoring and various measures of trust in LC strategies. Hence, the trust in LCs strategies can replace forest management rules [7]. It means that relying on LCs can represent a substitute for forest managers and personnel, reducing management costs if they are effectively utilized. LCs can themselves have a monitoring role in forestry practices.

Game Theory is a commonly used tool for analyzing community issues in order to attain a systematic approach and support for strategic situations. Game theory has been proposed for agents' decision making that is particularly relevant in conflict situations. This condition may occur when stakeholders are in competition with each other to attain certain goals [8,9]. CBFM is influenced by certain goals from forest managers and LCs. These goals can themselves be competing or conflicting. Hence, game theory is a suitable tool for analysis and understanding of forestry strategies. If used correctly, game theory can provide a refined analysis framework for successful policy implementation. Literature dealing with game theory models applied to forestry issues is relatively poor. For instance, Shahi and Kant [10] applied game theory to Joint Forest Management (JFM) in India, proposing a specific game for JFM modeling with four evolutionary strategy equilibriums: cooperators, defectors, defectors-enforcers, and cooperators-enforcers. The results from this study showed that this game has two asymptotically strong equilibriums: defectors-enforcers equilibrium, and cooperators-enforcers equilibrium. Mohammadi Limaei and Lohmander [11] adopted a dynamic game to analyze the timber market in northern Iran. The dynamic features of timber prices derived from the game model were finally applied to estimate the price chain in Iran. Mohammadi Limaei [12] applied a two-person non-zero-sum game to the pulpwood market in Iran.

Soltani et al. [13] applied the game theory between the LCs and state authorities in the Zagros forest (Iran). The results of this study show that a non-cooperative game exists between LCs and the executive management of that forest. Zandebasiri et al. [14] designed a Static Game of Complete Information (SGCI) for Zagros forest modeling the relationship between executive management and LCs. These authors found that the SGCI model has two Nash Equilibriums (NEs) in a composition of conservation of forest resources with a lack of corporation of the LCs and composition of precipitation storage with the corporation of the LCs. Many of the basic ideas for the game adopted in this paper are similar to the one of Zandebasiri et al. [14]. Also, a recently developed Static Game of Incomplete Information (SGII) method was used to analyze the community's forest management. In SGII, one of the players does not know the payoffs of the other players' strategies. Although game theory is an established method in forest management and policy analysis [10,12–14], to the best of our knowledge, our study is the first to apply SGII to forest management.

We applied this methodology to a particularly complex ecosystem in Zagros (Iran), regarded as an example of forest landscape has been heavily managed by executive management efforts [15,16], but it needs to develop the involvement of local communities in forest management decisions. The justification or significance of this study is providing principles consistent with local communities and social forestry. The results of this study can not only be useful for the management of local communities, but also for the management of drought and dieback in the forest. Everything in forest management depends on decisions [17,18], and the decisions determine the future of the forests. Based on the results of the game theory method in this research, we present the best options for forest managers.

The main socio-environmental issues in Zagros forests are the socioeconomic problems of local communities [19], the dependency of LC to forest resources [20], cattle overgrazing in woodland [21,22] and the consequent forest degradation [22,23]. For these reasons, the role of local communities in these forests is vital. The related ecosystem needs participatory management, enhancing opportunities and avoiding threats in forest utilization [24]. One of the main problems that LCs use to face is cattle overgrazing in forest land [22]. This issue conflicts with the technical demands of forest managers. The executive management will encircle these forests with barbed wire [25]. Hence, one of the main issues in this theme is forest conservation [16]. Additionally, for about eight years (from 2011 onwards), forest die-back has been the most widespread problem in Zagros [14], particularly in the aridest districts. To face these problems, the executive management of Zagros forests was increasingly oriented toward two strategies: (i) forest conservation and (ii) precipitation storage [15].

At the same time, local communities have employed informal responses to such problems: (i) cooperation, and (ii) non-cooperation with executive management strategies [14]. Based on these premises, it is clear how all executive management decisions and strategies are directly related to LCs response and to the social acceptance of such programs [16]. Local residents have traditional knowledge that may contribute significantly to effective forest management. In this context, the response of each local community's activity creates new information for the executive management of forests, and each of the executive management decisions causes new LC feedback [20]. According to such dynamics, a game is proposed here between the executive management and LCs using a Static Game of Incomplete Information (SGII) model. The model was built-up to specifically focus on two objectives of forest management in Zagros: (i) reducing the pressure of resident livestock on forest landscape and (ii) enhancing precipitation storage to cope with droughts and containing oak forest degradation.

2. Methodology

2.1. Game Theory

Game theory provides models for decision making for situations in which decision makers show interdependence and reciprocity. The condition when a decision-making process is affected by decision-makers where each decision relates to another decision is a process called the game [14]. In these situations, the suitable alternative for policy-making depends on other alternatives affecting decision-makers. For quantifying the policy-making, a payoff matrix is formed. Table 1 shows a payoff matrix for a two-player game.

Table 1. A payoff matrix for a 2-player game.

	S_{Y1}	S_{Y2}
S_{X1}	(x_{11}, y_{11})	(x_{12}, y_{12})
S_{X2}	(x_{21}, y_{21})	(x_{22}, y_{22})

S_{X1} is the first strategy and S_{X2} is the second strategy of the player. Strategies of player Y are defined in the same way. Numbers (x_{11}, y_{11}) denote that if the combination of S_{X1} and S_{Y1} strategies is applied by the two players, the benefit of players X and Y is equal respectively to x_{11} and y_{11}. In this game, problem answers are called Nash Equilibriums (NEs) of the game. The above matrix is valid for a two player game. In general, if a game has n players is presented as an n-player game. The above matrix has two strategies for each player. In general, when a game has n strategies for each player, it is represented by a non-matrix. The first step consists of presenting the game. A set of players in an N-player game can be represented by $N = \{1, 2, 3, \ldots, n\}$. Strategies of the i-th player can be represented by $S_i = \{A_1 * A_2 * A_3 * \ldots * A_n\}$, where A1, ... , An are n different strategies, $i \in N$. The rows and columns show the strategies of each player. The rows of this matrix depend on the number of strategies of one player and the column of this matrix depends on the number of strategies of the other player(s).

Games having simultaneous decisions among players are called Static Games (SGs), and games with the one-after-another decisions are called Dynamic Games (DGs). SGs are divided in Static Games of Complete Information (SGCI) and Static Games of Incomplete Information (SGII). When a player knows the payoffs of the other player(s), the game is called SGCI. On the contrary, in SGII, one (or more) player(s) does (do) not know the payoffs of strategies of the other players. Equilibriums of the SGCI and SGII are called, respectively, Nash Equilibrium (NE) and Nash Bayesian Equilibrium (NBE) [8,9].

2.2. Study Area

"Zagros forests" is the name for a vast system of forest landscapes in northwest, west, and southwest of Iran. This region contains 5 million hectares of forests (44 percent of Iran's forest area). The area of these forests had already contained 10 million hectares long ago. Due to the intense exploitation of word products, the forest area decreased to five million hectares in the present days. Brant's oak (*Quercus brantii* Lindl) is the dominant species in the landscape [26–28]. The study area includes the Watershed of Tang-e Solak in the province of Kohgiluyeh, South-western Iran. Tang-e Solak forest is located in Likak (Bahmaee) city in Kohgiluyeh province and includes fragmented cropland producing wheat and barley. In this area, natural cypress trees and oak stands form a unique stand of cypress-oak [29]. Local populations living in this region have progressively moved to urban areas in search for job, facilities, and higher income. Forest landscapes in Zagros are legally state-owned. Zagros forests were declared state forests by the law in 1962 [1]. However, the executive management of forestry projects had many conflicts with LCs in this forest. More specifically, there is a considerable dependency between LCs and the forest because of the traditional exploitation of forest production by LCs, with consequent degradation of forest structure through informal cuts of wood. As a consequence, the Zagros forest has been extensively coppiced [23,30]. In the last decade, a new issue was observed in the Zagros forest, since an extensive oak die-back occurred [31]. Arabic dust coming from neighboring Arab countries, climate change, and the intense summer heat, seem to be the main reason for this die-back [14]. This phenomenon requires specific planning to reduce the risk of forest die-back.

In the current situation, executive management has applied the Forest Management Plan (FMP) for the region with a focus on forest resources conservation [14]. By 1998, the management unit for Zagros forests concerned large watershed areas, but since then, the management unit in these forests changed to the level of customary rural development to include LCs [32]. In order to fix a sustainable utilization rate for local communities, the goal was to reach participatory planning decisions in line with the traditional forest related knowledge [33]. The investigated spatial level of the system was reduced to a specific area representative of the whole system: in our case, a watershed district was selected to model executive management strategies. The scores of each strategy were assumed to be variable in different local contexts depending on social and environmental conditions.

Although the intervention scale in Zagros forests can be broadened, it is a concern that the principle of unity of command—which is one of the principles of management [18]—may be lost at a larger scale. Accordingly, the present study was carried out with the aim to consider the level of decision-making at the level of the customary system rather than focusing on larger scales.

2.3. Selection of the Type of Game

This study is part of a larger research in Zagros forests of Iran on game theory methods and its application in participatory management. In this series of researches, we seek to explore the application of various types of game theory and its applications in the optimal management of local communities for forest conservation. To examine the relationship between local communities and executive management, for the first time in 2017, we used SGCI for this concept [14]. For this study, we selected the SGII method and in future researches; we will intend to use dynamic gaming models.

The questions that will be asked to experts to each method differ from the other method once each method needs separate data. Hence, in this study we can only use SGII method because the assumptions of the questions asked to the experts and the assumptions of the analyst team in this study were only assumptions of the SGII method.

2.4. Model Set-Up and Calculation

The most important issue in managing Zagros forests is the participation of local people. The most important reason for the failure of past projects in Zagros forests was the lack of social acceptance of forest planning [5,7,28,30]. Therefore, scenarios of social acceptance were defined by local communities. The reason for constructing scenarios is to focus on the issue of social acceptance in Zagros forestry projects. Based on these assumptions, two modeling scenarios were defined for the SGII: (1) a High social acceptance scenario (H) and (2) a Low social acceptance scenario (L). The high social acceptance scenario was used to illustrate forest sites that have appropriate social conditions. These forests may have few socioeconomic problems. Conversely, the low social acceptance scenario was applied to illustrate forest sites that have inappropriate social conditions.

These two scenarios have relevant consequences on computation. If social acceptance of projects increases, the success of the executive management programs will be highly probable, while if social acceptance of plans is reduced, the executive management plans will fail. Each of these scenarios may occur when executive management impacts FMPs. Consequently, SGCI cannot completely evaluate this game. Executive management can predict only the Stakeholder Analysis (SA) of the FMP. The SA of the FMP relates to land-use conflicts, the economic condition of local communities, and planning flexibility to the living standards of the area. Thus, model inputs were two scenarios with the SGII and prior SGCI. The purpose of applying the game theory of game in this section is to assess if the result of the implementation of executive management programs is in contrast to the local communities' practices in forest management.

2.5. Building Up the Payoff Matrix

A separate matrix was built-up for each scenario. Priority values were determined by asking three experts who were specifically selected for the estimation of payoffs of the game to answer the question of what privileges the expert had given to the emergence of different combinations of local community and executive management strategies. The higher the score, the more likely it will happen. Average values in each cell are considered as the number of matrices per cell. Likert scale numbers were calculated based on the mean of experts' opinions in each cell. The model's calculation was based on the SGII algorithm. Because of the complexity of the SGII model and combination of strategies in two conditions (i.e., high and low social acceptance), three experts were specifically selected for the estimation of payoffs of the game. The benefits of the different strategies for key players were evaluated using a Likert scale. Expert's opinions were used at this stage. The average score of each strategy was calculated based on the opinions of experts in the form of a Likert scale. Since the application of SGII is new, a short explanation of this theory is presented below. The normal form of the SGII for n players is as follows [8]:

$$N = \{1, 2, 3, \ldots, n\} \tag{1}$$

where n is the number of the total of players.

$$A_i = \{a_1, a_2, a_3, \ldots, a_n\} \tag{2}$$

where Ai is the collection of strategies of each player.

$$S = \{A_1 * A_2 * A_3 * \ldots * A_n\} \tag{3}$$

where S is the combination of strategies of all players. Strategies for each player are divided into Pooling Strategies (PSs) and Separating Strategies (SSs).

$$T_i = \{t_1, t_2, t_3, \ldots, t_n\} \tag{4}$$

where T_i is the set of conditions for each player.

$$T = \{T_1 * T_2 * T_3 * \ldots * T_n\} = \{t : (t_1, t_2, t_3, \ldots, t_n), t_1 \in T_1, \ldots, t_n \in T_n, t \in T\} \tag{5}$$

where T is the set of conditions for the game.

$$P_r = P_i(t_{-i}|t_i) \forall t_i \in T_i, \forall t_{-i} \in T_{-i}, \forall i \in N \tag{6}$$

where P_r is a prior belief for each player and T_{-i} is a set of conditions for competitor players.

$$u_1\left(q_1, q_2 \middle| (c, c_H)\right) = \left[\begin{matrix} (a - c - (q_1 + q_2(c_H)))q_1 \, if \, Q < a \\ -cq_1 \, if \, Q \geq a \end{matrix} \right] \tag{7}$$

where u_1 is the payoff of player 1, $q1$, and $q2$ are values of payoff matrix for player 1 and 2, respectively, $T_1 = \{c\}, T_2 = \{C_l, C_h\}$ and $q_2(c_H)$ is the value of payoff matrix for player 2 in H condition.

$$u_1\left(q_1, q_2 \middle| (c, c_L)\right) = \left[\begin{matrix} (a - c - (q_1 + q_2(c_L)))q_1 \, if \, Q < a \\ -cq_1 \, if \, Q \geq a \end{matrix} \right] \tag{8}$$

where $q_2(c_L)$ is the value of payoff matrix for player 2 under L condition.

$$u_2\left(q_1, q_2 \middle| (c, c_L)\right) = \left[\begin{matrix} (a - c_L - (q_1 + q_2)q_1 \, if \, Q < a \\ -c_L q_2 \, if \, Q \geq a \end{matrix} \right] \tag{9}$$

$$u_2\left(q_1, q_2 \middle| (c, c_H)\right) = \left[\begin{matrix} (a - c_H - (q_1 + q_2)q_1 \, if \, Q < a \\ c_H q_2 \, if \, Q \geq a \end{matrix} \right] \tag{10}$$

where u_2 is the payoff of player 2 in Equations (9) and (10).

The extensive form of SGII is called Harsanyi's Transformation (HT). HT is the first step in the resolution of SGII. An artificial player is defined in HT that will start the game by defining conditions of each player. The graphical model of HT is shown in Figure 1.

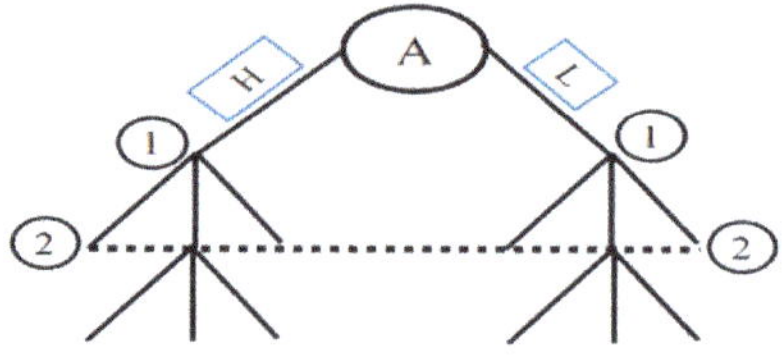

Figure 1. The graphical model of Harsanyi's Transformation (HT).

In Figure 1, signal A denotes the artificial environment to start the game, H and L are high and low SA conditions, and (1) and (2) are game players; player 2 does not know for sure that player 1 is H or L. HT was formed before Posterior belief (P_0) is calculated for players. P_0 is calculated from the below equation according to Bayes' rule:

$$P_0(t_{i-1}|t_i) = \frac{P(t_{-i}, t_i)}{P(t_i)} = \frac{P(t_i)}{\sum_{t_{-i} \in T_{-i}} p(t_{-i}, t_i)} \tag{11}$$

where P_0 is a posterior belief, $t = \{t_1, t_2, t_3, \ldots, t_n\} \forall t_i \in T_i$.

NBE is calculated from a combination of below strategies following the calculation of P_0:

$$NBE = \max_{s_i \in S_i} \sum u_i(s_1^*(t_1), \ldots, s_i, s_{i+1}^*(t_{i+1}), \ldots, s_n^*(t_n)|t_i, t_{-i})P(t_{-i}|t_i) \tag{12}$$

where $s^* = (s_1^*(t_1), \ldots, s_n^*(t_n)) \in S$

$$N = \{\text{Local resident, Executive management}\} \tag{13}$$

Executive management and the local community are the main stakeholders in participatory management of the Zagros forests [19].

$$S_1 = \{C, N\}, S_2 = \{E, S\} \tag{14}$$

C and N are acronyms of Cooperation and Non-cooperation of local community. E and S are acronyms of encirclement and saving rainfall strategies of executive management, respectively. Encirclement is applied for protection of forests with barbed wire. Saving rainfall is a strategy to preserve water under oak degradation.

$$S = S_1 * S_2 = \{(C, E), (C, S)(N, E), (N, S)\} \tag{15}$$

$$T_1 = \{H, L\}, T_2 = \{x\} \tag{16}$$

$$T = T_1 * T_2 = \{(H, x), (L, x)\} \tag{17}$$

$$P_1(x|H) = P_1(x|L) = 1, P_2(L|x) = 1 - P, P_2(H|x) = 1 - p \tag{18}$$

3. Results

Results for payoff matrixes for SGII are shown in Table 2.

Table 2. A payoff matrix for P probability and H condition (panel A) or L condition (panel B).

	E	S
C	(3,3)	(3,2)
N	(1,3)	(1,1)
(a)		
	E	S
C	(1,3)	(2,1)
N	(3,3)	(2,1)
(b)		

Table 2a shows that in the high social acceptance scenario, forest conservation can perform very well, and it acquires a high score. Table 2b shows that in the low social acceptance scenario, forest conservation cannot perform very well and score high; however, forest conservation is a key issue in the Zagros forests. The graphical model of HT for the game is shown in Figure 2.

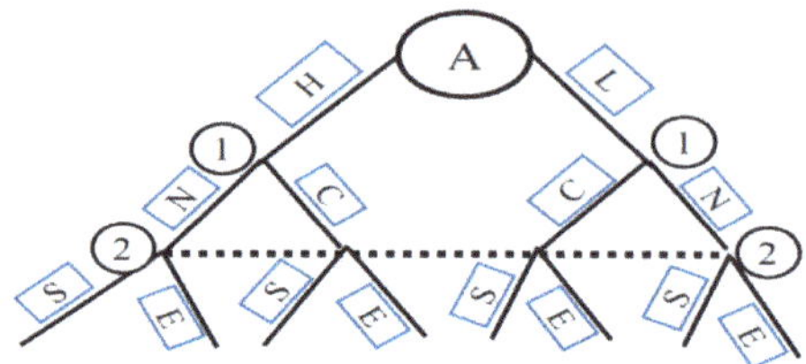

Figure 2. The graphical model of Harsanyi's Transformation for the selected game.

Interaction of strategies in the normal form of SGII is shown in Table 3.

Table 3. Interaction of strategies in the normal form of SGII.

	E	S
CC	$(2P+1,3)$	$(P+2,P+1)$
CN	$(3,3)$	$(P+2,P+1)$
NC	$(1,3)$	$(2-P,1)$
NN	$(-2P+3,3)$	$(-P+2,1)$

This table is the basis of the difference between the results of SGCI method and the SGII method in game theory that numbers in Table 3 were extracted from Equation (11).

$$U_1(CC,E) = 3P + (1-P) = 2P + 1 \tag{19}$$

The P number is a number for the probability of the H state in Figure 2 which occurs with high social acceptance for executive management programs.

$$U_2(CC,E) = 3P + 3(1-P) = 3 \tag{20}$$

The $(1-P)$ number is a number for the probability of the L state in Figure 2 which occurs with low social acceptance for executive management programs.

$$U_1(CC,S) = 3P + 2(1-P) = P + 2 \tag{21}$$

$$U_2(CC,S) = 2P + (1-P) = P + 1 \tag{22}$$

Number 3 represents the highest number among the results in these equations. The results of Equations (21) and (22) can only be approximated to the number 3 when the P number is equal to 1. These situations occur only when social acceptance is 100% probable. The S strategy (saving rainfall) is very important for the Zagros forests because these forests are facing drought today. If Zagros Forests Management wants to focus on the benefits of this strategy and shift the equilibrium of the game to this strategy, they must plan for social acceptance.

$$U_1(CN,E) = 3P + 3(1-P) = 3 \tag{23}$$

$$U_2(CN,E) = 3P + 3(1-P) = 3 \tag{24}$$

Equations (23) and (24) earned high scores for both executive management and local communities because number 3 represents the highest number among the results in these equations. When both executives and local communities have a high score, the equilibrium of the game can be here.

$$U_1(NC,E) = P + (1-P) = 1 \tag{25}$$

$$U_2(NC,E) = 3P + 3(1-P) = 3 \tag{26}$$

In Equation (25), the interests of local communities have been diminished, as only forest conservation reduces their interests. However, the score of management processes is raised in Equation (26).

$$U_1(NN, E) = P + 3(1 - P) = -2P + 3 \tag{27}$$

$$U_2(NN, E) = 3P + 3(1 - P) = 3 \tag{28}$$

Similar to Equations (25) and (26) the interests of local communities have been diminished in forest conservation.

$$U_1(CN, S) = 3P + 2(1 - P) = P + 2 \tag{29}$$

$$U_2(CN, S) = 2P + (1 - P) = P + 1 \tag{30}$$

Equations (29) and (30) show that if local communities do not cooperate, the scores for both executive management and local communities will decrease.

$$U_1(NC, S) = P + 2(1 - P) = -P + 2 \tag{31}$$

$$U_2(NC, S) = P + (1 - P) = 1 \tag{32}$$

The results of Equations (30) and (31) show low numbers $((-P + 2)$ numbers and 1 indicate this) because of the noncooperation approach for the first operation in the above equations.

$$U_1(NN, S) = P + 2(1 - P) = -P + 2 \tag{33}$$

$$U_2(NN, S) = P + (1 - P) = 1 \tag{34}$$

In the non-cooperative state of local communities, the benefits of both executive management and local communities will be reduced. $(-P + 2)$ numbers and 1 indicate this.

Accordingly, BNE is calculated after defeated strategies were erased.

$$NBE = (CN, E) \tag{35}$$

According to Equations (19)–(34), NBE strategies illustrated in Equation (35) were imported to SGII equilibrium for participatory management in Zagros forest. Participatory management can include different roles, such as executive roles and decision roles. NBE strategies demonstrate the importance of the E parameter and C and N strategies. The (CN, E) strategy is a strategy through which neither the local community nor the executive government wants to change because if the strategy changes, the payoff in the game will reduce. We found that this balance, which is achieved by integrating the demands of LCs and executive management-with a focus on the enclosure of forests-can be a source of conflicts in the implementation of Zagros forest management plans. Equation (19) and next ones until 35 allow calculating the payoff of other situations. Reaching a new equilibrium that would make precipitation storage more pronounced depends on further examination of native soil conservation methods. This result suggests that precipitation storage programs have not yet found a substantial role in FMPs, likely because of the lack of appropriate allocation of funds for the design and construction of this system. In an NBE state, LCs use a strategy that has the optimal response to selected strategies of the executive management and does not achieve the greatest outcome; instead, they use the best strategy related to decisions of another player [13].

4. Discussion

Compared to other game theory analyses, the model proposed in this study is new in forest management. Earlier studies have also emphasized the applications of this theory in forest economics [11,12]. For this reason, forest management knowledge needs basic definitions for the application of this theory to forest management. The static game theory developed in Zagros forests has practical applications, whose effectiveness depends on our understanding of mechanisms underlying the Nash equilibrium. More specifically, what forest manager's use in competitive situations depends on the Nash equilibrium? Representing theoretical work formulating a model and commenting on the basic characteristics of the model itself, our study takes advantage of game theory, and the conceptual expansion of this model can be a major innovation for ecosystem management. If developed in a wider model, this theory can be applied to forests under different socio-environmental conditions that can optimize forest management decisions. For instance, local communities in the study area have goals that usually relate to animal feeding, and especially goat feeding, while having no specific goals related to bark collection or cork production. In this landscape, there are very relevant social problems since forest products contribute to LCs subsistence, and thus local residents have an important role in forest management. It should be noted, however, that Zagros forest personals and managers have made great efforts to protect these forests [15,16], but a participatory management approach with local communities must be defined in the decision-making phase for these forests to prevent forest die-back in these landscapes.

Analysis of the NBE shows that E parameter (Encirclement or protection of forests by barbed wires) represents the most used strategy of executive management. The main responses of the local community to this strategy are cooperation in high SA conditions and noncooperation in low SA conditions. According to Shahi and Kant [10] and Soltani et al. [13], the cooperation of LCs has much influence on the game. NBE results present an unfavorable equilibrium in Zagros forests. The attention of stakeholders preferences [34], especially LCs [21], using the potential of forests, land-use planning, and site selection for sustainable development [35] can contribute to NBE strategies aimed at reaching sustainable forest management. In this study, we highlighted the role of the SA factor. SA is an adaptive goal with LCs in FMP. The core concept of forest management in Zagros forests emphasizes the goal-setting in FMP [21]. Goal setting is very important in the process of FMP formulation.

Examples of strategic conditions for forest management in Zagros include: (i) low tree production, (ii) forest structure degradation, (iii) overgrazing [36], (iv) soil erosion, and (v) oak decline [29,37]. The solution to most of these issues is grounded on the traditional agroforestry system existing in Zagros. The concept of agroforestry should be redefined for Zagros forests. Agroforestry is a combination of forestry components with other land-uses for the purpose of land-use planning [22]. In particular, grazing management plans can include organizing the number and timing of livestock in the forest. According to Table 3, SA represents the core of the Zagros forest management, outlining the importance of physical protection and encirclement for executive management. According to assumptions in Table 2, and considering few payoffs for saving rainfalls strategies it can be stated that precipitation saving strategies are not included in NBE. It is important to recognize that rainfall saving is not useful for executive management and has been little considered in its adoption strategies.

The results of this study are very sensitive to the early data collected using a Likert scale. In the current context, executive management has not yet clearly identified many aspects of the technical issues of local communities and their traditional knowledge, which can be seen as the main issue in the interaction between local communities and executive management programs. Contrary to earlier research in this field [14], an SGII method was used in this study. The SGII method is also a way of solving problems in game theory for maximizing the expected payoff of the players by considering the selective strategy of the other players. This is a difference between SGCI and SGII. SGII beliefs become quantities according to Bayes theorem (Equation (11) and Table 3), being an advantage for SGII methods. In SGII methods, the belief of the players indicates the probability of the type of strategy that players can play in the environment of the play (Tables 2 and 3). In a SGII strategy, an artificial player is defined (A symbol in Figure 2). This symbol actually defines the environment of chances and specifies the type or state of each player. SGII methods, players choose their strategies simultaneously, but at least one player does not have enough information about the payoff of the game. In forest management, this is much closer to the real world because, in forest management, uncertainty is always a very important issue for forest managers [32]. Until now, most of the uncertainty sources of forest management were ecological and economic [18]. To the best of our knowledge, this is the first time that social uncertainty is defined with an SGII model for forest management.

Our study illustrates an SGII method that is applied to forest management. Several types of game theory, such as an evolutionary game, dynamic game, and static game, including SGCI and SGII, are used to support the decision in forest management and economics, considering two types of agents (executive management and the local community). In this game, the local community has private information about something relevant to its decision making (social acceptance), but the executive management does not know that information. In this situation, it is necessary to model the game by SGII. The SGII contributes to clarify and open the information of a game that has already been hidden. Further approaches should implement a refined model of participatory management with multiple stakeholders such as research institutes, agricultural offices, NGOs, media, and other stakeholders. Further investigation of the performance of other games, like the one proposed by Mohammad Limaei in dynamic game [12], seems to be an appropriate research issue.

Hence, another issue that needs further investigation is the dynamics of the designed game. Dynamic game is another dimension of game theory in which players decide sequentially. In dynamic games, each player makes his own choice after the previous player. In this study, we adopted a static game because the game theory was only recently applied to the complex management of the world's forests by modeling its different branches. Examples of static games in forest management were recently provided [14]. Future studies should be deserved to model forest management under dynamic games. In the dynamic model, the following set of parameters, such as in a static game, are needed:

(1)   Players identification,
(2)   Set the strategies of each player
(3)   Players payoffs functions for each strategy combination

In dynamic games, there are a few additional points to consider [9]:

(1)   When does each player move?
(2)   What information does each player have on the move?
(3)   What can each player do when he moves?
(4)   What are the payoffs for players for each sequence of moves?

As such, we need the initial node, nature, decision nodes, terminal node, branches, predecessors, successors, and path of the game in the dynamic state. The initial node shows the start of the game. Decision nodes show the represent the starting point of each player's decision. The terminal node shows the final of the game. Multiple branches can originate from any non-terminal node. The nodes before a given node are called the predecessors, and the nodes after a given node are called the successors. Each path of the game is a sequence of nodes and branches. This creates the first difference in the graphical model for the dynamic game. The dynamic game model for Zagros forest management and their local communities can be designed as follows (Figure 3).

In this representation, the FMP management is assumed as the executive management of Zagros forests. So, logic dictates that the left panel in Figure 3 is the most reasonable representation of a dynamic game model in Zagros. Like static games, dynamic games can be a game with complete or incomplete information. As such, we face complex and wide-ranging situations in design that are best modeled individually. The main purpose is to exploit the benefits of this theory in forest management separately [14]. In dynamic games with complete information conditions, Backward Induction Nash equilibrium can be preferentially used, while in dynamic games with incomplete information conditions, perfect Bayesian Nash equilibrium can be used. Apart from these considerations, it is worth mentioning that forest systems are typically the most stable land-use, which is finally traced by static games representing complexity without providing excessively complex theory and controls for users.

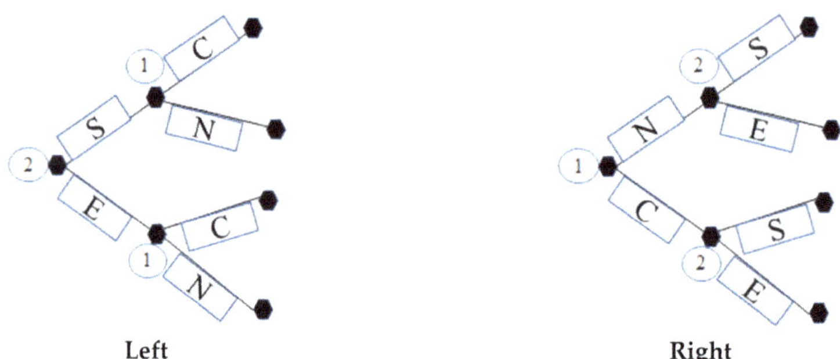

Figure 3. The graphical model of the Zagros forests' management of the dynamic game in a situation where (**left**) executive management or (**right**) local communities start the game.

5. Conclusions

Sustainable forest management benefited from the use of Multi-Criteria Decision Making (MCDM) methodologies in decision-making processes. In MCDM, what is important is decision-making by decision-makers. The real world of decision-making does not only depend on the decision-maker, but strictly depends on the other players' reaction to the decision situation. Game theories try to study these reactions. Many of the real-world games should be handled in incomplete information, especially in forest management. The performance of SGII is superior because of the high uncertainty in both climatic and social conditions. Our study finally demonstrates that the game between the executive management and the LCs in Zagros was not in the form of a cooperative game. Further examination of evolutionary and collaborative games may contribute greatly to a refined understanding of these deserving issues for forest management.

Author Contributions: Conceptualization, M.Z., J.A.F., J.S. and M.P.; methodology, M.Z. and M.N.M.; Validation, M.Z.; formal analysis, M.Z.; investigation, M.Z., J.S., M.P.; resources, M.Z and J.A.F.; data curation, M.Z.; writing—original draft preparation, L.S., J.A.F. and P.N.M.; writing—review and editing, L.S. J.A.F. and M.Z..; visualization, M.Z, J.S., M.P.; supervision, M.Z.; project administration, M.Z, J.S., M.P.; funding acquisition, J.A.F. and L.S., coordinate with correspondent author, P.N.M. and M.N.M. All authors have read and agreed to the published version of the manuscript.

Funding: Dr. Zandbassiri's doctoral thesis in Lorestan University, Iran was a static game with complete information (SGCI method) on oak decline. This article is out of his thesis, but more on that later a game with incomplete information (SGII method) is designed as an independent research in Behbahan Khatam Alanbia University of Technology (BKAUT). The charge for APC of this article was also paid by Isteki University, Portuguese.

Acknowledgments: The authors thank the Behbahan Khatam Alanbia University of Technology and Lorestan University, for their support. They also thank the Instituto Universitário de Lisboa and ISTAR-IUL, for their support. The authors acknowledge Zagros forest experts to their cooperation with this research. They finally thank of three anonymous reviewers for their valuable comments.

Conflicts of Interest: The authors declare no conflict of interest.

References

1. Ebrahimi Rostaghi, M. The role of policy-making and decision-making in protection of outside North forests. In Proceedings of the Conference on Protection of Forests in Sustainable Forest Management, Tehran, Iran, 11–13 October 2004; Iranian Society of Forestry: Tehran, Iran, 2005; pp. 137–151.
2. Anup, K.C.; Koirala, I.; Adhikari, N. Cost Benefit Analysis of a Community Forest in Nepal. *J. Sustain. For.* **2015**, *34*, 199–213.
3. Anup, K.C. Community Forest Management: A Success Story of Green Economy in Nepal. *J. Environ. Sci.* **2016**, *2*, 148–154.

4. Marasenia, T.R.; Bhattaraib, N.; Karkyb, B.S.; Cadmanc, T.; Timalsinab, N.; Bhandarib, T.S.; Apana, A.; Mad, H.O.; Rawate, R.S.; Vermae, N.; et al. An assessment of governance quality for community-based forest management systems in Asia: Prioritization of governance indicators at various scales. *Land Use Policy* **2019**, *81*, 750–761. [CrossRef]

5. Riggs, R.A.; Langston, J.D.; Margules, C.; Boedhihartono, A.K.; Lim, H.S.; Sari, D.A.; Sururi, Y.; Sayer, J. Governance Challenges in an Eastern Indonesian Forest Landscape. *Sustainability* **2018**, *10*, 169. [CrossRef]

6. Ellisa, E.A.; Monteroa, S.A.; Gómezb, I.U.H.; Monteroc, J.A.R.; Ellisd, P.W.; Rodríguez-Warde, D.; Reyesf, P.B.; Putzg, F.E. Reduced-impact logging practices reduce forest disturbance and carbon emissions in community managed forests on the Yucatán Peninsula, Mexico. *For. Ecol. Manag.* **2019**, *437*, 396–410. [CrossRef]

7. Kahsay, G.A.; Bulte, E. Trust, regulation and participatory forest management: Micro-level evidence on forest governance from Ethiopia. *World Dev.* **2019**, *120*, 118–132. [CrossRef]

8. Abdoli, G. *Game Theory and its Applications (Incomplete Information, Evolutionary and Cooperative Games)*; Samt Publication: Tehran, Iran, 2017. (In Persian)

9. Abdoli, G. *Game Theory and its Applications (Static and Dynamic Game with Complete Information)*; Jahad daneshgahi Publication: Tehran, Iran, 2019. (In Persian)

10. Shahi, C.; Kant, S. An evolutionary game-theoretic approach to the strategies community members under Joint Forest Management regimes. *For. Policy Econ.* **2007**, *9*, 763–775. [CrossRef]

11. Mohammadi Limaei, S.; Lohmander, P. A game theory approach to the Iranian forest industry raw material market. *Casp. J. Environ. Sci.* **2008**, *6*, 59–71.

12. Mohammadi Limaei, S. Mixed strategy game theory, application in forest industry. *For. Policy Econ.* **2010**, *12*, 527–531. [CrossRef]

13. Soltani, A.; Sanskhayan, P.L.; Hofstad, O. Playing forest governance games: State-village conflict in Iran. *For. Policy Econ.* **2016**, *73*, 251–261. [CrossRef]

14. Zandebasiri, M.; Soosani, J.; Pourhashemi, M. Evaluating existing strategies in environmental crisis of Zagros forests of Iran. *Appl. Ecol. Environ. Res.* **2017**, *15*, 621–631. [CrossRef]

15. Iranian Forests, Rangelands and Watershed Organization. *Guidelines for Sustainable Forest Ecosystem Management in Order to Prevent the Oak Mortality*; Forests, Rangelands and Watershed Organization Publication: Tehran, Iran, 2013. (In Persian)

16. Jazirei, M.H.; Ebrahimi Rostaghi, M. *Silviculture in Zagros Forests*; University in Tehran Press: Tehran, Iran, 2011. (In Persian)

17. Zandebasiri, M.; Pourhashemi, M. The place of AHP among the multi criteria decision making methods in forest management. *Int. J. Appl. Oper. Res.* **2016**, *6*, 75–89.

18. Zandebasiri, M.; Hoseini, S.M. *Sustainable Forest Management (SFM)*; Jahad Daneshgahi Press (Mazandaran Branch): Sari, Iran, 2019. (In Persian)

19. Imani Rastabi, M.; Jalilvand, H.; Zandebasiri, M. Assessment of socio-economic criteria and indicators in monitoring of Kalgachi Lordegan forest management plan. *Iran. J. For. Poplar Res.* **2005**, *23*, 1–8. (In Persian)

20. Ghazanfari, H.; Namiranian, M.; Sobhani, H.; Mohajer, R.M. Traditional forest management and its application to encourage public participation for sustainable forest management in the northern Zagros mountain of Kurdistan province, Iran. *Scand. J. For. Res.* **2004**, *19*, 65–71. [CrossRef]

21. Soltani, A.; Sanskhayan, P.L.; Hofstad, O. A dynamic bio-economic model for community management of goat and oak forests in Zagros, Iran. *Ecol. Econ.* **2014**, *106*, 174–185. [CrossRef]

22. Valipour, A.; Plieninger, T.; Shakeri, Z.; Ghazanfari, H.; Namiranian, M.; Lexer, M.J. Traditional silvopastoral management and its effects on forest stand structure in northern Zagros, Iran. *For. Ecol. Manag.* **2014**, *327*, 221–230. [CrossRef]

23. Pourhashemi, M.; Zandebasiri, M.; Panahi, P. Structural characteristics of oak coppice stands of Marivan Forests. *Iran. J. Plant Res.* **2004**, *27*, 766–776. (In Persian)

24. Zandebasiri, M.; Ghazanfari, H. The main consequences of affecting factors on forest management of local settlers in the Zagros forests (Case study: Ghalegol watershed in Lorestan province). *Iran. J. For.* **2010**, *2*, 127–138. (In Persian)

25. Zandebasiri, M.; Soosani, J.; Pourhashemi, M. Evaluation of the crisis severity in forests of Kohgiluye and Boyerahmad province (Case study: Tang-e Solak). *Iran. J. For. Poplar Res.* **2015**, *24*, 665–674. (In Persian)

26. Pourhashemi, M.; Dey, D.C.; Mehdifar, D.; Panahi, P.; Zandebasiri, M. Evaluating acorn crops in an oak-dominated stand to identify good acorn producers. *Austrian J. For. Sci.* **2018**, *35*, 213–234.

27. Pourhashemi, M.; Panahi, P.; Zandebasiri, M. Application of visual surveys to estimate acorn production of Brant's oak (*Quercus brantii* Lindl.) in northern Zagros Forests of Iran. *Casp. J. Environ. Sci.* **2013**, *11*, 85–95.

28. Sagheb Talebi, K.; Sajedi, T.; Pourhashemi, M. *Forests of Iran: A Treasure from the Past, a Hope for the Future*; Springer: Berlin/Heidelberg, Germany, 2014.

29. Zandebasiri, M.; Parvin, T. Investigation on importance of Near East Process's criteria and indicators on sustainable management of Zagros forests (case study: Tange Solak Water Catchment, Kohgiloye and Boyer Ahmad province). *Iran. J. For. Poplar Res.* **2012**, *20*, 204–216.

30. Pourhashemi, M.; Zandebasiri, M.; Panahi, P. Estimation of acorn production of gall oak (*Quercus infectoria* Olivier) in Baneh forests by Koenig visual method. *Iran. J. For. Poplar Res.* **2011**, *19*, 194–205. (In Persian)

31. Hoseinzadeh, J.; Pourhashemi, M. The study of crown indicators in *Quercus brantii* tress in relationship with mortality phenomenon in Ilam forests. *Iran. J. For.* **2005**, *7*, 57–66. (In Persian)

32. Zandebasiri, M.; Ghazanfari, H.; Sepahvand, A.; Fatehi, P. Presentation of decision making pattern for Forest management Unit under uncertainty conditions (case study: Taf local area-Lorestan). *Iran. J. For.* **2011**, *3*, 109–120. (In Persian)

33. Zandebasiri, M.; Pourhashemi, M. Traditional forest related knowledge, Part 4: Management unit in Zagros forests. *Iran. J. Nat.* **2018**, *12*, 8–11. (In Persian)

34. Paletto, A.; Giacovelli, G.; Grilli, G.; Balest, J.; De Meo, I. Stakeholders' preferences and the assessment of forest ecosystem services: a comparative analysis in Italy. *J. For. Sci.* **2014**, *60*, 472–483. [CrossRef]

35. Salehnasab, A.; Feghhi, J.; Danehkar, A.; Soosani, J.; Dastranj, A. Forest park site selection based on a Fuzzy analytic hierarchy process framework (Case study: the Galegol Basin, Lorestan province, Iran). *J. For. Sci.* **2016**, *62*, 253–263. [CrossRef]

36. Soltani, A.; Angelsen, A.; Eid, T.; Noori Naieni, M.S.; Shamekhi, T. Poverty, sustainability, and household livelihood strategies in Zagros, Iran. *Ecol. Econ.* **2012**, *79*, 60–70. [CrossRef]

37. Zandebasiri, M.; Vacik, H.; Etongo, D.; Dorfstetter, Y.; Soosani, J.; Pourhashemi, M. Application of time-cost trade-off model in forest management projects. *J. For. Sci.* **2019**, *65*, 481–492. [CrossRef]

Article

Managing Open Innovation Project Risks Based on a Social Network Analysis Perspective

Marco Nunes [1,*] and António Abreu [2,3,*]

[1] Department of Industrial Engineering, University of Beira Interior, 6201-001 Covilhã, Portugal
[2] Department of Mechanical Engineering, Polytechnic Institute of Lisbon, 1959-007 Lisbon, Portugal
[3] CTS Uninova, 2829-516 Caparica, Portugal
* Correspondence: marco.nunes@tetrapak.com or D2317@ubi.pt (M.N.); ajfa@dem.isel.ipl.pt (A.A.)

Received: 30 March 2020; Accepted: 9 April 2020; Published: 13 April 2020

Abstract: In today's business environment, it is often argued, that if organizations want to achieve a sustainable competitive advantage, they must be able to innovate, so that they can meet complex market demands as they deliver products, solutions, or services. However, organizations alone do not always have the necessary resources (brilliant minds, technologies, know-how, and so on) to match those market demands. To overcome this constraint, organizations usually engage in collaborative network models—such as the open innovation model—with other business partners, public institutions, universities, and development centers. Nonetheless, it is frequently argued that the lack of models that support such collaborative models is still perceived as a major constraint for organizations to more frequently engage in it. In this work, a heuristic model is proposed, to provide support in managing open innovation projects, by, first, identifying project collaborative critical success factors (CSFs) analyzing four interactive collaborative dimensions (4-ICD) that usually occur in such projects—(1) key project organization communication and insight degree, (2) organizational control degree, (3) project information dependency degree, (3) and (4) feedback readiness degree—and, second, using those identified CSFs to estimate the outcome likelihood (success, or failure) of ongoing open innovation projects.

Keywords: risk management; project management; sustainability; social network analysis; collaborative networks; project lifecycle; project critical success factors; open innovation; predictive model; project outcome likelihood; organizational competencies

---

1. Introduction

In today's complex, and unpredictable business landscape, if organizations want to achieve sustainable competitive advantages, they need to develop strategies that enable them to enhance performance and innovation to meet actual market needs and demands [1,2]. Innovation and performance strongly depend on how an organization´s top management drives and motivates the organization´s employees to overcome barriers such as different geographic locations, time-zones, cultures, and functions [3], as well as to have the capacity to acquire the necessary resources (human, technological) and to adopt an effective innovation model. For example, some authors argue that an organizational ambidextrous leadership style enhances the chances of gaining and holding sustainable competitive advantage. Such a style is essentially characterized by the exploitation of the present conditions in order to optimize the current business models operation and, at the same time, exploring the opportunities that contribute to redefining the business model by making decisions in a pioneering risky way [4,5]. However, most organizations do not contain, on their own, all the resources, such as brilliant minds, technologies, and know-how, just to name a few, necessary to be able to respond today´s market complex and dynamic demands. In an attempt to overcome such constraints, organizations engage in collaborative or trade-off partnerships, whereby in interacting with other organizations or individuals such as business partners, customers, universities, scientific institutes, public institutions, and even inventors, they hope to find the best methods for supporting innovation and improving organizational performance. Such collaborative partnerships are essentially characterized by an exchange of ideas, resources, and technologies in a controlled environment—enhancing synergy. The aim of these partnerships is to provide organizations with benefits such as reduced innovation costs, faster innovative processes, increased differentiation, easier access to the market, creation of new revenues streams, more diverse R&D investments, and the sharing of innovation development risks [6]. The description mentioned above—however not without criticism regarding its benefits [6]—fits one of the most popular and adopted innovation models [7–9], with an adoption rate reaching up to 78 percent of companies in North America and Europe. We can credit to Henry Chesbrough, for having coined the term "open innovation" [10] to describe the above phenomena. In fact, the literature shows that one of the key factors—if not the major key factor—in today's business landscape, including performance and innovation, is how well organizations are able to work in networks of collaboration [11–13]. Furthermore, research shows that working in networks of collaboration that are fueled with diversity and inclusion [14,15] efficiently distributed across the different organizational functions, geographies, and technical expertise domains [16] strongly contributes to the achievement of competitive advantages [11,12] and also boosts innovation and performance [17,18]. In fact, several authors and researchers argue that the network factor—working in networks of collaboration—is a higher success predictor, regarding innovation and organizational performance than individual competencies and know-how, especially if those networks of collaboration are built with positive energy, diverse problem-solving skills and reach [13,19,20]. While the benefits of working in networks of collaboration are well documented throughout the literature, in order that organizations might efficiently and effectively profit from them, these networks, must be effectively managed [21]. In fact, research shows [22] that organizations that engage in networks of collaboration and that adopt a more hands-off approach (less control from the management team) have considerably lower success rates for their projects than do organizations that adopt a hands-on approach (more control from the management team). However, it is often argued that the lack of effective models to support the management of collaborative networks—such as the open innovation model—is one of the major factors preventing organizations from engaging with more frequency in collaborative networks models such as the open innovation one [23]. Among the potential challenges that organizations may need to address when working in networks of collaboration, several authors argue that the management of three different collaborative risks dimensions—(1) behavioral risks, (2) the risk of assigning tasks to partners, and (3) the risk of uncooperative partners—represents by far the greatest challenge that organizations need to deal with [24]. The reason behind this challenge, has to do with the nature of how work in most organizations is accomplished. In most organizations work is done through internal and external networks that are usually a mix of formal and informal networks of collaboration [25]. While there do exist formal organizational hierarchies that determine how work and collaboration should be done, these alone, due their structural rigidity, seem unable to effectively respond to the actual needs of organizations, namely those regarding innovation and performance [25]. Research shows that as organizations engage in networks of collaboration, more work within and between organizations will be done through informal networks of relationships, that will emerge as the collaboration evolves, reducing to a certain extent the role of the formal organizational structure [13,26]. While organizational formal networks consists of a designed chain of authority where often are ruled by the rational-legal authority system based in universalistic principles that are understood as fair, informal networks are usually hidden behind an organizational formal chart, very hard to see with a naked eye [27], and very often not ruled by the rational-legal authority system but rather by unfair and particularistic principles, such as friendship, propinquity, homophily, dependency, trust, and so on, which are characteristic of personal and social needs of individuals [25]. In fact, it is often argued that informal networks are almost entirely responsible for how organizations find relevant information, solve problems, capitalize opportunities, and generate satisfaction well-being and retention [17,25,26]. However, several authors argue that in an organizational context, it is very difficult to distinguish whether relations between organization´s entities (employees) and between different organizations are informal or formal [28,29]. Moreover, informal relationship networks may become formal and vice-versa [29], which shows that there is a blurred line between informal and formal networks in an organizational context. It can be concluded that formal and informal networks of relationship simultaneously influence and are influenced by the behaviors of the different entities that comprise a social network. This phenomenon, in an organizational context, will influence how project tasks or activities will be executed, which to a certain extent may result in collaborative risks, such as (1) behavioral risks, (2) risks in assigning tasks to partners, and (3) risks in selecting critical partners. Several researches also show that if formal and informal organizational networks of collaboration are not effectively managed, they may strongly hinder the performance and innovation capacity of an organization and ultimately can evolve either to an overload collaborative status or to an inefficient organizational collaboration status [12,26,27]. Efficiently managing networks of collaboration has been pointed out by renowned authors, researchers, and institutes, as a major factor that influences results as for example project outcomes [26,27]. The most effective way to study, analyze, and quantitatively measure the dynamic interactions, which mirror existing and forge future behaviors of social entities that occur throughout formal and informal networks of relationships between entities in a social network, is through the application of social network analysis (SNA) centrality metrics [16,26,27,30]. Social network analysis centrality metrics or measures are developed based on graph theory—a branch of discrete mathematics structure, used to model pairwise relationships between entities such as persons, organizations, and others—which is the only effective method that enables the mapping, analysis, and quantification of relationship data between entities in a dynamic environment, contributing thus to explain how social structures evolve across time and how they impact the environment where they do exist [30]. However, due the complex nature of the subject (the application of social network analysis metrics in the organizational context)—characterized by a non-straightforward linear process regarding the understanding and the practical application in organizations, namely in the organizational managerial field, regarding the practical application of the SNA concepts and how their benefits are understood—still most organizations have not done the shift towards the integration of relational data (data from networks of collaboration analyzed and measure by the application of SNA centrality metrics) into their organizational strategic management processes [12,17,20].

In this work, a heuristic model, Open Innovation Risk Management model (OI-RM), developed on the basis of four scientific fields (Figure 1)—(1) collaborative networks, (2) project management, (3) risk management, and (4) social network analysis—has as main goal the identification of critical success factors regarding the formal and informal dynamic interactions (behaviors) between the different entities (organizations) that participated in projects characterized by a collaborative network approach, such as the open innovation one, so that they can be replicated and used as guidance in future projects characterized by a collaborative network approach.

In order to uncover and quantitatively measure dynamic interactional behaviors that will be used to identify and quantify project critical success factors, the proposed model in this work will analyze a set of successfully and unsuccessfully delivered projects that were executed under a collaborative network approach—such as the open innovation one—searching for unique dynamic behaviors between entities that participated in those delivered projects, that characterize each of the two project outcomes (success, and failure). The data that the proposed model in this work requires in order to quantitatively measure project critical success factors is to be collected through two different methods. They are (1) project meetings, where essentially the number of meetings and participating organizations in each meeting, characterized with their respective project competencies is recorded and (2) project emails, where essentially the number of exchanged emails between the participating entities, characterized according to their content and temporal timeline, is recorded. The data collected in project meetings and project emails, will then be analyzed and quantified by the application of social network analysis centrality metrics, which in turn will be used to characterize four different interactional collaborative dimensions

(4-ICD) that usually occur in collaborative network projects. They are (1) key project organization communication and insight degree, which has as objective to measure how the presence of important key (function of their specific project competence) entities (organizations) in project meetings and email communication network, triggers communication dynamics (communication proactivity) between the participating entities, (2) organizational control degree, which has as objective to measure how a given organization controls the email communication network, in terms of send/receive project information related, (3) project information dependency degree, which has as objective to measure the dependency degree of a given organization or organizations, regarding project information to execute their project tasks or activities, and finally (4) feedback readiness degree, which has as objective to measure the speed of answering/replying to project information requests through the email communication network. In Table 1, a comprehensive description is illustrated that translates the integration of the individual contributions of the four scientific fields that build the foundations of the proposed model designed to quantitatively measure open innovation projects critical success factors across the different phases of their lifecycles.

Figure 1. The four scientific fields, which support the development of the Open Innovation Risk Management model (OI-RM) model.

Table 1. OI-RM Model individual integration contributions.

	Four Scientific Fields	Individual Contributions for Proposed Model
	Project management	Provides the definitions and structure of an open innovation project where the application of the proposed model will be deployed.
	Collaborative networks	Provides the definitions of the different dimensions of collaboration (networking, coordination, cooperation, and collaboration) that are used to define the four different informal collaboration dimensions (4-ICD).
OI Project Lifecycle	Social network analysis	Provides the tools and techniques to uncover and quantitatively measure the four informal collaborative dimensions (4-ICD), between organizations across an open innovation project lifecycle, where the 4-ICD are (1) key project organization communication and insight degree, (2) organizational control degree, (3) project information dependency degree and (4) feedback readiness degree.
	Risk management	Provides definitions, approach, and standard risk management process, to be adopted throughout the process of identifying, analyzing, measuring, treating, monitoring, and updating (continuous improvement cycle) dynamic collaborative risks, in other words, project critical success factors.

1.1. Relevance and Novelty of the Research

The proposed model in this work addresses three different collaborative risks dimensions—(1) behavioral risks, (2) risks of assigning tasks to partners, and (3) risks of selecting critical partners—that emerge as organizations engage in collaborative network models, such as the open innovation one. By doing so, the proposed model in this work aims to provide meaningful insight that contributes to answer the following three questions that together form its research question. They are

Question 1: To which extent do dynamic interactions (dynamic behaviors) between different entities (organizations) that participate in collaborative projects across a project lifecycle influence project outcome (failure or success)?

Question 2: Are there critical successful factors that can be identified associated with project success outcome?

Question 3: If there are critical success factors, can they be quantitatively measured and replicated in future projects?

By addressing the mentioned research questions, the proposed model in this work is contributing with valuable insight in three different dimensions.

First and as main objective of the proposed work, the proposed model in this work provides organizations a heuristic model that helps to manage in a holistic way (formal and informal) the networks of collaboration that emerge between the different entities that engage in collaborative networks working models, as they participate in innovation initiatives. The management of collaborative networks in innovation initiatives supported by effective management models provides organizations benefits as proved in the latest research [12,22], which argues that organizations that adopt a more hands-off approach management style (less control from the management team in the negotiation phase, as different organizations define the collaborative guidelines processes and procedures) leaving strategic, operational, or cultural incoming issues or differences to be managed as the collaboration evolves across time—a fix-it-as-you-go issue resolution approach—have considerable lower success rates than organizations that adopt more a hands-on approach (where the active involvement of top management to anticipate and resolve issues before any collaboration between different organizations begins). The proposed model in this work is in line with what is mentioned above, in the sense that it provides organizations with a structure to control innovation initiative developments, identifying project critical success factors based on behavioral patterns of collaboration uncovered in past successfully delivered collaborative projects.

Second, provides organizations a way to quantify and verify how much does a more, or a less centralized collaborative network structure—regarding communication, control, dependency and feedback degrees, measured by the application of social network analysis centrality metrics as result of the formal and informal dynamic collaborative interactions—positively, or negatively influences collaborative project´s outcome. This will enable to argument in a more data-driven way, research that defends the fundamental role for innovation and performance, of formal and informal networks in organizations [13,17,18,31] and research that points in other direction, defending that other factors—rather formal and informal organizational networks of collaboration—are of more importance for innovation and performance in organizations [2].

Third, the implementation of the proposed model in organizations is aligned with the organizational digital transformation strategy and industry 4.0 [32] to the extent that it automatizes the collection, processing, and analysis of behavioral data, associated to successful and unsuccessful project outcomes, continuously refining the process of identifying project critical successful factors through an automatized continuous improvement cycle, characterizing the proposed model in this work as a machine learning model.

1.2. The Importance of Organizational Sustainability in the Global Sustainability Context

Sustainability in organizations can be seen as holistic, consistent, and incremental growth processes (economic, social, and environmental), rather than non-constant growth processes over time. This means that these holistic and consistent growth processes over time focuses not only in the immediate important challenges of an organization but also in the long-term challenges. As organizations engage in open innovation projects, effective collaboration becomes the "neuralgic center" that strongly affects the outcome of collaborative development between the different organizations that participate in open innovation initiatives. Understanding and efficiently managing the different collaborative networks dimensions that emerge between the different organizations that engage in open innovation, such as communication, information-sharing, feedback, just to name a few, is a critical factor to better and more accurately develop efficient collaborative planning and take corrective or optimization measures, in a timely manner. By acting in this way, one is optimizing the necessary different resources associated to such actions. The application of social network analysis has in this scenario a key role regarding the identification (essentially quantitative) of such different collaborative networks dimensions that emerge as organizations engage in open innovation initiatives. Social network analysis enables to identify and quantify risks associated to dynamic interactive collaboration between the different organizations that participate in open innovation initiatives in a holistic way, such as collaborative overload and poor or lack of collaboration or information sharing between the different organizations according to the following value-chain: The identification of open innovation collaborative risks in a timely manner contributes to a better understanding of the actual and future collaborative developments, which in turn will enable a more accurate planning of effective responses to upcoming challenges. This in turn will enable optimization regarding the allocation of necessary resources to plan responses, which in turn strongly contributes to a leaner organizational and societal approach, providing organizations sustainable competitive advantages (by essentially saving resources, time, and money), which in turn, strongly contributes to the three fundamental pillars of sustainability known as economic, social, and environmental.

1.3. Structure of this Work

In this work a heuristic model is proposed to manage collaborative networks in open innovation projects, by identifying project collaborative critical success factors. The present work is divided into four distinct but interrelated chapters.

In chapter 1 (Introduction) a brief introduction to the architecture of the proposed model in this work is presented, highlighting the linkage between individual contributions to the different scientific fields that support the development of the proposed model and the motivation (reasons for the need of such model in the organizational context, namely in organizational collaborative networks perspective) to develop it.

In chapter 2 (Literature review) an extensive literature of the four scientific fields (collaborative networks, risk management, social network analysis, and project management) that support the development of the proposed model is undertaken, highlighting the most important contributions in the organizational collaborative networks context, to the development of the proposed model in this work.

In chapter 3 (Model development and implementation) the research methodology, function principles—supported by an application case—, important concepts, development process, implementation process across a project lifecycle, and ethical considerations of the proposed model are illustrated in an extensive and detailed approach.

In chapter 4 (Conclusions), the managerial implications, further developments, and benefits and limitations of the proposed model are presented and discussed in an open innovation context. The chapter finalizes by suggesting a number of aspects for further development and research.

2. Literature Review

2.1. Collaborative Networks

Collaborative networks (CN) can be defined as networks that comprise a variety of entities such as organizations, people, and so on, which are geographically distributed, autonomous, and heterogeneous regarding their operating environment, culture, social capital, and goals but collaborate to better achieve common or compatible goals, where the interactions are supported by computer network and can take several forms, such as service-oriented organizations, virtual organizations, dynamic supply chains, industry clusters, virtual communities, virtual laboratories, and so on [33]. Organizations that are members of long-term networked structures, usually result in an increase in circulation and production of knowledge within the network, contributing that to a more effective way of working of organizations, as they pursuit their objectives [18]. However, it is argued that the lack of performance indicators that identify and measure the production and circulation of knowledge in a collaborative environment make difficult to prove its relevance [18]. The concept of collaboration may often be confused with cooperation and gains a higher level of ambiguity, when the concepts such as networking, communication, and coordination are considered [34]. All the concepts mentioned above are components of collaboration but have different dimensions in the organizational field regarding the value that they offer. Therefore, it is important to identify what each one of them really represents. Their unique contribution is described as follows [33]: (1) Networking is the exchange of communication, information, and experiences for mutual benefit, where usually there is no common goal or structure that regulates timing and individual contributions. (2) Coordination, in addition to the characteristics of networking, involves aligning/altering activities for mutual benefit in order to achieve results in a more efficient way. (3) Cooperation involves not only information exchange and adjustments of activities (Networking and coordination) but also sharing resources and division of some labor among participants in order to achieve compatible goals. The sum of the individual contributions by the various participants in an independent manner, forms the aggregate value of this collaborative level. (4) Collaboration involves all the other three mentioned before and includes jointly planning, implementing, and evaluating a program of activities in order to achieve a common goal. It is a "working together" collaborative level, which requires mutual engagement in problem solving activities from the participating entities, trust, effort, and dedication. It still implies the sharing of resources, responsibilities, rewards, risks. Ultimately, it enhances the capabilities of the participating entities.

2.1.1. Open Innovation

Open innovation is an organizational collaborative model type that has been gaining increasing attention in the past years, essentially due its measurable benefits in enhancing the innovation capacity of organizations [23]. The term "open innovation" is credited to Henry Chesbrough [10] and can be defined as a model that uses a wide range of external entities (actors and sources) to help organizations to achieve a sustainable innovation behavior [23]. Essentially, open innovation means that organizations should work together in networks of collaboration, sharing ideas, experiences, know-how, and technologies, to generate value [35] that otherwise could not be achieved if organizations work in an isolated mode. Some authors argue that there are essentially three different levels of collaborative innovation, which to a certain extent are dependent from strategic, operational and structural, legal, and cultural issues or challenges [23]. They are (1) management of interorganizational collaboration process (managing the interactions between the different organizations that participate in an open innovation project), (2) management of the overall innovation process (managing all the processes, phases, and innovative breakthroughs across the timeline evolution that defines the duration of the collaboration), and (3) creation of a new collaborative knowledge (organizing, documenting, and making available critical information regarding the innovation process evolution). To better understand how the open innovation model functions, Chesbrough [10] proposes the comparison between a traditional approach to innovation initiatives (the blockbuster business model type or closed innovation model) and a new approach (open innovation model). Chesbrough [10] argues that the blockbuster business model type—also known as the closed innovation model type—is no longer economically sustainable and that organizations need to engage in open innovation working approaches if they want to survive. He argues that the open innovation model, by contrast with the traditional closed innovation model, is a new way of creating value in an organization [10]. Essentially, the open innovation model is characterized by having two different flow-types of knowledge and resources [10].

They are (1) outside-in flow type (occurs when an organization brings external knowledge or/and resources from business partners, customers, universities, scientific institutes, and public institutions to improve its innovative performance. By acting so, innovation initiatives, costs, and time can be reduced by acquiring, buying, or borrowing only those resources that are really needed. (2) inside-out flow type (occurs when organizations search for possibilities to share already available in-house knowledge or/and resources with the external environment in a way which will add value to the sharing organization such as for example out-licensing and transfers of rights, promoting spinoffs, turning to open source, just to name a few. By sharing their ideas with the external environment, organizations can create value chains (downstream chains) with other organizations or earn a royalty when others use their ideas, which very likely organizations on their own could not use, develop, or even implement. If both flow types are simultaneously adopted by an organization, they can be named as coupled flow type, where the exchange of resources, ideas, and technologies occurs trough collaborative partnerships in the form of joint researches, consortiums, and joint ventures just to name a few. As already mentioned before, the literature shows that there is a lack of effective models to support co-innovation collaborative models. Nevertheless, some models can be found in the literature [23,36], where two of them are very popular among practitioners of open innovation projects, also known as crowdsourcing [23]. The first model, the InnoCentive [36], created in 2001, runs in a web based platform and consists of six steps that start with the identification of problems and ideas, formulation of a challenge, specification of intellectual property agreements, publication of the challenge, evaluation of the solutions, and a price to end with the transfer of intellectual property. The second model, developed by Procter & Gamble (P&G), called Connect + Develop, works inwards and outwards spanning from packaging to trademarks, marketing to engineering, and commercial services to design.

2.1.2. Open Innovation Model Benefits and Challenges

One of the biggest benefits in engaging in well-managed open innovation projects, is the positive impact in economic, social, and environmental sustainability [37]. However, despite the benefits, that can easily be found in the literature [38], according to a survey conducted by Accenture in 2015 (a consultancy company) about 50% of surveyed organizations, said that open innovation don't seem to be yielding as many new products or other benefits as they had hoped [38]. Several studies show that the reason for the low adoption rate is essentially linked to two factors [22,38]. They are (1) political, and (2) cultural, but surprisingly not technical. Factors like, different cultures and different attitudes regarding sharing intellectual property, different concerns about risk sharing, multiple gatekeepers, relationship between large and small organizations, skepticism regarding anything "not invented here," [22] and disputes between organizational rival groups over organizational "territory", hold back adoption. Research in the field of open innovation, suggests three major risks that are likely to be experienced by organizations as they engage in open innovation projects [24]. They are (1) pure risk or uncertainty (related to the probability of an event occurrence that puts at stake the success of the innovation project), (2) risk of an innovation project (related to the fact that there is a substantial portion of the risks associated with estimates such as resources, duration of the task and costs. Includes also business, political or regulation, and operational risk sub types), and finally (3) collaborative risks (related to the fact that a collaborative ecosystem can be characterized by a set of relationships that are established between several entities, such as companies/organizations, knowledge, resources and tasks, and contains the risk of collaborative management, behavioral risks, risk of assigning tasks to partners, and the risk of selecting critical enterprises). Other challenges that may be an obstacle, or even hinder pursuing in to these collaborative project types include, but not only, finding creative ways to exploit internal innovation, incorporating external innovation into internal development, and motivating outsiders to supply an ongoing stream of external innovations processing ideas quickly and efficient, establishing an efficient internal structure, proper management of intellectual property issues and other legal risks, the fear of failure, the lack of incentives and critical creative resources [39]. Crafting effective strategies and models to properly address the above-mentioned risks, represents a huge challenge for organizations. Despite the benefits that are credit to open innovation, the model is not without critics or downsides. High costs in people coordination and loss of control and power over innovative processes, are just a few mentioned in the literature [6]. To better understand the advantages and the disadvantages of open innovation, in Table 2 are illustrated the differences between closed, and open innovation, which represent the benefits and limitations of both models [6,37].

Table 2. Benefits and Limitations of Open Innovation and Closed Innovation.

	Benefits	Limitations
Closed Innovation Model	• Full overall control on the innovative process and intellectual property (IP) • Less or non-dependence on external knowledge • No risk of leak of confidential information • Less faults on routine works • What one organization discover it will get it to market first	• Not all the smart people in the necessary fields to innovate, work for *us* • Higher levels of investments to supply the R&D departments • Development performs at a slower pace • Gains limited market share • Higher risk, because developed ideas may end not being supported by the organization
Open Innovation Model	• Allows to knowledge, ideas, technology flow in and out between organizations • Diversification of R&D investments • Easier market entry • Resource acquisitions advantages • Development performs at a higher pace • Broader base of ideas • Technological synergy effects • Increase of the learning capacity • Use intellectual non-own property as strategic asset • Reduced costs of innovation initiatives • Share innovation investments risks with other partners • Increase differentiation and the creative process • Create new revenues streams (Copyright- royalties)	• Increase in process coordination and implementation costs • More faults in routine workflows • Strong dependence on external knowledge • Loss of key knowledge control and flexibility, creativity, and strategic power • Lack in legacy for additional tasks • Risk of leak, of confidential information • Loss of overall control over the innovative process and intellectual property (IP)

As it can be seen in Table 2, the benefits of the open innovation outweigh by far the limitations, namely when compared with the benefits of a closed innovation system. Essentially, the major benefits can be traced in financial outcomes in two different ways. First the exchange of resources and technologies enables faster innovation processes and opens more doors to existing and different markets, wherefrom more revenues may arrive. Second, by sharing the risk of failed innovation initiatives with innovation initiative partners, may spare organizations from bankruptcy or even extinction.

2.2. Risk Management and Critical Success Factors in Project Management

A project, according to the PMI (the Project Management Institute) is a temporary endeavor with a defined start and end, that aims the creation of a unique product or service or result [40].

In order organizations successfully deliver a project, they should have support of standard structured approaches (so-called best practices standards). Such best practices on how to efficiently manage projects, are usually provided by the project management scientific field [40]. Project management can be defined as the application of knowledge, skills, tools, and techniques to project activities, to meet project requirements, across all the different project phases of a project lifecycle [40]. Very often, as organizations deliver projects, challenges in the form of risks (threats or opportunities), arise across the different phases of a project's lifecycle. Such risks, if not properly managed (usually threats), very likely will drastically reduce the chances of achieving a successful project outcome, which in other words means, deliver a project within the planned scope, quality, schedule, costs and resources (the so-called project constraints [40]). Project risk management expert David Hillson defines project risks, as the uncertainty that matters [41]. He argues that this definition aims to divide what represent real project risks, from what not represents real project risks. Hillson suggests four types of risks that may outbreak in project management. They are [41]: (1) Event risk, (2) variability risk, (3) ambiguity risk, and (4) emergent risk. In Table 3 a comprehensive description regarding the four types of risks and the respective uncertainties is presented.

Table 3. Risks and respective uncertainties types.

Risk Types	Uncertainty Types	Description	Management Approach
Event Risk	*stochastic uncertainty*	Also called event risks, are risks related to something that has not yet happened, and it may not happen at all, but if it does happen, it will impact on one or more project objectives.	There is a set of well-established techniques for identifying, assessing and managing them, based on risk management standards and best practices.
Variability Risk	*aleatoric uncertainty*	Are a set number of possible known outcomes, but one does not exactly know, which one will really occur.	Advanced analysis models such as the Monte Carlo simulation, are the actual solution to model and manage these risk types.
Ambiguity Risk	*epistemic uncertainty*	Are uncertainties, arising from lack of knowledge or understanding. Also called of know-how and know-what risks, comprise the use of new technology, market conditions, competitor capability or intentions, and so on.	Learning from experience from past, or others–lessons learned. Prototyping and simulating, before taking real action.
Emergent Risk	*ontological uncertainty*	Known as "Black Swans", these risks are unable to be seen because they are outside a person´s experience or mindset, so one doesn't know that he should be looking for it at all. Usually they arise from game-changers and paradigm shifters, such as the release of disruptive inventions or products.	Contingency planning, is the key to manage such risk types.

The model proposed in this work, addresses the ambiguity risk type, characterized as an epistemic uncertainty (according to Table 3) which essentially is characterized by a lack of knowledge and understanding regarding a given subject, which in the case of the proposed model is the extent to which the dynamic behavioral interaction—characterized by the 4-ICD—of entities in an open innovation context, influence a project outcome (success or failure). Such type of risks can be properly addressed by learning from past experiences and simulating future events (according to Table 3) which is exactly what the proposed model in this work offers, as it identifies project critical success factors from closed projects and replicate those critical success factors in future projects. Other authors and researchers in the area of project risk management, argue that project challenges (project risks) can be instead called as project critical success factors [42]. Factors such as experience of project teams, project manager's ability to solve problems, management level during the strategy formulation stage [43] are just a few of them. Notable is the work in this area done by Pinto and Slevin back in 1988 [42], as they identified a set ten project critical success factors, that change importance function of a given project phase. Such critical factors, are considered major project risks, that if not properly managed, will jeopardize the chances of a successful project outcome. They are [42]: (1) project mission not properly defined, (2) lack of top management support, (3) undetailed project schedule, (4) poor client consultation, (5) lack of necessary and proper technology, and expertise, (6) poor team skills and experience, (7) ambiguity client acceptance, (8) lack of proper monitoring and feedback of project activities, (9) poor communication, (10) non-readiness to handle crises and deviations from plan -contingency plans. In order to manage project risks, organizations have risk management standards that they can incorporate in their project management activities, provided by institutes or body of knowledge such as the PMI [40] and ISO [44], that essentially recommend— based on experience and best practices principles—a set of problem-solving strategies an methods, supported by ad hoc tools and techniques. A very popular approach to manage risk in organizations is proposed by the ISO (the International Organization for Standardization), in their standard–31000:2018 [44]. In this standard, a set of principles aim the creation of value in organizations by effectively assessing and treating risks. The standard consists in a set of well-structured six steps that essentially aim the identification, treatment, and monitoring of risk. They are [44]: (1) Establishing scope (defining the scope of the risk management activities), context (defining the external and internal context, which is the environment in which the organization seeks to define and achieve its objectives), and criteria (specifying the amount and type of risk that an organization may or may not take, relative to objectives), (2) perform a risk identification (consists in finding, recognize and describe risks that might help or prevent an organization achieving its objectives), (3) perform a risk analysis (comprehend the nature of risk, uncertainties, risk sources, consequences, likelihood, events, scenarios, controls and their effectiveness), (4) do a risk evaluation (comparing the results of the risk analysis with the established risk criteria to determine where additional action is required, (5) proceed to risk treatment (specify how the chosen treatment options will be implemented, so that arrangements are understood by those involved, and progress against the plan can be monitored), and finally (6) record and report the previous steps (continuously keep monitoring, and reviewing the evolution of identified risks and the efficacy of the applied controlled measures).

2.3. Social Network Analysis in Organizations

The beginning of the use of graph theory in analyzing dynamic relationships between entities (persons) is credit to the Romanian American psychiatrist, psychosociologist, and educator Jacob Levy Moreno (1889–1974), as his work "Who Shall Survive" was published in 1934 [45–47]. Nowadays the application of social network analysis covers a wide range of different areas such as organizational science industry, management and leadership [48], political science [49], behavioral sciences [50], communication, learning and media [51], law, national safety, criminology and terrorism [52], just to name a few. Social network analysis (SNA) is the process of studying and analyzing social structures data with a variety of measures developed based on graph theory, that contributes to explain how social structures evolve across time, and how they impact the environment where they do exist [53]. It can be more simply defined as "a specific set of linkages among a defined set of persons, with the additional property that the characteristics of these linkages as a whole may be used to interpret the social behavior of the persons involved" [54]. SNA plays an important role in bringing light to the social capital challenges [55] and has been incorporated into traditional Risk Management processes of Organizations, as a supportive tool for decision making and risk analysis [56] and simultaneously being used to study subjects such as talent shortages and retention, incompetence, innovation, network collaboration, collective and individual performance, cultural fit, values, unethical behavior, low morale, employee wellness, noncompliance with industry, and fraud [57]. The application of social network analysis to study social structures has achieved high popularity essentially due the desire to understand to which extent people´s behaviors and relationships influence others and outcomes such as performance, innovation, social cohesion, information diffusion, and so on [47,58]. Such relationships are complex by nature and cannot be entirely explained trough traditional social theory and data analysis methods but rather by methods that are based in sociology, because they consider the individual´s social context in the process of making choices [59]. In 1979, Noel Tichy, Michael L. Tushman and Charles Fombrun [60] enumerated some benefits of applying SNA in organizations, namely the potential offered by the study and understanding of organizational theory and the dynamic component of it.

2.4. The Application of Social Network Analysis in Project Management

The application of social network analysis has been expanding to several diverse scientific areas such as project management, for example, although it remains so far at a very initial stage [61]. Essentially, in project management, the identification of critical success factors concerning the dynamic of project informal networks that may contribute for a success project outcome is the principal reason for the application of SNA [62,63]. In the past years, several works that evidence the successful application of SNA in projects have been published. In 1988 Pinto and Slevin identified that a defined set of critical factors were changing the importance degree and function of the project phase. Among those identified, some are related with the interaction between entities across a project lifecycle such as top management support, client consultation, and communicating network [42]. These findings done by Pinto and Slevin [42] were revalidated by the latest research conducted in 2012 [64]. In 1993, professor David Krackhardt and Jeffrey R. Hanson, pointed out the importance of managers uncovering their informal organizational networks as being a key contributor to success. According to them, three collaborative networks are crucial to be mapped in an organizational context. They are [13] (1) advice network, which reveals the people to whom others turn to get the job done; (2) the trust network, which identifies who shares delicate information with whom; and (3) the communication network, which shows who talks to whom about work-related matters. One of the most cited ever works regarding the application of social network analysis in project management [65] is the work done by Stephen Mead in 2001 [66]. He applied SNA to visualize project teams, namely the analysis of the informal project stakeholder´s communication network [66], and after having identified central and isolated stakeholders regarding the informal communication network, he created a corrective plan to improve the performance of those that were not so well integrated in the project network. One of the most notable works regarding he application of SNA in organizations has been published by Professor Rob Cross, who in one of his most known books, "The Hidden Power of Social Networks", published in 2004, illustrates a ten-year case-study collection of the benefits of the application of SNA in organizations, namely in managing project collaborative networks [67]. Cross identified that in every organization, there is an informal type of network that is responsible for how the work gets done and coined a set of specific actors based on their location within a social structure [26,67]. They are The Central Connectors, Boundary Spanners, Information Brokers, Peripheral Experts, Peripheral Intentionally and the Energizers. In 2009, Prell et al. [68] applied SNA to identify and analyze stakeholder networks in natural resource management using the results in the selection of the important stakeholders. Toomey [69] identified four key subjects of SNA theory that play a major role in project development. They are centrality, structural holes, boundary management, and tie strength. In 2017, Mok et al. [70] applied SNA network centrality measures to identify key challenges in major engineering projects (MEPs) based on interdependencies between important stakeholder concerns, resulting in the identification of a number of key challenges that occur in those MEPs and helped to develop a set of good practices to release those challenges that could be used in future MEPs. In 2018, Michael Arena, former chief talent officer for General Motors, concluded, after years of investigation in several different organizations, that successful organizations operate in a more networked way, enjoying what he called as adaptative space [12], which is essentially a virtual place that enables proper connection between the operational and entrepreneurship pockets of an organization, were employees explore new ideas, and the most creative ones are empowered to propagate their ideas across the organization, enabling it to work in a more agile way. This adaptive space is built, managed, and maintained using SNA (social network analysis).

SNA Centrality Measures in Project Management

Centrality, in a social collaborative network, refers to the structural location of an entity, rather to the entity own inherent attributes such as age, tender, or gender. SNA researchers suggest that centrality is a measure of importance, influence, prestige, control and prominence [71,72] and that such structural locations can be quantitatively measured by the application of graph theory centrality metrics such as degree, betweenness, and closeness. According to Freeman [71], for each of these centrality metrics, a respective social direct implication exists as follows: activity (degree can be an index of potential for the network's activity), control (betweenness is an index of communication control by serving as bridge between two different subgroups of an network), and independence (closeness is an index of the potential independence from network control), respectively. Essentially, network centrality is associated to informal power in collaborative social networks [26,27,73] that will influence coordination and decision-making in project management and ultimately influence project outcome—success or failure [74,75].

3. Model Development and Implementation

3.1. The Proposed Model: The OI-RM (Open Innovation Risk Management) Model

The proposed model in this work is called OI-RM Open Innovation Risk Management and has as objective to contribute to answer the following research question:

To which extent does the dynamic collaborative interaction between different organizations that participate in an open innovation project across all the different phases of a project lifecycle influence a project outcome (success or failure)?

The proposed model in this work is divided into two parts. In part 1, the model will identify project outcome critical success factors regarding the dynamic interaction/collaboration between organizations that participated across a project lifecycle in open innovation environment projects, in closed (successfully and unsuccessfully delivered) projects, by analyzing project related data collected in project E-mail-exchange network and project meetings, throughout all the different phases of a project lifecycle. The model will analyze the collected project data, where essentially it will be looking for repeatable behaviors (RBs) regarding the dynamic interaction/collaboration between the different organizations that worked in open innovation project environment from successfully and unsuccessfully delivered projects, by analyzing four interactional collaborative dimensions (4-ICD) that usually occur as a project is being delivered. They are (1) key project organization, communication, and insight degree; (2) organizational control degree; (3) project information dependency degree; (3) and (4) feedback readiness degree. If the proposed model identifies unique RBs associated to successfully and unsuccessfully delivered open innovation projects, they are considered the critical success factors. In part two, the model will provide guidance, and estimate an outcome likelihood of an ongoing open innovation project type, by comparing the deviation between the real ongoing project evolution and the desired evolution (which is defined by the identified critical success factors in part 1 of the model) of the ongoing project. By providing answer to the above-mentioned research question, the proposed model in this work is essentially addressing two risk types previous mentioned in chapter 2 that may occur when organizations work in open innovation projects. The first one, proposed by Abreu et al. [24], concerns the collaborative risks, which are delayered into critical enterprises risk, assigning tasks to partners risk, behavioral risk, and collaborative network management risk. The second is proposed by Hillson [41], essentially addressing the ambiguity risk type, also known as ambiguity risk—also known as "epistemic uncertainty"—where uncertainties (risks) arise from lack of knowledge or understanding, and its efficient management is done with lessons learned, prototyping, and simulations. Before introducing how the model was developed and working functioning principles, the model's key concepts must be introduced. They are presented in Table 4.

Table 4. OI-RM model key concepts.

OI-RM Key Concept	Description
Open Innovation Project	For the proposed model in this work, an open innovation project is a temporary endeavor undertaken to create a unique product, service, or result.
Open Innovation Project Outcome	The proposed model assumes only two types of project outcomes. They are successful and unsuccessful project outcomes. The criterion that defines both types is not given by the proposed model.
Number of Open Innovation Projects	The model does not preview a maximum number of open innovation projects to be analyzed. However, at least two projects—one with a successful and one with an unsuccessful outcome—are required as input.
Open Innovation Project Organizations	Project organization is any organization that participated in a project, across its lifecycle, or/and is officially assigned to participate in an open innovation project. This means it is any organization that has participated in project meetings and email project information-related exchange.
Open Innovation Project Organization-Competencies	Project organization competencies are the different competencies that different organizations play as they participate in delivering open innovation projects. They can be from the most diverse fields such as engineering, marketing, sales, human resources, and so on.
Open Innovation Project phases and Lifecycle	Every project used as input for the model has a finite number of project-phases. Usually four generic phases can be used (but not necessarily four only), (phase 1—Starting the project, phase 2—Organizing and preparing, phase 3—Carrying out the work, phase 4—Ending the project). The sum of all project phases of a project is the so-called project lifecycle.
Collaborative Interaction	The dynamic collaborative interaction of project organizations, which is characterized by the four interactive collaborative dimensions (4-ICD) that usually occur as a project is being delivered, comprises the formal and informal networks of collaboration.

3.2. OI-RM Model Function Principle

3.2.1. Research Methodology

The proposed model in this work is the result of an extensive literature review and consulted case studies that are illustrated in the previous chapter, regarding the already mentioned four scientific fields that lay the foundations of the development of the proposed model. The proposed model in this work integrates the proven individual benefits of each of the four scientific fields (project management, risk management, collaborative networks, and social network analysis) in organizations, in a network collaborative context—without neglecting its limitations—which gives to the proposed model strong trustworthy basis regarding the success of its application in an organizational context. The fundamental reason for the development of the proposed model in this work essentially relies on the countless studies and researches presented in the previous chapter in the fields of the social sciences and organizations that argue and prove that there is a relationship between the dynamic behavioral interaction between entities of an organizational social network and organizational outcomes essentially translated into performance and innovation. However, there is still by far a higher number of application cases in organizations that perform to a certain extent bureaucratic or the so-called true office work such as call centers, R&D, maintenance, operations, or factory work, rather than in project environments. The proposed model in this work is aimed to project environments, namely collaborative projects such as the open innovation ones. The methodological approach follows a well-defined sequential approach based on the literature review and case studies where the application of social network analysis has been one of the key pillars that enables the identification of the influence of dynamic behaviors in organizational outcomes. Regarding the research methodology process of the proposed model in this work, the following steps have been undertaken:

The first was defining the physical and spatial environment where the action is to occur; a project (with a well-defined start and end) with a typical structure (typical project lifecycle) and a variety of different entities (organizations) undertaken in an open innovation working model.

The second was the definition of the different levels of collaboration (communication and information exchange) between the different entities that are designed to participate in the project.

The third was the selection of the collection data methods (project meetings and project email exchange) that will enable to visualize and quantify the dynamic interactions that mirror a certain dynamic behavior.

The fourth was the selection of the most effective tools and techniques (according to literature research in the application of SNA in projects) to analyze in a quantitatively way the collected data, based social network analysis centrality measures such as in-degree, out-degree, total-degree, density, and reciprocity.

Finally, the fifth was the association of results obtained to project outcomes, enabling to draw conclusions regarding the relationship between the different dynamic behaviors across a project lifecycle and project success or failure outcomes.

3.2.2. Introduction to the Functioning Principles of the Proposed Model

To understand the working principles of the proposed model in this work, a demonstrative example based on Figure 2, is presented. The presented projects A and B, in Figure 2, that aim to the demonstration of the functioning principles of the proposed model in this work were carefully selected in order to show its potentiality for the identification of a project's critical success factors regarding the dynamic behavioral interaction of different entities (organizations) across a project lifecycle. Essentially, this has to do with the process of identifying critical success factors that must obey the criteria of being unique and repeatable to one of the two possible project outcomes—failure or success. This is, for example, to be seen in two different examples. First, in phase 1 of both projects in Figure 2, the architecture of the green lines that connect the different organizations are different. In project A, which was successfully delivered, all entities are connected through the green lines, and in project B, which was unsuccessfully delivered, not all entities are connected by the green lines. The difference in the number of connecting lines between projects A and B in phase 1 is captured quantitatively by the proposed model through the application of social network analysis' centrality metrics, which enables to characterize the difference between the project that was successfully delivered (project A) and the project that was unsuccessfully delivered (project B), regarding the number of connecting lines in a uniquely measurable way. On the other hand, in phase 4 for both projects in Figure 2, although project A was successfully delivered, the number of connection lines between the entities are the same from both projects. Here, the proposed model in this work cannot find a measurable unique difference regarding the number of connecting lines, which ultimately means that the number of collecting lines in phase 4 of the analyzed project do not influence any of the two project outcomes. This phenomenon is also one of the reasons why the proposed model in this work does not only operate with one unique centrality metric but rather with five different ones, in order to capture other dynamic behaviors that may not be able to be captured by a given centrality measure.

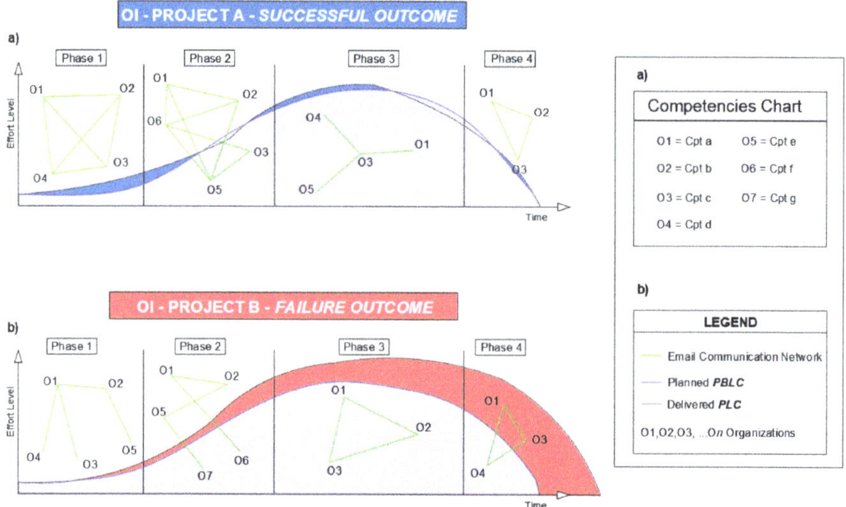

Figure 2. Project lifecycles for delivered Open Innovation Projects A and B.

3.2.3. Functioning Principle of the Proposed Model—an Application Case

In Figure 2, the lifecycles of two delivered open innovation projects are represented (projects A and B). Project A was successfully delivered, and B was unsuccessfully delivered. Both projects serve only as a demonstrative example of how the proposed model in this work functions. Organizations O1, O2, O3, O4, O5, O6, and O7 participated in both open innovative projects A and B, each with its own specific competency (a, ... , g) according to (a) in Figure 2. The blue lines across the project lifecycles represent how the projects were planned to be delivered. The grey dashed lines represent how the projects were delivered. In each phase, on both projects, the green lines represent the email communication network direct or indirect channels between the project organizations, which result from the data collected from project-related emails exchanged. For example, in project A at phase 1, there has been an email communication channel between project organizations O1 and O2. Analyzing the email communication at the first two phases (phase 1 and phase 2) of both projects (Project A and B), it is clear to see with a naked eye that the email communication network of the project that was successfully delivered (Project A) is by far denser than the email communication network in project B. In other words, there are more email communication channels between the organizations. At this point, considering only this factor, one can conclude that a denser email communication in phases 1 and 2 of an open innovation project is associated with a project success outcome.

However, it still needs to be quantitatively measured (the difference between a denser and a less denser email network communication channel). It is at this point that the application of Social Network Analysis (SNA) turns into a powerful tool. SNA uses the graph theory metrics that can be used to measure any graph-structure like the one that represents the email communication network illustrated in Figure 2. For this case, a centrality measure, *network average total degree (degree centralization)* [45] metric, has been chosen to quantify the email communication network in the proposed model. The network average total degree is the ratio between the sum of all links attached to one organization (total degree) and is given by the formula (1) adapted from [45]:

$$ATD(pha) = \frac{\sum_{o=1}^{n}(x_o)}{n} \tag{1}$$

where:

ATD = average organizational network total degree
x = number of existing links attached to one organization o
n = total number of project organizations ($o = 1, \ldots , n$)

Applying (1) for phase 1 of project A:

$$ATD(A,\ 1) = \frac{3+3+3+3}{4} = 3 \tag{2}$$

Maximum for $(A,\ 1) = 3$
Applying (1) for phase 2 of project A:

$$ATD(A,\ 2) = \frac{3+3+2+4+4}{5} = 3,2 \tag{3}$$

Maximum for $(A,\ 2) = 4$
Applying (1) for phase 3 of project A:

$$ATD(A,\ 3) = \frac{1+1+1+3}{4} = 1,5 \tag{4}$$

Maximum for $(A,\ 3) = 3$
Applying (1) for phase 3 of project A:

$$ATD(A,\ 4) = \frac{2+2+2}{3} = 2 \tag{5}$$

Maximum for $(A,\ 4) = 2$
Applying (1) for phase 1 of project B:

$$ATD(B,\ 1) = \frac{1+1+1+2+3}{5} = 1,6 \tag{6}$$

Maximum for $(B,\ 1) = 4$
Applying (1) for phase 1 of project B:

$$ATD(B,\ 2) = \frac{1+1+2+2+2}{5} = 1,6 \tag{7}$$

Maximum for $(B,\ 2) = 4$
Applying (1) for phase 2 of project B:

$$ATD(B,\ 3) = \frac{2+2+2}{3} = 2 \tag{8}$$

Maximum for $(B,\ 3) = 2$
Applying (1) for phase 3 of project B:

$$ATD(B,\ 4) = \frac{2+2+2}{3} = 2 \tag{9}$$

Maximum for $(B,\ 4) = 2$

3.2.4. Conclusions and Interpretation of Results

After applying (1), to all the open innovation project phases of Figure 2, it can be concluded that for the first two project phases, the ATD value of project A is almost twice the value of project B. This means that the organizations that participated in project A at the first two phases were more connected (through the email communication network) than the organizations that participated in project B, in the same project phases.

For example, in phase 1 of project B, the email communication network was much more centralized (in this case O1 has a disproportional number of links in relation to the other organizations) than in project A of the same project phase. In other words, attending only to the first two phases of projects A and B, one can conclude that a more centralized email communication network, between the organizations that participated in open innovation projects, is associated with project failure outcome.

However, when analyzing phase 4, the same conclusion cannot be taken. In fact, there is no difference between the ATD value for both projects A and B, but project B failed, and project A succeeded. This means that this metric (ATD) is no longer suitable for identifying differences regarding the interactional collaborative dimensions (the search of repeatable behaviors) between projects that were successfully delivered and projects that were unsuccessfully delivered. In this case, a new metric, based on centrality SNA metrics, should be applied.

Throughout this brief application case regarding Projects A and B illustrated in Figure 2, it is clearly demonstrated, first, how dynamic relationships (networks of collaboration) between interacting entities organized in any network form can be quantitatively measured, regardless of the size of the network, and second, how conclusions can be outdrawn regarding the association between quantitative dynamic behaviors and project outcomes (success or failure).

Furthermore, the proposed model in this work is aligned with the findings from Pinto and Slevin [42], as they stated that project critical success factors change the importance degree, according to the project phase of a project lifecycle.

3.3. OI-RM Project Success Profile and Project Failure Profile

Continuing with the example illustrated in Figure 2, if the project lifecycle of both open innovation projects A and B were the average result regarding the number of organizations, and the number of email communication channels that existed in each phase of a set of analyzed open innovation successfully and unsuccessfully delivered projects, one could say that repeatable behaviors would have been found, regarding the email communication network. In other words, the resulting project lifecycles presented in Figure 2, would not represent the email communication of one project, rather of a set of projects delivered, which would be called as OI-PSP (open innovation project successful profile—for the project lifecycle (a) of Figure 2), and OI-PFP (open innovation project failure profile—for the project lifecycle (b) of Figure 2). In this case, one could say that critical success has been identified regarding the email communication network, with a measurable value associated to them. Furthermore, function of the number of analyzed delivered projects, one could talk about a certain working open innovation collaborative culture.

3.4. OI-RM Model Application Span

The proposed model in this work (OI-RM) is not limited to a certain number of *project phases* of a project lifecycle. For the example of Figure 2, a four-phase model project lifecycle was adopted. However, the number of project phases of projects to be analyzed by the proposed model must be the same for both successfully and unsuccessfully delivered open innovation projects. The proposed model was designed to be applicable, regardless of *project size and complexity*. The *project organization-competencies* are not limited to those mentioned in Table 4, as long they are well defined for both successfully and unsuccessfully delivered projects. Still, the model does not preview a maximum number of project *organizations* that take part in in the different phases of open innovation project lifecycles.

3.5. OI-RM Model Part 1 and Part 2

The proposed model in this work has two parts (Figure 3). In part 1, the model aims to identify open innovation project critical success factors, and in part 2, the model provides guidance and estimates an outcome likelihood of an ongoing open innovation project, by comparing the deviation between the real, and the and the desired evolution of the ongoing project, based on identified project critical success factor in pat 1. To be able to run part 2 of the model, part 1 of the model needs to be previously run. This means, critical success factors must have been previously found, otherwise, part two of the model has no effect so ever.

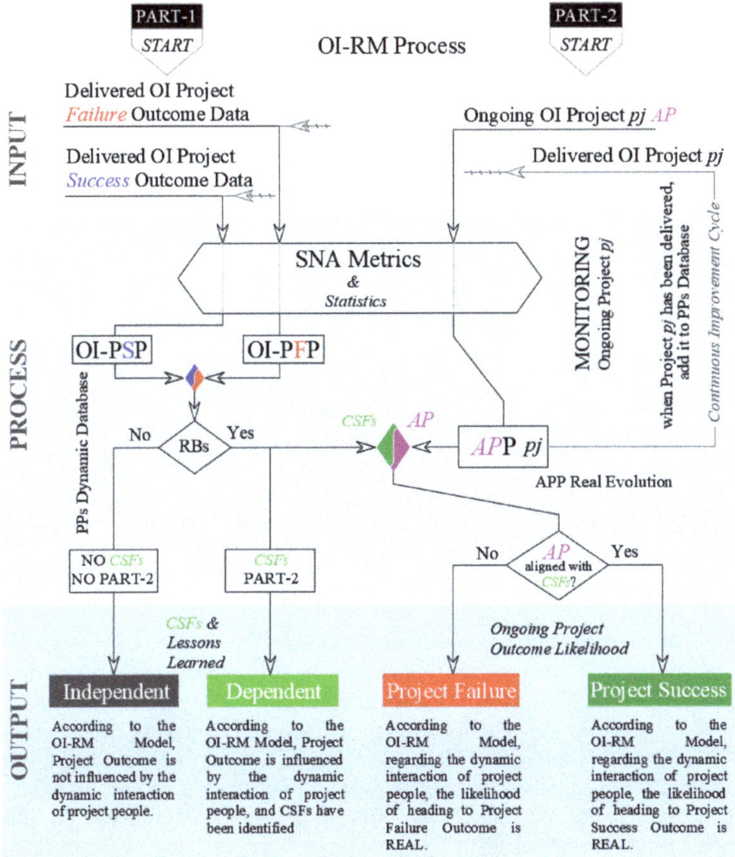

Figure 3. OI-RM model process for both Part 1 and Part 2.

A framework for both part 1 and part 2 of the proposed model is illustrated in Figure 3. To properly understand how the proposed model functions, Figures 3 and 4 should be interpreted in parallel. In Figure 3, the model process of the OI-RM model is illustrated, divided in three main blocks. They are Input (which relates to the necessary data that will be analyzed by the proposed model), Process (which relates to the mathematical and statistical operations of the proposed model), and Output (which relates to the quantitative results and conclusion provided by the proposed model). In Figure 3 on the left side (PART-1 START), the process for part 1 of the proposed model is illustrated. First, delivered OI (open innovation) failure and success outcome data—arising from project email information-related and project meetings in each phase of delivered project lifecycles (Figure 4)—is collected from a set of delivered projects and undergo a process of analysis by the application of social network analysis metrics and statistics (Figure 3). After all introduced projects have been quantitatively analyzed, they follow the average process of creating a OI-PSP (open innovation project success profile), and a OI-PFP (open innovation process failure profile), which represent in average, the repeatable behaviors of all analysis of successfully and unsuccessfully delivered projects.

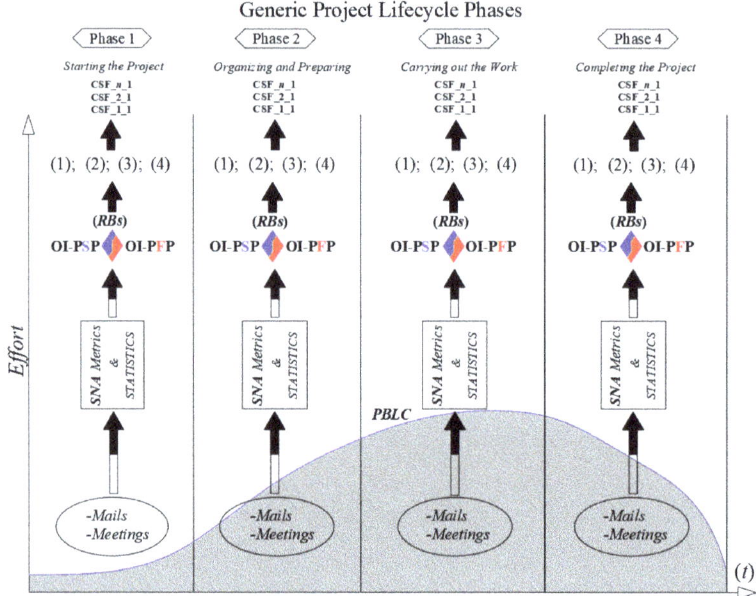

Figure 4. OI-RM Model—Part 1 and 2 Process, for Success, and Failure open innovation project outcomes. Source: Adapted from PMI, 2017 [40].

Next, the averaged repeatable behaviors from all successfully delivered projects will be compared with the averaged repeatable behaviors from all unsuccessfully delivered projects, and if unique differences between successfully and unsuccessfully delivered projects, regarding the four interactional collaborative dimensions have been found, then open innovation project critical success factors have been identified (Figures 3 and 4). If not, the conclusion to be taken is that, according to the proposed model, the dynamic collaborative interaction between the different organizations that participate across all the different phases of an open innovation project lifecycle do not influence a project outcome. In other words, they are independent. At this point, part 1 of the model is concluded. On the right side of Figure 3 (PART-2 START) part 2 of the proposed model is illustrated. At this point, it is only meaningful to go further, if previously critical success factors have been identified. In part 2 of the proposed model, the objective is to use the identified critical success factors identified in part 1 of the model to provide guidance across the evolution of an ongoing open innovation project.

First, at an ongoing project *pj*, one must to define in which project phase the ongoing project is, run the model for the collected data until the actual point that defined the present status of the ongoing project (AP), and run the comparison between the critical success factors for the respective project phase to be analyzed, and the results obtained for the AP point of the ongoing project. By the same analogy, the resulting measurable outputs for the AP point, represent an actual ongoing project profile (APP). If the results of the comparison show, that the evolution of the ongoing project at AP point, regarding the four interactional collaborative dimensions, are aligned with the critical success factor previously identified, that the conclusion to be taken is that the likelihood of heading to a successful project outcome is real (Figure 3). If not, then the likelihood of heading to an unsuccessful outcome is real, by opposition. After the project *pj*, is delivered, it undergoes once more through the all analysis process, and the results will be added to the project profiles dynamic database (PPS dynamic database), contributing for the refinement of the identification of critical success factors.

This last step is previewed by the model as the continuous improvement cycle (Figure 3). Although in the present work is not objective to demonstrate the process of estimating an open innovation project outcome likelihood, the outcome likelihood will be estimated by applying a simple averaging mathematical process, which essentially is based on the highest percentage of metrics indicating success or failure outcome. In other words, the more metric results–for an ongoing open innovation project–are aligned with the critical success factors identified in part 1 of the model, the highest is the success outcome likelihood for that ongoing project.

3.6. OI-RM Model Requirements

In Table 5, are illustrated the required open innovation projects information to be collected regarding mails and meetings projects, that are needed as input to the proposed model. Project meetings refer to any type of F2F meetings, that occur in each phase of a project lifecycle. Project mails, refers to all the project email information-related data, that was exchanged between the different project organizations, in each phase of a project lifecycle.

Table 5. Required information for input to the IO-RM model.

Open Innovation Projects Information	
Project Meetings	- Total number of project meetings occurred in each open innovation project phase, across a project's lifecycle - Total number of participating organizations in each open innovation project meeting, in each project phase, across a project's lifecycle - Organization Project Official Competency, from each of the participating organizations across an open innovation project lifecycle

Project Mails	- Total number of emails sent/received in each phase of an open innovation project, that relate to project information data related to.
	 - Organization Project Official Competency, from each participating organization that sent/received emails project related information.
	 - Chronological Mail Exchange Time (send/received/answered)
	 - Categorize emails according *:
	   ○ Mails sent in seeking for help, or advice regarding project information related matter
	   ○ Mails sent, providing help, or advice, regarding project information related matter

* Mails need to be identified and characterized, either by their subject or content, as being seeking mails type, or providing help mails type.

3.7. OI-RM Model four Interactive Collaborative Dimensions (4-ICD) and Respective Metrics

As previously mentioned, the proposed model in this work, will look for repeatable behaviors that are associated with successful, and unsuccessfully delivered open innovation projects, in four different dimensions (Table 6). These dimensions, named as interactive collaborative dimensions (4-ICD), that usually occur between organizations across a project lifecycle, are described in Table 6, as well as which SNA centrality metrics will be applied.

Table 6. The four interactive collaborative dimensions (4-ICD) of the IO-RM model.

(a)	**Key Project Organization Communication and Insight Degree**	Description and Objective: How does important project organizations (function of their competency across the accomplishment of an open innovation project) present at the in-project meetings and emails networks, and to which extent their presence influences a certain project outcome. Regarding Meetings: How the presence of those important project organizations in project meetings, triggers communication and insight on what is ongoing throughout the different phases of a project lifecycle, namely at the transitional phase of the different project phases. SNA Metric: For this case, the *total-degree* (C_{DT}) [45] SNA metric will be applied, to first measure the project meetings participation degree of each participating organization in each open innovation project phase. $$C_{DT}(n_i) = \sum_j x_{ji}$$ Where: C_{DT} = total degree of an entity within a graph n = total number of entities within a network (graph) for $i = 1 \ldots ,n$ x_{ji} = number of links from entity j to entity i, where $i \neq j$, and vice-versa. After having all the total degrees for each participating organization, a linear regression will be applied to characterize the evolution within a given project phase. There are three possible outcomes. They are 1- Negative evolution: characterizes a decreasing participation degree as a given project phase heads to its end. 2- Positive evolution: characterizes an increasing participation degree as a given project phase heads to its end. 3- Neutral evolution: characterizes a stable (continuously) participation degree as a given project phase heads to its end. Regarding Mails: How cohesively is the mail communication network? Are email communication channels open to all the organizations that participate in project activities across all different phases of a project lifecycle? SNA Metric: For this case, the *density* (**Ds**) [45] will be used, to characterize the amount of existing email communication channels that exist between the different organizations that participate in open innovation projects. $$Ds = \frac{N_{L\ REAL}}{N_{LMAX}}$$ Where: Nr of Maximum Possible ties = $N_{LMAX} = \frac{n(n-1)}{2}$ n = number of entities within a graph The outcome for this metric is: (a) Full density: characterized by an email communication network that reaches all the organizations that participate in a project (b) Relative density: ratio between all possible email communication channels, and existing email communication channels.

Table 6. *Cont.*

		Description and Objective: To which extent does a given organization controls and holds "power" over the email communication network, in terms of send/receive project information related. Regarding Mails: How is the volume of mail communication between the different participating organizations in open innovation projects? Who holds the most volume of email communication related to project information data? SNA Metric: For this case, the *In-degree*, and *Out-degree* (C_{OT}) [45] will be applied in order to identify which organization holds control over the email communication network.
(b)	**Organizational Control Degree**	$$C_{OT}(n_i) = \sum_j x_{ji}$$ Where: C_{OT} = total out-degree of an entity within a graph n = total number of entities within a network (graph) for $i = 1 \ldots ,$n x_{ji} = number of links from, only entity j to entity i, where $i \neq j$. The output for this metric, is: (a) Full control: organization holds completely control over the email communication network across a project phase. (b) Average control: several organizations hold control, over the email communication network across a project phase.
		Description and Objective: To which extent, does the project-related information, provided by one organization to other organization is recognized as important and decisive to enable evolution in project activities? What is the degree of dependency of a given organization regarding to another organization in order to accomplish project activities or tasks? Regarding Mails: How is the volume of emails sent seeking vs providing vital information to project activities? SNA Metric: For this case, will be used the *Out-degree* (C_{OT}) (4) and the *Average degree* (A_D) [45], will be applied, which will characterize how much a given organization is dependent on other organization to accomplish project activities.
(c)	**Project Information Dependency Degree**	Average Degree: $$A_D(n_i) = \frac{\sum_j x_{ji}}{n} = \frac{\sum_{i=1}^{n} C_{OT}(n_i)}{n}$$ Where: A_D = Average degree n = total number of entities within a network (graph) The output for this metric is: 1- Total dependency: characterized by an organization that receives input from other organizations for all activities that respects an open innovation project. 2- Shared dependency: characterized by the existence of several organizations that receives input from other several organizations for all activities that respects an open innovation project.

Table 6. *Cont.*

		Description and Objective: To which extent, does the speed of answering emails by providing or seeking project information related, influences project outcome?
		Regarding Mails: How fast or how slow is the speed of answering a request from an organization to other organization? Analyze the volume of emails sent/received crossing them with the chronological time.
		SNA Metric: For this case, first the *reciprocity* (**R**) [45] will be used to analyze which emails were answered providing project information related, and second, the chronologic time associated to each pair sent/received.
(d)	**Feedback Readiness Degree**	$R = \dfrac{L^{<->}}{L}$ Where: $L^{<->}$ = Number of links pointing in both directions L = *total number of links within a network* The output for this metric is: • An average value, in hours, ranging from "1" (meaning an instantaneous answer reply has been made in less than 1 h period of time) up to a maximum of the project time duration "0", for cases where feedback is not to be found, across the lifetime of a project.

3.8. IO-RM Model Implementation Process

In Figure 5, is illustrated the implementation process framework of the proposed model for both, parts 1 and 2. The framework illustrated in Figure 5, is details the implementation process for project phase 1, however the process is not exclusively of phase 1 of a project, rather is to be fully replicable for all different project lifecycle phases across the PBLC (project baseline curve), which represents the planned project evolution. In Figure 5 are represented the project meetings chronology, and emails that were exchanged within phase 1 of project-phase 1. Project meetings and emails must be documented, as previously seen, according to Table 5. At the beginning of each project phase, a formal Competencies Chart (displayed at the left side, above the first project meeting (E1) in Figure 5), must be defined, where all the organizations that are assigned to take part of the activities of an open innovation project, have well defined project competencies and responsibilities. Organizations can have different project competencies, as previously seen, such as those in Table 4. In Figure 5, are illustrated six open innovation project meetings, that did occur across phase 1 of the illustrated project. These project meetings, or events (E) are represented by E1, E2, E3, E4, E5, and Et, where Et–for the case of Figure 5–can represents the sixth project meeting.

Project Meeting E1, was the first meeting that did occur within phase 1 of the project, and project meting Et represents the last project meeting of phase 1. For example, at meeting E1 (left red marked dot in Figure 5), three organizations were present. They are O1, O2, and O3, which represent competencies a, b, and c respectively. The lines that connect the three organizations, inside each box above each project meeting (Meetings), represent the relationship degree between them, regarding the pairwise meetings participation degree across the phase 1 of the illustrate project. For example, in project meeting 1, as it is the first time that O1, O2, and O3 are together in any project meeting or the project illustrated in Figure 5, they have a line with value 1× (relationships degree box). In project meeting E4, organizations O1, O2, and O3 have degree 4×, because is the fourth project meeting that they participate together. The boxes above the Meetings boxes (Σ *Emails*), represent all the project information related exchanged emails between the period of any two project meetings. The lines represent the email communication channels. For example, between project meetings E1, and E2, organizations O1, O3, O4 and O6, are almost all connected through the email communication network, except for O1 and O6.

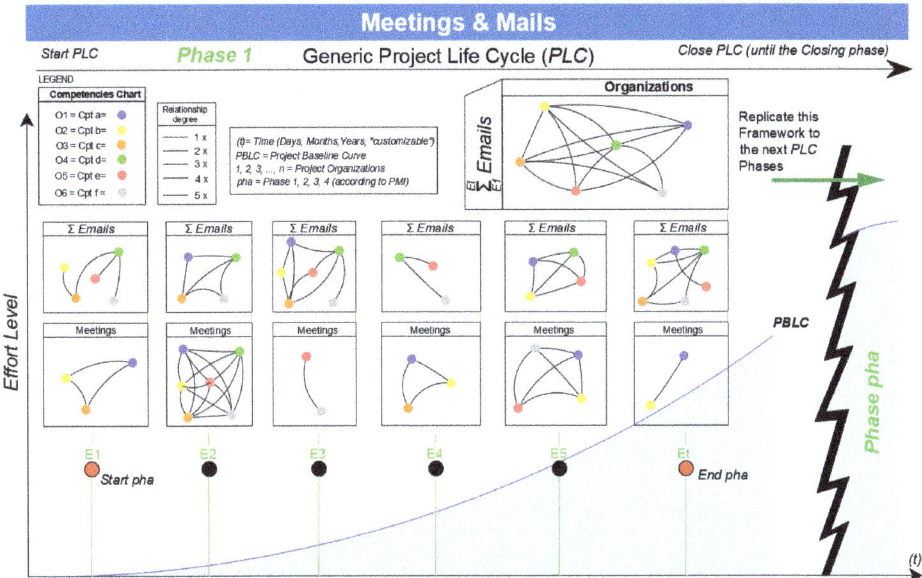

Figure 5. Implementation Timeline Framework of the OI-RM model.

At the end of each phase of a project lifecycle, all the project information related exchanged emails must be collected ($\sum$ *Emails*), as it is illustrated in Figure 5 at the box above the last project meeting E6 (Et). The end of a project phase may be determined by a last project meeting or by an open innovation project milestone. The process above described–related to Figure 5–is to be replicated throughout all the other project phases of a project lifecycle.

3.9. IO-RM Legal and Ethical Considerations

The proposed model in this work, accesses, and analysis open innovation project related information that flows across the different project organizations throughout the different phases of a project lifecycle. Such project information is very often considered confidential and therefore not desired to be accessible and exposed to the exterior of organizations. This aspect may represent a constraint to the implementation of the proposed model. Therefore, the effective implementation of the proposed model in this work may be dependent on the acceptance of the competent authorities at the organizational and nation level that manage the legal and ethical respective data protection issues. Nevertheless, it is expected that all the project organizations that participate in an open innovation project that will be supported by the proposed model should be informed of it, before the collaborative project starts.

4. Conclusions, Implications, and Further Developments

The proposed model in this work, aims to quantitatively identify project critical success factors regarding the dynamic behavioral interactions between the different entities that participate across the different phases of a project lifecycle in a collaborative network context such as the open innovation model. The proposed model in this work was developed based on four scientific fields: (1) Collaborative Networks, (2) Project Management, (3) Risk Management, and (4) Social Network Analysis, integrating in a holistic way the individual proven benefits for organizations from each of the four scientific fields, essentially regarding performance and innovation. Concretely the proposed model has two parts. In part 1, the model will analyze and quantitatively measure those dynamic behavioral interactions through the application of social networks analysis centrality measures, using data arriving form project meetings and project email exchange, from successfully and unsuccessfully delivered collaborative projects, searching for repeatable behaviors associated to each of the outcomes (success or failure). In part 2, the proposed model offers a framework that provides guidance to an ongoing collaborative project. Here, the proposed model measures the deviation between the real evolution of an ongoing project against a desired (planned) evolution of the ongoing project, having a criteria the identified critical success factors in part 1 of the model.

4.1. Proposed Model and Researched Literature

The proposed model in this work has as main aim to provide organizations with a model to support the management of collaborative network projects, such as the open innovation system, where the lack of such models is pointed out as the major obstacle [23] for organizations to more often engage in collaborative network project models. The model addresses two of the most important risks in collaborative network projects as proposed by [24] and [41], which are detailed in chapter 2 of this work. They are (1) collaborative risks [24] (comprised by a subset of risks such as critical enterprises risk, assigning tasks to partners risk, and behavioral risk) and (2) ambiguity risks [41], also known as "epistemic uncertainty", where uncertainties essentially arise from lack of knowledge or understanding. According to [24] and [41], both risks (collaborative and ambiguity) can effectively be managed by consulting project lessons learned and undertaking simulations to systems before implementation on the field.

The proposed model in this work provides valuable and unique insight into those dimensions by, first, identifying in a quantitatively way invisible dynamic behavioral interactions that cannot be fully understood by traditional statistical tools and techniques, which will later enable to monitor and simulate the evolution of a system, which in this case is the project social network.

Furthermore, the proposed model also addresses one of the most trendy subjects that organizations currently face—the organizational transformation through digitalization—defended by several authors [7,12,20,31] and known as well by industry 4.0 [32], which argues that organizations need to change the way they think and do work, addressing organizational processes, procedures, and mindsets transforming themselves into adaptable machine learning systems, through formal and informal networks of collaboration.

The proposed model in this work also addresses this subject as—once properly implemented in organizations—a continuous improvement cycle (Figure 3, Process) takes place, regarding the refinement of the identification of critical success factors process in collaborative projects, as it is demonstrated in chapter 3 of the present work.

4.2. Managerial Implications

In a nutshell, the key findings in this work, essentially regard the demonstration of the applicability of the proposed model in detecting (in a quantitatively way) dynamic interactive behavioral patterns associated to unsuccessful and successfully delivered projects, run under a collaborative network model approach such as the open innovation model, by essentially measuring communication and information flow exchange between the entities (organizations) that comprise a collaborative project social network, across the different phases a project lifecycle.

The proposed model in this work, provides organizations with a valuable historic picture regarding how collaboration between the different organizations that did participate in collaborative network projects occurred, across the different phases of a project lifecycle. In other words, the model provides organizations with a dynamic lesson-learned knowledge- base, which enables them to learn from past experiences (failures and successes) regarding which dynamic behavioral interactions are associated with success or failure project outcomes.

From a macro perspective, the continuous application of the proposed model in this work in organizational collaborative network projects, such as open innovation, will help organizations to identify and quantify different collaborative working cultures that emerge as they work in collaborative network models, enabling thus to identify which collaborative working culture is more effective regarding organization performance and innovation.

There are still other advantages that the proposed model offers from a micro perspective, compared with other HR people analytics models, where the data to model and analyze organization performance is usually collected through pulse survey or 360° questionnaire approaches. In this dimension, the data collecting method of the proposed model in this work is almost completely bias-free and eliminates organizational down-time as organization members do not need to answer performance and quality pulse surveys regarding how collaboration occurred when they worked in an open innovation project environment.

The identification of project critical success factors is a mathematical process, with quantitative outcomes. From a management perspective, where managers need to take decisions most of the time based on quantitative approaches expecting to improve the quality of results (performance and innovation), the proposed model in this work, by outputting quantitative results regarding the interactive dynamic behavior of the different entities that participate in collaborative network projects, enables organizations to quantitatively understand the effect of such dynamic behavior in organizational outcomes and to craft strategies and take actions in a more data-driven way, rather than traditional approaches essentially based on gut feeling and key influencers' opinions.

Still, from a managerial perspective, the model provides the organization with another benefit related to the actual trend, working through collaborative networks. By quantitatively analyzing the influence that the blur of formal and informal networks of collaboration in successfully and unsuccessfully delivered projects, it provides managers with a unique insight into the real importance of essential informal networks of collaboration, enabling them to take appropriate action in order to support, maintain, or even foster collaborative network dynamics that are associated with successful outcomes.

Finally, the proposed model in this work aims to provide organizations with a much clearer insight regarding how organizations can benefit from the integration of relational data (data that quantitatively mirrors the dynamic network relationships between entities in collaborative network projects) into their organizational strategic management processes or frameworks, where an effective implementation of the proposed model in organizations will enable them to do more with less, thus contributing to the achievement of sustainable competitive advantages.

This means that, as the proposed model in this work enables organizations to better plan and manage their organizational collaborative networks (by understanding and identifying the critical factors that drive successful project outcomes), organizations reduce or eliminate risks associated with collaborative projects (collaborative risks), which in turn optimizes resource usage, orienting organizations towards being leaner, which ultimately will contribute positively to economic, social, and environmental sustainability.

4.3. Suggestions for Future Research

The implementation of the proposed model in organizations can be a challenge to them. This may occur essentially because organizations need to access to the right technology and the employment of a working culture, as they work through collaborative networks, that enables necessary data to the proposed model in this work, to be recorded as previewed in Table 5. Creating and implementing an automated process that collects the necessary data to the proposed model demands appropriate technology that may not be at reach for most organizations.

Here, further research should be undertaken, in order to develop a system that can be efficiently implemented in organizations where data can be properly collected and the impact to the working culture be minimized as much as possible so that the collected data mirrors as much as possible how really collaboration occurs.

The proposed model in this work only collects data from project meetings and emails. However, much project related information flows through other communication and collaboration channels such as phone calls, instant messaging, corridor meetings, and so on. Essentially, due legal constraints, the collaboration done thought these other mentioned channels is not captured by the actual proposed model. Here, further research should be undertaken to develop new data collection methods in order to be able to filter personal from professional interactions as entities participate in collaborative projects, so that collaboration done through those other channels mentioned should be able to be captured and analyzed.

Finally, it is suggested that research should be continuously undertaken regarding the development of new social network analysis metrics that can complement existing metrics to better identify, measure, and understand dynamic behavioral patterns that occur as organizations engage in collaborative network projects.

Author Contributions: Author M.N. carried out the investigation methodology, writing—original draft preparation, conceptualization, the formal analysis, and collected resources. Author A.A. contributed with the conceptualization, writing—review and editing, the supervision, and final validation. All authors have read and agreed to the published version of the manuscript.

Funding: This research received no external funding.

Conflicts of Interest: The authors declare no conflict of interest.

References

1. Nuryakin, M. Competitive Advantage and Product Innovation: Key Success of Batik SMEs Marketing Performance in Indonesia. *Acad. Strat. Manag. J.* **2018**, *17*, 1544–1558. Available online: https://www.abacademies.org/articles/competitive-advantage-and-product-innovation-key-success-of-batik-smes-marketing-performance-in-indonesia-7164.html (accessed on 1 January 2020).

2. Ng, D.; Law, K. Impacts of informal networks on innovation performance: Evidence in Shanghai. *Chin. Manag. Stud.* **2015**, *9*, 56–72. [CrossRef]

3. Hansen, M. *Collaboration: How Leaders Avoid the Traps, Create Unity, and Reap Big Results*; Harvard Business School Press: Cambridge, MA, USA, 2009.

4. Lutfihak, A.; Evrim, G. Disruption and ambidexterity: How innovation strategies evolve? *Soc. Behav. Sci.* **2016**, *235*, 782–787.

5. Lee, K.; Woo, H.; Joshi, K. Pro-innovation culture, ambidexterity and new product development performance: Polynomial regression and response surface analysis. *Eur. Manag. J.* **2017**, *35*, 249–260. [CrossRef]

6. Ullrich, A.; Vladova, G. Weighing the Pros and Cons of Engaging in Open Innovation. *Technol. Innov. Manag. Rev.* **2016**, *6*, 34–40. [CrossRef]

7. Chesbrough, H. How to Capture All the Advantages of Open Innovation. 2020. Available online: https://hbr.org/podcast/2020/01/how-to-capture-all-the-advantages-of-open-innovation (accessed on 27 February 2020).

8. KPMG European Innovation Survey. 2016. Available online: https://assets.kpmg/content/dam/kpmg/nl/pdf/2016/advisory/KPMG-European-Innovation-Survey.pdf (accessed on 3 April 2020).

9. Sabine, B.; Henry, C. The Adoption of Open Innovation in Large Firms. *Res.-Technol. Manag.* **2018**, *61*, 35–45. [CrossRef]

10. Chesbrough, H. The Era of Open Innovation. *MIT Sloan Manag. Rev.* **2003**, *44*, 35–41.

11. Pertusa-Ortega, E.; Molina-Azorin, J. A joint analysis of determinants and performance consequences of ambidexterity. *BRQ Bus. Res. Q.* **2018**, *21*, 84–98. [CrossRef]

12. Arena, M. *Adaptive Space: How GM and Other Companies are Positively Disrupting Themselves and Transforming into Agile Organizations*; McGraw Hill Education: New York, NY, USA, 2018.

13. Krackhardt, D.; Hanson, J. *Informal Networks the Company behind the Charts*; Harvard College Review: Cambridge, MA, USA, 1993; Available online: https://hbr.org/1993/07/informal-networks-the-company-behind-the-chart (accessed on 5 March 2018).

14. Bouncken, R.; Brem, A.; Kraus, S. Multi-cultural teams as sources for creativity and innovation: The role of cultural diversity on team performance. *Int. J. Innov. Manag.* **2015**, *20*, 1650012. [CrossRef]

15. Hewlett, S.; Marshall, M.; Sherbin, L. How Diversity Can Drive Innovation—From the December 2013 Issue—Harvard Business Review. 2013. Available online: https://hbr.org/2013/12/how-diversity-can-drive-innovation (accessed on 1 February 2020).

16. Cross, R. Networks and Innovation: Rob Cross, Mack Institute for Innovation Management. 2013. Available online: https://www.youtube.com/watch?v=-ECSFbgWSFo (accessed on 2 March 2019).

17. Arena, M.; Cross, R.; Sims, J.; Uhl-Bien, M. How to Catalyze Innovation in Your Organization. MIT, 2017. Available online: https://www.robcross.org/wp-content/uploads/2017/10/how-to-catalyze-innovation-in-your-organization-connected-commons.pdf (accessed on 1 January 2019).

18. Urze, P.; Abreu, A. Mapping Patterns of Co-innovation Networks. In *Collaboration in a Hyperconnected World. PRO-VE. IFIP Advances in Information and Communication Technology*; Afsarmanesh, H., Camarinha-Matos, L., Lucas Soares, A., Eds.; Springer: Berlin/Heidelberg, Germany, 2016; Volume 480.

19. Cowen, A.; Vertucci, L.; Thomas, T.; Cross, R. Leading in a connected world: How effective leaders drive results through networks. The network roundtable at the University of Virginia. 2017. Available online: https://www.robcross.org/wp-content/uploads/2016/10/org_dynamics_leading_in_a_connected_world.pdf (accessed on 1 August 2019).

20. Workday Studios, In Good Company—Michael Arena, Chris Ernst, Greg Pryor: Organizational Networks. 2018. Available online: https://www.youtube.com/watch?v=6faV0v0yVFU (accessed on 12 July 2019).

21. Dutton, W.H. Collaborative Network Organizations: New Technical, Managerial and Social Infrastructures to Capture the Value of Distributed Intelligence (November 17, 2008). OII DPSN Working Paper No. 5. Available online: https://ssrn.com/abstract=1302893 (accessed on 4 March 2020). [CrossRef]

22. Narsalay, R.; Kavathekar, J.; Light, D. A Hands-Off Approach to Open Innovation Doesn't Work. 2016. Available online: https://hbr.org/2016/05/a-hands-off-approach-to-open-innovation-doesnt-work (accessed on 4 March 2020).

23. Santos, R.; Abreu, A.; Anes, V. Developing a Green Product-Based in an Open Innovation Environment. Case Study: Electrical Vehicle. In *Collaborative Networks and Digital Transformation*; Camarinha-Matos, L., Afsarmanesh, H., Antonelli, D., Eds.; IFIP Advances in Information and Communication Technology; Springer: Cham, Switzerland, 2019; Volume 568.

24. Abreu, A.; Martins Moleiro, J.D.; Calado, J.M.F. Fuzzy Logic Model to Support Risk Assessment in Innovation Ecosystems. In Proceedings of the 13th APCA International Conference on Automatic Control and Soft Computing (CONTROLO), Ponta Delgada, Portugal, 4–6 June 2018; pp. 104–109. [CrossRef]

25. Kadushin, C. *Understanding Social Networks Theories, Concepts, and Findings*; Oxford University Press: New York, NY, USA, 2012.

26. Cross, R.; Prusak, L. The People Who Make Organizations Go—Or Stop. June 2002 Issue of Harvard Business Review. 2012. Available online: https://hbr.org/2002/06/the-people-who-make-organizations-go-or-stop (accessed on 15 March 2020).

27. Cross, R.; Rebele, R.; Grant, A. Collaborative Overload. *Harv. Bus. Rev.* **2016**, *94*, 74–79.

28. Björkman, I.; Kock, S. Social relationships and business networks: The case of Western companies in China. *Int. Bus. Rev.* **1995**, *4*, 519–535. [CrossRef]

29. Kontinen, T.; Ojala, A. Network ties in the international opportunity recognition of family SMEs. *Int. Bus. Rev.* **2011**, *20*, 440–453. [CrossRef]

30. Nunes, M.; Abreu, A. Applying Social Network Analysis to Identify Project Critical Success Factors. *Sustainability* **2020**, *12*, 1503. [CrossRef]

31. Müller, J.; Buliga, O.; Voigt, K. The Role of Absorptive Capacity and Innovation Strategy in the Design of Industry 4.0 Business Models-A Comparison between SMEs and Large Enterprises. *Eur. Manag. J.* **2020**. [CrossRef]

32. Digital Transformation Monitor of the European Commission. Germany: Industrie 4.0. 2017. Available online: https://ec.europa.eu/growth/tools-databases/dem/monitor/sites/default/files/DTM_Industrie%204.0.pdf (accessed on 3 April 2020).

33. Camarinha-Matos, L.; Afsarmanesh, H. Collaborative Networks: Value creation in a knowledge society. In *Knowledge Enterprise: Intelligent Strategies in Product Design, Manufacturing, and Management*; International Federation for Information Processing Digital Library; Springer: Boston, MA, USA, 2006; Volume 207. [CrossRef]

34. Denise, L. Collaboration vs. C-Three (Cooperation, Coordination, and Communication). *INNOVATING Reprint* **1999**, *7*, 3.

35. Abreu, A.; Camarinha-Matos, L. On the Role of Value Systems and Reciprocity in Collaborative Environments. In *Network-Centric Collaboration and Supporting Frameworks*; IFIP International Federation for Information Processing; Springer: Boston, MA, USA, 2006; Volume 224, pp. 273–284. [CrossRef]

36. Innocentive. Available online: https://www.innocentive.com/ (accessed on 1 January 2019).

37. Rauter, R.; Globocnik, D.; Perl-Vorbach, E.; Baumgartner, R. Open innovation and its effects on economic and sustainability innovation performance. *J. Innov. Knowl.* **2018**, *4*, 226–233. [CrossRef]

38. Deichmann, D.; Rozentale, I.; Barnhoorn, R. Open Innovation Generates Great Ideas, So Why Aren't Companies Adopting Them? Available online: https://hbr.org/2017/12/open-innovation-generates-great-ideas-so-why-arent-companies-adopting-them (accessed on 1 January 2019).

39. West, J.; Gallagher, S. Challenges of Open Innovation: The Paradox of Firm Investment in Open-Source Software. *Fac. Publ.* **2006**, *36*, 319–331. [CrossRef]

40. PMI® (Project Management Institute). *Project Management Body of Knowledge (PMBOK® Guide)*, 6th ed.; Project Management Institute, Inc.: Newtown Square, PA, USA, 2017.

41. Hillson, D. *How to Manage the Risks You Didn't Know You Were Taking*; Phoenix, A.Z., Ed.; Paper presented at PMI® Global Congress 2014—North America; Project Management Institute: Newtown Square, PA, USA, 2014.

42. Pinto, J.K.; Slevin, D.P. Critical success factors across the project life cycle: Definitions and measurement techniques. *Proj. Manag. J.* **1988**, *19*, 67–75.

43. Li, L.; Li, Z.; Wu, G.; Li, X. Critical Success Factors for Project Planning and Control in Prefabrication Housing Production: A China Study. *Sustainability* **2018**, *10*, 836. [CrossRef]

44. ISO—The International Organization for Standardization. Available online: https://www.iso.org/home.html (accessed on 1 January 2019).

45. Wasserman, S.; Faust, K. Social Network Analysis in the Social and Behavioral Sciences. In *Social Network Analysis: Methods and Applications*; Cambridge University Press: Cambridge, MA, USA, 1994; pp. 1–27. ISBN 9780521387071.

46. Scott, J. *Social Network Analysis: A Handbook*, 4th ed.; SAGE: London, UK, 2017.

47. Borgatti, S. Introduction to Social Network Analysis Stephen, University of Kentucky. 2016. Available online: https://statisticalhorizons.com/wp-content/uploads/SNA-Sample-Materials.pdf (accessed on 15 January 2019).

48. Kacanski, S.; Lusher, D. The Application of Social Network Analysis to Accounting and Auditing. *Int. J. Acad. Res. Account. Financ. Mang. Sci.* **2017**, *7*, 182–197. [CrossRef]

49. Ward, M.; Stovel, K.; Sacks, A. Network Analysis and Political Science. *Annu. Rev. Political Sci.* **2011**, *14*, 245–264. [CrossRef]

50. Clifton, A.; Webster, G. An Introduction to Social Network Analysis for Personality and Social Psychologists. *Soc. Psychol. Pers. Sci.* **2017**, *8*. [CrossRef]

51. Jarman, D.; Eleni, T.; Hazel, H.; Ali-Knight, J. Social network analysis and festival cities: An exploration of concepts, literature and methods. *Int. J. Event Festiv. Manag.* **2014**, *5*, 311–322. [CrossRef]

52. Maghraoui, A.; Richman, B.; Siegel, D.; Zanalda, G. Networks of Cooperation and Conflict in the Middle East (2017–2018). 2019. Available online: https://bassconnections.duke.edu/project-teams/networks-cooperation-and-conflict-middle-east-2017-2018 (accessed on 4 January 2020).

53. Durland, M.; Fredericks, K. An Introduction to Social Network Analysis. *New Dir. Eval.* **2006**, *2005*, 5–13. [CrossRef]

54. Mitchell, J.C. *Social Network Analysis for Organizations*, 4th ed.; Tichy, N.M., Tushman, M.L., Fombrun, C., Eds.; Academy of Management Review: Briarcliff Manor, NY, USA, 1979; pp. 507–519.

55. Abreu, A.; Camarinha-Matos, L.M. Understanding Social Capital in Collaborative Networks. In *Balanced Automation Systems for Future Manufacturing Networks. BASYS 2010. IFIP Advances in Information and Communication Technology*; Ortiz, Á., Franco, R.D., Gasquet, P.G., Eds.; Springer: Berlin/Heidelberg, Germany, 2010.

56. Organizational Network Analysis Gain Insight Drive Smart. 2016. Copyright 2016 Deloitte Development LLC. Available online: https://www2.deloitte.com/us/en/pages/human-capital/articles/organizational-network-analysis.html (accessed on 1 February 2019).

57. Meyer, M.; Roodt, G.; Robbins, M. Human resources risk management: Governing people risks for improved performance. *SA J. Hum. Resour. Manag.* **2011**, *9*, 366. [CrossRef]
58. Harary, F. *Graph Theory*; CRC—Westview Press Taylor and Francis: Reading, MA, USA, 1969.
59. Borgatti, S.P.; Everett, M.; Johnson, J. *Analyzing Social Networks*, 2nd ed.; Sage Publications Ltd.: London, UK, 2017.
60. Tichy, M.; Tushman, M.; Fombrun, C. Social Network Analysis for Organizations. *Acad. Manag. Rev.* **1979**, *4*, 507–519. [CrossRef]
61. Ruan, X.; Ochieng, P.; Price, A. The Evaluation of Social Network Analysis Application's in the UK Construction Industry. In Proceedings of the 27th Annual ARCOM Conference, Bristol, UK, 5–7 September 2011; Association of Researchers in Construction Management, ARCOM: London, UK, 2011.
62. Carlsson, L.; Sandström, A. Network governance of the commons. *Int. J. Commons* **2008**, *2*, 33–54. [CrossRef]
63. Newig, J.; Günther, D.; Pahl-Wostl, C. Synapses in the network: Learning in governance networks in the context of environmental management. *Ecol. Soc.* **2010**, *15*, 24. [CrossRef]
64. Müller, R.; Jugdev, K. Critical success factors in projects: Pinto, Slevin, and Prescott—The elucidation of project success. *Int. J. Manag. Proj. Bus.* **2012**, *5*, 757–775. [CrossRef]
65. Zheng, X.; Le, Y.; Chan, A.; Hu, Y.; Li, Y. Review of the application of social network analysis (SNA) in construction project management research. *Int. J. Proj. Manag.* **2016**, *34*, 1214–1225. [CrossRef]
66. Mead, S.P. Using social network analysis to visualize project teams. *Proj. Manag. J.* **2001**, *32*, 32–38. [CrossRef]
67. Cross, R.; Parker, A. *The Hidden Power of Social Networks: Understanding How Work Really Gets Done in Organizations*; Harvard Business School Press: Boston, MA, USA, 2004.
68. Prell, C.; Hubacek, K.; Reed, M. Stakeholder analysis and social network analysis in natural resource management. *Soc. Nat. Resour.* **2009**, *22*, 501–518. [CrossRef]
69. Toomey, L. *Social Networks and Project Management Performance: How Do Social Networks Contribute to Project Management Performance? Paper Presented at PMI® Research and Education Conference, Limerick, Munster, Ireland*; Project Management Institute: Newtown Square, PA, USA, 2012.
70. Mok, K.Y.; Shen, G.Q.; Yang, R.J.; Li, C.Z. Investigating key challenges in major public engineering projects by a network-theory based analysis of stakeholder concerns: A case study. *Int. J. Proj. Manag.* **2017**, *35*, 78–94. [CrossRef]
71. Freeman, L. Centrality in social networks conceptual clarification. *Soc. Netw.* **1979**, *1*, 215–239. [CrossRef]
72. Ove, F. Using centrality modeling in network surveys. *Soc. Netw.* **2002**, *24*, 385.
73. Monge, P.R.; Eisenberg, E.M. Emergent Communication Networks. In *Handbook of Organizational Communication*; Jablin, F.M., Ed.; Sage: Newbury Park, CA, USA, 1987; pp. 305–342.
74. Dogan, S.; Arditi, D.; Gunhan, S.; Erbasaranoglu, B. Assessing Coordination Performance Based on Centrality in an E-mail Communication Network. *J. Manag. Eng.* **2013**, *31*, 04014047. [CrossRef]
75. Wen, Q.; Qiang, M.; Gloor, P. Speeding up decision-making in project environment: The effects of decision makers' collaboration network dynamics. *Int. J. Proj. Manag.* **2018**, *36*, 819–831. [CrossRef]

 *sustainability*

Article

Toward an Evolutionary and Sustainability Perspective of the Innovation Ecosystem: Revisiting the Panarchy Model

James Boyer

Lille Economy and Management, HÉMiSF4iRE Design School, Catholic University of Lille, 59800 Lille, France; James.Boyer@univ-catholille.fr

Received: 25 March 2020; Accepted: 14 April 2020; Published: 16 April 2020

Abstract: This paper proposes an evolutionary and sustainability perspective of the innovation ecosystem. This study revisits the Panarchy model in order to generate new perspectives on the innovation ecosystem. The Panarchy model describes the evolutionary nature of complex adaptive systems relying on four phases, without, however, being deterministic: exploitation, conservation, decline, and reorganization. When ecosystems face important shocks, adaptive mechanisms and properties within the ecosystem lead the ecosystem to a new reorganization phase, which gives birth to another exploitation phase. In this perspective, the innovation ecosystem allows the avoidance of technology lock-ins and structural and organizational rigidity by providing mechanisms to enhance both resilience and competitiveness. Innovation ecosystem sustainability relies on two major dual forces: the exploitative function and the generative or autopoiesis function. Therefore, evolutionary and sustainability perspectives remain the "natural home" for developing works and models about the innovation ecosystem, and instrumental for policy-makers and practitioners involved in innovation management.

Keywords: innovation ecosystem; sustainability; evolutionary economics; Panarchy; resilience; adaptation; competitiveness

1. Introduction

During the past two decades, a great deal of interest in ecosystem research in management and economic fields has developed [1–3]. A growing number of scientific studies have popularized the concept of the innovation ecosystem as a new framework for academia, policy-makers, and practitioners [4–6]. Thus, more and more policymakers and economic agents are resorting to the innovation ecosystem framework, which highlights the role of open innovation, and actors' collaboration and co-evolution, as well as knowledge production and transfer.

An innovation ecosystem can be defined as a dynamic and adaptive system characterized by complex (formal, informal, organic, or institutional) relationships between a set of heterogeneous actors, performing distinct activities, playing different roles and having various motivations and capabilities, which contribute to the development of innovation processes or technologies [7–9]. From an evolutionary and ecological perspective, this notion remains strongly committed to the original biological metaphor. Moore introduced the concept of the "business ecosystem" to develop the thesis of firms' coevolution in a strategic context, in opposition to blind competition [1]. Frosch and Gallopoulos, on the other hand, developed the "industrial ecosystem" concept to advance the idea that a strategic context with interdependencies between heterogeneous actors could make it possible to develop innovative logics of circularity and environmentally friendly recycled products within the manufacturing sector [10].

The approach that developing innovation processes and competitive advantage relies on a business-friendly environment with complex relationships between actors is not new. The Marshallian hypothesis on "industrial districts" have already described the importance of geographic and relational proximity characterized by complementarities and interdependences between a diversity of economic actors regarding the competitiveness of a given territory [11]. Becattini, in the 1980s, expanded the concept of Industrial districts to characterize Italian districts' competitiveness [12]. Porter delved deeper into this concept and popularized, in the 1990s, the cluster concept, which is defined as a concentration of small, medium, or large firms, organizations, and institutions, which are in synergy in a particular technological field within a geographic area [13]. At the same time, evolutionary economists have developed the concept of the "innovation system" as incorporating the role of institutions in the structuring (at different scales: sector, territorial, or topological) of this strategic and interactionist environment favoring the development of innovation processes [14–17].

Given such an abundant theoretical corpus, many are critical of and skeptical about the potential contribution of the innovation ecosystem concept. According to Oh et al., the added value this concept generates is so low that its mobilization exposes the researcher much more to questionable scientific rigor and invalid knowledge production. Its use can even lead to harmful and dangerous political and strategic choices [18]. Faced with these criticisms, the ecosystem concept has become a more and more contested concept, calling for more robust conceptual and theoretical foundations. In addition, these criticisms highlight the need for concrete operationalization of this concept for policy-makers and practitioners.

This article proposes an evolutionary and sustainability perspective of the innovation ecosystem in order to offer an answer to the call for conceptual rigor to this concept. Moreover, this paper addresses the main implications for performance measures and the sustainability conditions of innovation ecosystems.

This study uses an evolutionary and sustainability perspective of innovation ecosystems. We revisit the adaptive cycle, known as "Panarchy", developed by Gunderson and Holling [19]. The Panarchy model was developed to describe the evolutionary nature of complex adaptive systems and their sustainability. It has been applied mainly to natural ecosystems (forests, meadows, lakes, rivers, and seas), socio-economic systems (territorial governance structures), as well as socio-ecological systems.

Because the ecosystem of innovation concept was originally an ecological metaphor, and since many scholars use it as a complex adaptive system [20]), it seems relevant to know to what extent the Panarchy model can be applied to the innovation ecosystem concept and how this can reveal specific features of this concept.

The first part of this paper highlights the theoretical foundations and main approaches to innovation ecosystems, as several bibliometric works have referenced them. The second part explains our methodological framework. The third part tests the adaptive cycle main hypotheses on two emblematic cases of innovation ecosystems. Finally, we discuss propositions arising from this analysis to bring out the properties of innovation ecosystems and their implications in terms of performance, sustainability, collective strategies, and public policies.

2. Innovation Ecosystems, Theoretical Foundations, and Main Approaches

A cross-analysis of two bibliometric works carried out on the innovation ecosystem concept in economic and management fields allows us to identify five main theoretical corpora, with a major influence on the development of the innovation ecosystem framework [5,6].

The first corpus is the Open Innovation framework [21,22]. From this viewpoint, the innovation ecosystem is a system of complex relationships between various actors in the *context of open innovation* [23,24].

The second corpus is the Strategic Management field [25–27]. With that in mind, the ecosystem is seen as a *strategic context* on which companies' performances depend. It provides a framework that fosters the emergence and development of innovations and technologies in order to improve value creation and competitiveness [28,29].

The third corpus is the Organization Studies field [30,31]. The innovation ecosystem is, therefore, a *collaborative organizational and institutional arrangement* in which companies and other economic players combine their knowledge, know-how, and methods consistently to foster the development of innovation processes [8,32].

The fourth corpus is that of Evolutionary Economics [33]. From this perspective, the innovation ecosystem is considered as a *complex adaptive system*, characterized by permanent interactions between various actors, which allows the combination of top-down and bottom-up initiatives that contribute to developing innovations. Actors' coevolution and ecosystem dynamics are then critical.

The fifth corpus is the Industrial Ecology field (this has currently been identified by the bibliometric work of Tsujimoto et al.) [5,10,34]. From this viewpoint, the concept used is the "industrial ecosystem", which is a much deeper analogy of the natural ecosystem. The main hypothesis relies on the transformation of the industrial system through limiting the industrial impacts on environmental and natural resources by optimizing the production of goods and materials besides energy flows.

The three main approaches structure the work on innovation ecosystems: the *platform based-ecosystem* or *digital ecosystem*, the *regional/local ecosystem*, and the *industrial ecosystem*. In the literature, the difference between the innovation ecosystem and the business ecosystem is not clear. Augusto et al., explain the difference: the innovation ecosystem focuses on value creation, whereas the business ecosystem is about capturing value [6]. Our study uses a cross-cutting approach that integrates both value creation and value capture.

The *platform-based ecosystem* approach highlights mainly the role of digital platforms [35–37]. The ecosystem is, therefore, an open and collaborative space built from various actors' interactions around a pivotal company or a *keystone* [38–40]. This ecosystem is structured around a multi-sided platform with at least two main faces. The first one enables the leader, stakeholders, and peripheral players or developers communities to design and develop complementary innovative products and services. The second one facilitates the management, sale, or monetization of new products and services between platform stakeholders and users or customers. Both sides favor exploiting data on the evolution of customers' behaviors, practices, and preferences. They ease up the continuous improvement of the digital platform's products and services. Digital platforms, therefore, foster complex and dynamic interactions, complementarities, and automated transactions between a variety of players (customers, developers, users, and suppliers).

The *Regional/Local Ecosystem* approach is an extension of Saxenian's work, highlighting the territorial or regional dimension in the dynamics of the innovation process [41]. These works point out the central role of geographic proximity and interactions between actors and institutions in the development of innovation processes [42,43]. Works on creative cities or creative ecosystems have made one of the most relevant contributions. They place the same level of emphasis on formal and informal relationships, institutions' dynamics, and epistemic communities. These entities foster as well as exploration and creativity or exploitation and development of new products or services [44–47]. According to these works, the innovation ecosystem has three main components, which are in organic interaction:

(i) The upperground. It is made up of stable, formal entities with a great capacity for exploitation, development, and standardization (e.g., companies, universities, innovation agencies, public institutions);

(ii) The underground: It is made up of talented, marginal or alternative elements, artists, informal collectives, which essentially play a major role in the exploration and generation of new ideas;

(iii) The middleground, i.e., spaces, places, actors, and communities that connect the actors of the underground and those of the upperground.

The performance of the innovation ecosystem therefore depends on the quality of the middleground or the quality of the organic relationships between formal and informal entities.

The third approach is the industrial ecosystem connected to the industrial ecology field, as we explained above [10]. The industrial ecosystem approach is the one that best embodies the ecological dimension and legitimizes the prefix "eco". This approach is linked to a broader perspective of sustainable development. This approach opposes a linear vision of material and energy flows. It relies on the assumption of a "perfect industrial system", which minimizes its impact on the environment and natural resources through cyclic or circular processes [48]. Within complex collaboration, heterogeneous actors develop then innovation processes to

(i)   promote circular economy renewable energies, recycling products;

(ii)   reduce drastically industrial, natural resources, and energy waste;

(iii) exploit local resources, intangible and cultural elements while taking into account their limits;

(iv)   enhance the reproductive capacity and the conditions of the ecosystem renewal.

The diversity, cooperation, complementarities, and interdependence of the economics agents are sources of flexibility, renewal and adaptability of this industrial system.

Torre and Zimmermann emphasized the ecological dimension as the main specificities to the innovation framework [49]. The ecological dimension does not refer automatically to sustainable development, but rather to sustainability. In a generic way, it is the taking into account of living beings in their environment, their interactions, their adaptability, and their coevolution.

We, therefore, position our consideration on innovation ecosystems in an evolutionary sustainability perspective. These are our research questions: From an evolutionary perspective, what theoretical foundation(s) could justify the development of research on the innovation ecosystem concept face to traditional theoretical frameworks on innovation systems and clusters? What are the main implications for performance measures and sustainability conditions of innovation ecosystems? To answer these questions, we revisit the "Panarchy" model, which describes the evolutionary nature of complex adaptive systems and their sustainability [19].

3. Methodological Framework

In this paper, we follow in the footsteps of the innovation ecosystem as a complex adaptive system characterized by complex (formal, informal, organic, or institutional) relationships between heterogeneous actors, and by a mixed of top-down, bottom-up, and individual and collective initiatives that promote and foster the development of new products and services, as well as the co-evolution of heterogeneous players [20].

The adaptive cycle, Panarchy (Figure 1), describes the evolutionary process of complex adaptive systems and ecosystems [19]. This adaptive cycle includes four recursive phases, without, however, being deterministic. Namely, the *reorganization phase*, which will give rise to an *exploitation phase*, then the *conservation phase* and *the decline phase,* due to shocks or external variability. Adaptive mechanisms within the ecosystem will lead the ecosystem from a decline to a new *reorganization phase*.

These phases are different from phases describing the innovation ecosystem life cycle within the management field [28]. However, the Panarchy model offers an original perspective for the innovation ecosystem. Works on Panarchy rely on ecosystems sustainability, which is based on the ability to continuously adapt so as to reach their equilibrium state or several equilibrium states when faced with severe shocks and external variability [19].

The cycle is based on several important observations on natural ecosystems that echo the innovation ecosystems. In this paper, we highlight three of them:

1-   An ecosystem lifecycle is characterized by one or more equilibrium states. This (or these) state(s) of equilibrium correspond(s) to the optimal situation when the functioning and outputs generation of this ecosystem are maximal.

2-   An ecosystem lifecycle is interrupted by episodes of shocks and unpredictable disturbances with varying magnitudes that compromise these equilibrium states. These episodes of shock induce uncertainties for the ecosystem's future and constrain the ecosystem to activate adaptive mechanisms so as to launch the reorganization phase. The reorganization phase will be followed by the exploitation phase, which leads the ecosystem to regain its own, or another, equilibrium state.

3-   A duality determines an ecosystem's sustainability. On the one hand, the stabilization forces maintain productivity, the accumulation process, and the ecosystem efficiency. On the other hand, the destabilizing forces are essential to foster resilience and generativity.

In this study, we apply the adaptive cycle hypotheses on innovation ecosystem dynamics. We use the case of the dynamics of forests as natural ecosystems, as described in Gunderson and Holling's work [19]. It shows how these phases happen in practice when forests are submitted to forest fires. After highlighting the key elements, we test these phases on the characteristics of two emblematic cases of innovation ecosystems commonly studied in the literature: regional or local ecosystems and platform-based ecosystems (Figure 2).

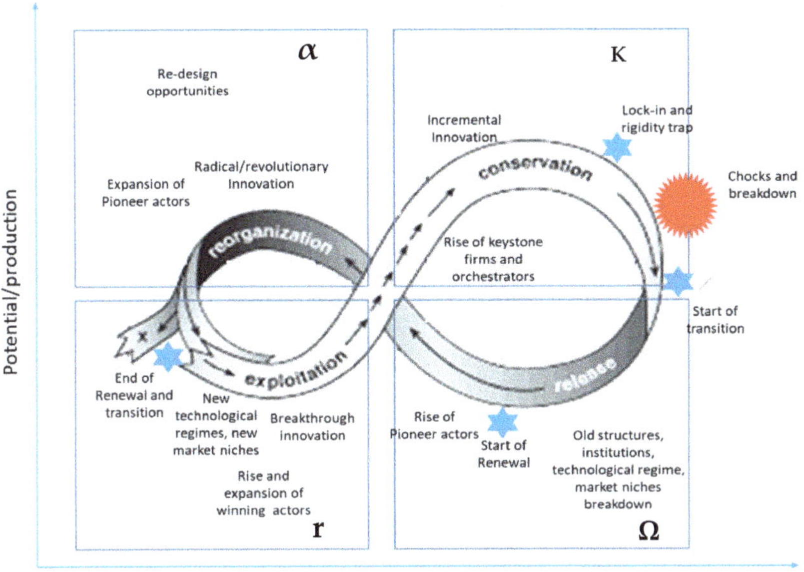

Figure 1. Panarchy applied to Innovation Ecosystem (adapted from Gunderson and Holling [19]).

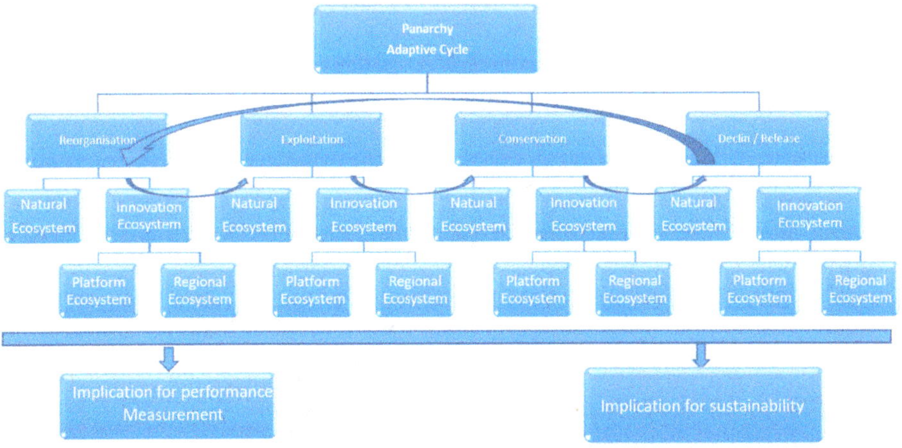

Figure 2. Method of the Panarchy model application to main cases of Innovation Ecosystems.

Finally, we highlight the implications for innovation ecosystem performance measures and sustainability conditions.

4. Applying the Panarchy Model to the Main Cases of Innovation Ecosystems

In this paper, we start with the *conservation phase* because we believe it is more relevant to highlight adaptation mechanisms. We do not describe the creation process or the innovation ecosystem, a topic already much studied in the literature [28].

4.1. Conservation Phase

The so-called *"conservation phase"*, noted as **K** (Figure 1) [19], corresponds to the optimal situation in the ecosystem dynamic in terms of performance, accumulation process, population size, as well as materials and energy flows. In this phase, the growth rate is initially very low and eventually becomes zero or declines. Connectivity between the different components is very high, which causes some rigidity in the ecosystem. At this stage, the ecosystem becomes more and more fragile and vulnerable to severe accidents and shocks because its resilience is weak.

In the case of a forest, trees reach maturity and the forest becomes very dense because of high connectivity and proximity between forest components. Production and energy flows are at the top. Nevertheless, this ecosystem is fragile. Dead leaves fall and some parts of trees and other plants become sclerotic. The conjunction of this situation and external factors such as rising temperatures during the summer period (drought leading to the evaporation of the water contained in plant tissues and to twigs and grass drying up) exposes this ecosystem to fire risks.

In the lifecycle of an innovation ecosystem, this phase relates to the *maturity or leadership phase* [28].

In the case of a *regional innovation ecosystem,* this phase is characterized by the maximum exploitation of a technological cycle, hyper-specialization, a high level of connectivity between actors, and an appropriate institutional environment. Those conditions foster the development and performance of this innovation ecosystem. The ecosystem then gets a competitive advantage over its competitors through cost or differentiation advantages. However, at the end of this phase, the technological waves that have supported the competitiveness of the innovation ecosystem are running out of steam. Even though incremental innovations are important, they can no longer generate permanent value creation for this innovation ecosystem. Relationships are becoming more and more formal and hierarchical. Firms and organizations might become bureaucratic and rigid or "locked into the patterns of traditional and vertically-integrated industrial structure" [41]. While the logic of specialization leads to exploiting a unique set of skills and technologies, they can, over time, cause a lock-in phenomenon. Institutions that fostered a business-friendly environment for exploiting the current technological waves might become inappropriate for generating a new technological paradigm. As a result, the innovation ecosystem may find itself in a hyperspecialization trap, institutional and structural rigidity, or lock-in. While this ecosystem resilience is low, a wave of creative destruction or market changes can, at any time, disrupt this ecosystem and lead it towards a phase of decline, as Schumpeter pointed out [50].

In the conservation phase, the platform-based ecosystem is in a leadership and competitive advantage position over competitors. Orchestration and complementarities between stakeholders are optimal. The platform has reached its critical mass, value creation, and growth [51]. The two main faces are very well structured, and the platform is running at full speed to design and manufacture new products and services. At the end of this phase, however, growth slows down. The influence of the leader or the *keystone firm* is weakening due to the wide range of actors whose interactions are increasingly complex. To optimize its influence, the *keystone firm* repositions its activity on its core business and develops more and more formal relationships and institutionalized interactions in order to safeguard its own growth. These strategies will have negative consequences for the growth of the whole platform ecosystem, thereby increasing its vulnerability (see Table 1) [52].

Table 1. Conservation phase.

Ecosystems	Equilibrum Situation	Main Characteristics	Internal Vulnerabilities	Indicators of Efficiency
Forest ecosystem	Maximum accumulation Maximum population Optimal functioning of material and energy flows	Production ++++ Connectedness ++++ Complementarities ++++ Resilience +	Tree maturity Dead leaves tanks Sclerosis of fragments of trees and other plants	Optimal functioning of material and energy flows
Regional ecosystem	Maximum exploitation of a technological cycle Adequate institutional framework Very strong connectivity between actors with formal and institutionalized relationships Optimal allocation of resources Hyper-specialization	Competitiveness ++++ Connectedness ++++ Complementarities ++++ Resilience +	Rigid, hierarchical organization Bureaucracy Hyper-specialization Path dependency	Competitiveness Attractiveness
Platform-based ecosystem	Market leader position Optimal layout around Keystone firms and stakeholders High complementarities between actors High attractiveness		Too many stakeholders Decrease of Keystone firms influence Optimization and refocusing on core activities and profitable niches	Value creation and value-capture Competitiveness Attractiveness

+ Weak   ++ Medium   +++ High   ++++ Very High.

4.2. Decline Phase

The decline phase is characterized by the situation where an external (or endogenous) event disturbs the initial equilibrium state of the ecosystem. It is noted as Ω (Figure 1) [19].

In a forest ecosystem, the trigger event could be accidental: the occurrence of lightning, dry thunderstorms, or anthropic fires. [19]. This situation occurs when the conditions within this forest enable a small local fire to spread quickly enough to start a forest fire that can devastate hundreds or thousands of hectares (as was the case in Australia and the Amazon in 2019). Actually, strong destabilizing feedbacks occur between disturbing elements (lightning, dry storms, or small anthropic fire), established aggregates (mature forest trees, leaves, and dry grass), and other surrounding conditions (the oxygen in the air). The conjunction of these elements is sufficient to provide enough activation energy to torch the entire forest. This fire could hardly have spread if the forest had been very sparse. Forest vulnerability is due to population density and close proximity between the elements. The ecosystem then gets into a crisis situation where connections are broken and regulatory mechanisms weakened. This is the phase when the conditions for chaotic behavior are met.

This phase is related to the first step of the "renewal" phase for innovation ecosystems [28]. Consequently, the main disruptive elements are new technological and market changes, components of the creative destruction process.

In the case of a *regional innovation ecosystem*, this phase corresponds to the moment when industrial change and disruptive innovation disturb the technical status quo of existing technological and economic paradigms to generate new ones. This disruptive transformation process is inherent to the lifecycle of the regional innovation ecosystem and clusters. As a result, it destroys the value capture of well-established firms and organizations generated by a monopoly or oligopoly position associated with previous technological paradigms. This process submits the innovation ecosystem to uncertainties. This situation questions both orchestration and the position of dominants and periphery actors [53]. Control and regulation mechanisms are weakened, institutional and social boundaries dividing firms, as well as trust and confidence, break down. The survival of the ecosystem relies on adaptation mechanisms, new combinations, and organizational and institutional change [41]. Here are the most frequent indicators of these crises: delocalization of industrial activities and firms, sharp declines of foreign investment, loss of competitiveness.

During the decline phase, the stakeholders of *platform-based ecosystems* experience a structural drop in their sales and in the capture of economic value. This situation may be due to the emergence of their market niches and core business of other platforms using better digital technologies or offering higher quality services at lower costs. It could also be explained by the value aspiration created by actors in very peripheral positions who are in connection with the core of a competing or complementary platform. This structural drop in profits can also result from the capture of value by a set of platforms on neighboring market niches. The platform-based ecosystem then risks disintegration. The actors and stakeholders who provided complementary services and products can leave the platform to create their own platforms or be plugged in other competing or complementary platforms [51]. A large proportion of users or customers have gone. The platform is in a hemorrhage and great-uncertainty situation. The adaptation of this ecosystem relies on the leader's and the remaining stakeholders' ability to find other development pathways. It depends on their capacity to appropriate new digital technologies that could make their offer more attractive. Adaptation could also mean to completely change their market niches or their initial core business. It is also the ability to stop the hemorrhage caused by the ongoing departure of users and stakeholders and to build confidence and a healthy environment. Even the leader's position could change.

To cope with these shocks and ensure their survival, a reorganization process must be initiated, whether in a platform-based ecosystem or a regional innovation ecosystem. This is the reorganization phase of the adaptive cycle (see Table 2).

Table 2. Decline phase

Ecosystems	Decline Situation	Main Characteristics	Shock Activators	Indicators of Efficiency
Forest ecosystem	Forest fire	Production + Connectedness + Complementarities + Resilience +++	Temperature rises Drought Fire from natural or anthropic origin	Resilience: Resistance to chocks
Regional ecosystem	Technological disruption Drop of competitiveness Mass layoffs Relocation of companies	Competitiveness + Connectedness + Complementarities + Resilience +++	Creative destruction due to the rise of new technological waves Evolution of the competitive context	Resilience: Resistance to drop of competitiveness and value creation
Platform-based ecosystem	Technological disruption Decline in value creation		Rise of competing platforms Aspiration of value by complementary platforms Emergence of more efficient digital technologies	

+ Weak   ++ Medium   +++High   ++++ Very High.

4.3. Reorganization Phase

The reorganization phase, noted as α (Figure 1) [19], relates to the phase when the ecosystem turns to these residual resources and its regenerative capacity. This adaptive process relies on species that were not destroyed during the previous shock phase and can, therefore, survive within these new and harsh environmental conditions. This reorganization phase also relies on the emergence or expansion of so-called pioneer species or organisms that are beginning to take advantage of the uncertain environment.

As regards the forest ecosystem, the main adaptation actors are species coming from germinating seeds stored in seed banks accumulated from the past, as well as branches and tree trunks not consumed by the forest fire. The impact of the wind or birds or other animals can also bring them in [54]. They can come from the nutrients unconfined by organic matter decomposition or seed banks established in the soil. Diversity is essential for the reorganization process. Pioneer species might be varieties that were not widely represented or actively involved in previous conservation phases. They could even lay dormant and re-emerge after the decline phase to allow the ecosystem to adapt to future unforeseen external changes. As internal connectivity to the ecosystem is weak, pioneer species are very much influenced by external variability, both as opportunities to exploit and as constraints to bear. As a result, individuals and communities adapt to live in and exploit the opportunities of a variable environment with harsh and extreme conditions. They develop new combinations and relationships in order to reinforce each other. The future begins to be more predictable and less guided by uncertain forces independent from ecosystem control. This phase lays the foundations for the exploitation phase.

The reorganization phase is equivalent to the end of the renewal phase in an innovation ecosystem lifecycle. This phase is characterized mainly by the exploration, co-creation, regeneration, and restructuring of an innovation ecosystem. Works describing this phase are not many in the literature on innovation ecosystems.

In the case of a *regional innovation ecosystem*, this phase highlights the need for a creative class and diverse, pioneer actors within this ecosystem [55]. Individual or collective actions to explore and experiment novelties and new concepts become essential. The regulation processes and economic policies should be adapted to this transition and transformation period to achieve better change management. Disruptive innovations are developing from pre-existing technological niches [56]. Actors, companies, and organizations that will expand in this phase are mainly pioneer actors and risk-takers, who are able to deal with external variability (appropriating the new technological waves, adapting to the new competitive context) and the new conditions prevailing within the innovation ecosystem. This is a business-friendly environment for start-ups, spin-offs, and business angels development driven by a pioneering spirit. As a result, reorganizing the innovation ecosystem fosters the development of new products and services, new business models, new markets designed to create and capture economic value. Successful initiatives and success stories will recreate the confidence necessary for effective cooperation, collaboration, and the clustering process. The whole environment will gradually become more accommodating and more predictable. These pioneering behaviors foster new combinations and more complex relationships around new technologies or the exploitation of new markets. As the environment is more reliable, the *exploitation phase* can follow.

During the *reorganization phase*, a *platform-based ecosystem* is involved in a realignment and repositioning process in order to generate new innovative products or services and new complementary goods. The leader and stakeholders re-design or reinvent their value chain, attract new developer communities, and identify new business opportunities. This could happen through the appropriation of new digital technologies. Peripheral actors and users and developers' communities can make up a stock of new ideas or new concepts, which could allow reorganizing the platform. Routines are changing. Actors' positions in the ecosystem are also changing. Actors in the periphery position could move up at the heart of the platform or even become a leader. In this phase, former customers or users are reassured and find it worth staying. New stakeholders and new communities rejoin the former ones to create communities of interest, sharing a common vision. Coherence emerges, although not yet stabilized, the platform-based ecosystem is sufficient to enable stakeholders to move forward together in a new innovation process, a new trajectory, or the exploitation of a new business model.

This renewal process depends on the ability of this ecosystem to take into account changes in behaviors, to integrate new knowledge, to appropriate new technologies, or to adapt to new market conditions. A co-evolution process then begins, which will shape the final reconfiguration of the ecosystem (see Table 3).

4.4. Exploitation Phase

This phase, denoted as **r** (Figure 1) [19], relates to the phase when the growth-rate of the ecosystem is higher. The *winning species* are growing. The *pioneer species* may not necessarily become the *winning species*, but they create conditions for the exploitation phase by the *winning species*. New players emerge to strengthen the ecosystem. The environment is healthier, and the future becomes more predictable. Uncertain forces less guide the future of the ecosystem. Connectivity between species is intensifying. A complex relationship between cooperation and competition takes place, as well as self-organized relationships.

Table 3. Reorganization phase.

Ecosystems	Situation	Main Characteristics	Internal and External Elements of Re-Generativity	Indicators of Efficiency
Forest ecosystem	Mobilization of residual resources Development and expansion of pioneer species	Production ++ New Combinations ++++ New Complementarities ++++ Resilience++++	Expansion of so-called pioneer organisms or species Ability of pioneer species to live and exploit opportunities in a harsh and extreme environment Arrivals of other exogenous species	Resilience: Regeneration speed
Regional ecosystem	Reorganization, restructuration Mobilization of technological niches Exploitation of creative class and underground actors Rise of pioneer actors	Competitiveness ++ New Combinations ++++ New Complementarities ++++ Resilience++++	Incubation system Creative class Middleground artefact and Underground actors Exploration and experimentation initiatives Development of the start-up spirit Arrivals of exogenous actors and risk-takers New innovation policies	Resilience: Speed in the appropriation of emerging technological waves (digital), the deployment of technological niches, the rearrangement of players and the development of new markets
Platform-based ecosystem	Realignment and repositioning of actors Mobilization of communities of users and peripheral or external developers communities		Re-design of new products and services Reinventing the value proposition and the market target Attraction of new developer communities Reconfiguration of the platform and repositioning of stakeholders	

+ Weak   ++ Medium   +++High   ++++ Very High.

In a forest ecosystem, at the exploitation phase, shrubs are in full development, and so is biodiversity. This ecosystem attracts exogenous species (e.g., insects, birds) that find favorable conditions to their development. Soils are improving, thanks to the decomposition of organic matter. Microclimatic variability is moderated by vegetation. Material and energy flows are becoming more and more secure. This results in a system of complex relationships, cooperation, mutualization, symbiosis, but also competition. As phase **r** progresses, the accumulation of nutrients and biomass becomes more and more closely linked to the existing vegetation, preventing other competitors from using them. Ecosystem connectivity increases with clustering processes.

In a *regional innovation ecosystem*, the exploitation phase relates to the phase generating increasing returns to scale, the exploitation of one or more new technological field(s), or new markets. The ecosystem is running at full speed. The institutional context fosters more and more innovations. Even a small incremental innovation leads to strong value creation. During this phase, the return on investment is very high. The development of complex relationships leads to a more efficient ecosystem by minimizing transaction costs and operations rationalization. Orchestrators emerge to improve the coherency of the ecosystem. Confidence is at its highest level, technical skills are developing, and markets are growing. This ecosystem is becoming very attractive for investors, talent, and entrepreneurs. Over time, competitive processes will create new monopoly and oligopoly situations, which harm the ecosystem diversity. Indeed, this situation will have consequences for newcomers and innovations: fewer opportunities to emerge, despite their potential superiority [57,58]. This is the start of the conservation phase.

Regarding the platform-based ecosystem; at this phase, the platform is almost completely reformatted with the two main faces and the *keystone firm*. The core business and target market are very well-identified. On the innovation side, the first partners and developer groups are developing organic relationships and complementarities with the leader or new leaders. However, strategies to attract other developer communities are still operating. Old and new communities constitute a new community of destiny [1,28]. Identifying the core business provides immense opportunities for developing complementary and interdependent products and services. On the management or sales sides, the early loyal customers and the first new users will form the new market target. The evolution of preferences is captured by users' and consumers' data, which become the main factor of the innovation process. Now, it is all about attracting other customers and users. Harmonizing faces, once stabilized, becomes the main driver of the whole ecosystem. The platform can then develop expansion strategies and new value propositions through the development of a number of features. It can be a system of customer recommendations or rating, decentralized quality control, search engines, user-friendly access to the platform and a notification system that allows reaching other users who are connected to their customers on social networks [41].

The exploitation phase can lead to a new phase of maturity or conservation (see Table 4).

Table 4. Exploitation phase.

Ecosystems	Situation	Main Characteristics	Exploitive Elements	Indicators of Efficiency
Forest ecosystem	Expansion of so-called winning species Development of networks and complex cooperation between species Harmonization of energy and material flows	Production +++ Connectedness +++ Complementarities +++ Resilience ++	Expansion of so-called winning species Development of complex relationships, cooperation, mutualization, and symbiosis process	Growth rate
Regional ecosystem	Exploitation of new technological waves and new market conditions Development of complex networks and cooperation Emergence of orchestrating actors/organization	Competitiveness +++ Connectedness +++ Complementarities +++ Resilience ++	Business incubation and acceleration system Innovation policies for strengthening business clusterization process and complex relationships between actors Development of unicorns and firms with strong capacity to exploit new technological paradigm	Growth rate Increasing returns of Innovation
Platform-based ecosystem	Exploitation of new business model, new digital technologies, new pools of customers or users Harmonization of the multi-faces of the platform Strong feed-back between Communities of developers and communities of users		Development of Common destiny community between keystone firms, stakeholders, peripheral developers, users and consumers of the renewed platform	

5. Discussion

The adaptive cycle, called Panarchy, provides an evolutionary and sustainability perspective, which is truly relevant to the innovation ecosystem concept and its main application. Operationally, this evolutionary vision already provides valuable tools for policy-makers and economic actors for better innovation policy implementation and innovation-related collective actions, or individual initiatives, as well.

First, this perspective shows that the dynamics of an innovation ecosystem relies on its adaptation capacity, which is mainly governed by two main dual functions. The first is the exploitation function, which maximizes value creation and value capture from a competitiveness perspective. The second is the generativity function, which maximizes creativity, invention, and exploration in a resilience and adaptation perspective.

The most obvious manifestation of the exploitation function could be seen during two phases of the Panarchy model: the exploitation phase and the conservation phase (maturity or leadership). These are characterized by cooperative and competitive strategies between a set of diverse actors in a business-friendly institutional context. They foster innovation processes or the exploitation of technological waves or market opportunities [20]. These phases prioritize growth and competitiveness due to the mastery of existing technological paradigms and market conditions. Here are the main performance indicators: growing returns from innovation, attractiveness, and value creation or value-capture [6]. This performance results from optimal allocation of resources, business-friendly institutions, complementarities, and complex relationships between actors, as well as maximum exploitation of one or more current technological cycles. The co-evolutionary trend depends, therefore, on a predictable context and a probable future.

Nevertheless, works on innovation ecosystems, as well as works on clusters and innovation systems, point out such favorable conditions that could foster innovation processes leading to competitiveness [5,6,13,16,41,59–61]. As a result, while fundamental, these ideas could not have a critical added value in the literature on the innovation process because they have already been studied and deepened by works on innovation systems and clusters. However, interactions between heterogeneous actors in a given institutional context meant to exploit technological waves and market conditions, even through the innovation process, could lead to the lock-in phenomenon [62].

This lock-in phenomenon may be the result of hyper-specialization, institutional and structural rigidity, and path-dependency [63,64]. In this situation, policies for clusters reinforcement and strategies for strengthening cooperation between actors are no longer sufficient to ensure sustainable performance. This ecosystem sustainability, therefore, depends on another force, which is part of its adaptive capacity: autopoiesis or generativity.

Here, we point out one of the main criticisms of the clusters works carried out by the advocates of the innovation ecosystem concept, which at the same time sheds light on the contribution of the ecosystem works resulting from ecological and sustainability perspective: the quasi-absence of cluster-specific mechanisms to avoid the lock-in phenomenon, as well as institutional and structural rigidity [65]. Without regeneration mechanisms, lock-in and institutional and structural rigidity could turn into a trap that prevents cluster renewal and could prove fatal to the cluster or the regional innovation system. This is what Saxenian observed in her study on the Route 128 regional innovation system [41]. Furthermore, while works on innovation systems dynamics do exist, their conceptual frameworks do not ontologically incorporate regenerative mechanisms. They are based essentially on institutionalist and interactionist points of view on innovation processes [56,66].

Generativity or autopoiesis is the second major function that governs the adaptive capacity of innovation ecosystems as a social system [67]. It is particularly obvious during the decline and reorganization phases. These phases are characterized by a situation of great uncertainty and unforeseen crises. The objective of the innovation ecosystem is no longer to increase productivity gains or competitiveness. The ecosystem's performance is, therefore, its resilience. It relies on its speed in appropriating emerging technological waves and in adjusting actors' behaviors to market changes [19]. It also depends on the ability to build new value creation patterns. Reinforcing innovation ecosystem resilience requires creativity, invention, exploration, and exploiting diversity.

Cohendet et al. explain that this autopoiesis is essentially based on the *underground*, namely, talented elements, artists, peripheral actors, and informal communities [45]. Florida describes the crucial role of the creative class in exploring alternative pathways of value creation [55]. For platform-based ecosystems, generativity relies mainly on substantial feedbacks between *keystone firms*, main stakeholders, and communities of users or peripheral developers. Within the research system, this could arouse alternative research to the dominant paradigm. For the technological system, this could give rise to technological niches or communities of engineers working in their garages or third-places [56,68]. These peripheral or alternative elements are important for the resilience of the innovation ecosystem. In addition to diversity, the pioneering actors foster new combinations as well as experimenting and testing new pathways. Epistemic communities can be the bearers of collective action in this situation of weak ties and informal relationships [47]. Business incubators can facilitate individual high-risk initiatives by fostering start-up development. Nevertheless, firms' absorptive capacity also remains fundamental [69].

During the decline and reorganization phases, actors' coevolution becomes more complex due to the highly turbulent context, which even destabilizes established institutions. The reorganization of the ecosystem, a consequence of its generativity or autopoiesis, also relies on artefacts or "actants" that facilitate the circulation, testing, and exploitation of new ideas, new inventions … [70,71]. These artefacts or "actants" also foster the appropriation of a new exogenous technological wave. They can prove to be places, platforms, spaces, and events that enable organic connections and interaction between research and industry, actors in the underground and those in the upperground, *keystone firms*, the main stakeholders of the platform, and users or peripheral developers communities [45,47,72]. These meetings are decisive for transitioning from ideas to concepts and from concepts to the success stories of pioneering initiatives that support the emergence of new innovation pathways.

These two major functions, exploitation and generativity or autopoiesis are obviously in tension, but determine both the resilience and competitiveness of the innovation ecosystem [73,74]. The performance of an innovation ecosystem is based on its ability to cope with these opposing or contradictory functions. It depends on the coexistence and best use, at an appropriate time, of the generative capacity and the exploitative capacity. It also depends on the capability of promoting tight and loose relationships between actors, collective (co-creation, co-design, co-development, and open innovation) and individual initiatives, openness logic (control of external variability) and use of its singularity, coherence and chaos initiatives, well-planned strategies, together with improvisation or spontaneity, as well as top-down and bottom-up initiatives. Furthermore, the performance of an innovation ecosystem relies on the coexistence of actors with different profiles: pioneers, consolidators, non-conformists, revolutionaries, leaders, and orchestrators [45].

While the coexistence of opposing forces is decisive for an innovation ecosystem performance and sustainability, the two functions cannot be maximized simultaneously. Actually, maximizing all of these forces together could damage the existence of this innovation ecosystem. These functions can only be maximized sequentially, as happens in natural ecosystems [19]. The optimization of a function depends on this ecosystem context and dynamics. In decline and reorganization phases, the obvious step is to optimize the generativity function. However, during the exploitation and conservation phases, the innovation ecosystem performance could result from the optimization of the exploitation function. Nevertheless, from a time perspective, the performance of an ecosystem depends on its ability to adapt quickly to an exogenous or endogenous shock and on remaining in an exploitation phase as long as possible. For an ecosystem, the risk is to lose one of these two pivots at the core of its ambidexterity [73]. Both functions determine the innovation ecosystem sustainability.

This analysis shows the advantage of revisiting the Panarchy model to consolidate the framework on innovation ecosystems and to highlight its relevance. The adaptive cycle developed within the framework of the "Panarchy" then becomes a relevant theoretical point if one is to understand—from an evolutionary and sustainability perspective—the ontological singularity of the framework of "innovation ecosystems" compared to more traditional theoretical frameworks on innovation systems and clusters.

6. Conclusions

While technological cycles get shorter in the digital era, volatile world economy, and globalization, adaptation issues are becoming increasingly critical. Today, economic agents find themselves at the crossroads of major challenges: the digital revolution (e.g., AI, blockchain, Industry 4.0), climate change, ecological and energy transitions, and the demographic transition. Furthermore, in a context of uncertainties and accelerating transformations, it becomes crucial to be able to build a path in uncertainty. Adaptation is, therefore, key for any firm and organization. It is obvious that, as a result, policy-makers and economic agents are looking for relevant frameworks or tools to enable them to be both competitive and resilient. Therefore, the innovation ecosystem framework proves to be relevant because it offers conceptual and operational tools, able to address both resilience and competitiveness through adaptation capacity. Adaptation implies the co-evolution of actors, institutions, networks, and knowledge. Revisiting the Panarchy model has enabled us to depict these fundamental properties and features of innovation ecosystems. The whole point of the innovation ecosystem framework is to provide economic actors and policy-makers who will avoid institutional and structural rigidity and technological lock-in. Moreover, the innovation ecosystem framework equips policymakers and practitioners to shorten decline situations caused by technological and market changes and to move toward reorganization and new exploitation phases. However, we do not assert that the four phases of the adaptive cycle are predetermined. An exploitative phase can even give rise to another exploitation phase. We only highlight properties and mechanisms, which foster the permanent adaptation of the innovation ecosystem to technological and market changes. We also point out which public policies and strategies are relevant according to the dynamics of an innovation ecosystem.

Future works on innovation ecosystems would be well advised to focus on dynamic, ecological, and evolutionary perspectives, which make it possible to model and capture the adaptation mechanisms of innovation ecosystems. Indeed, the innovation ecosystem framework and its contribution to the scientific literature on innovation processes cannot be fully understood without an evolutionary and sustainability perspective.

Funding: This research was funded by Hauts-de-France regional Concil. and "The APC was funded by Catholic Institute of Lille- HEMISF4IRE.

Conflicts of Interest: The authors declare no conflict of interest. The funders had no role in the design of the study; in the collection, analyses, or interpretation of data; in the writing of the manuscript, or in the decision to publish the results.

References

1. Moore, J.F. Predators and prey: A new ecology of competition. *Harv. Bus. Rev.* **1993**, *71*, 75–86. [PubMed]
2. Bogers, M.; Sims, J.; West, J. What is an ecosystem? Incorporating 25 years of ecosystem research. Presented at the 2019 Meeting of the Academy of Management, Boston, MA, USA, 9–13 August 2019; Available online: https://ssrn.com/abstract=3437014 (accessed on 14 January 2020).
3. Adner, R. Match your innovation strategy to your innovation ecosystem. *Harv. Bus. Rev.* **2006**, *84*, 98–107. [PubMed]
4. Kapoor, R. Ecosystems: Broadening the locus of value creation. *J. Organ. Des.* **2018**, *7*, 12. [CrossRef]
5. Tsujimoto, M.; Kajikawa, Y.; Tomita, J.; Matsumoto, Y. A review of the ecosystem concept—Towards coherent ecosystem design. *Technol. Forecast. Soc. Chang.* **2018**, *136*, 49–58. [CrossRef]
6. De Vasconcelos Gomes, L.A.; Facin, A.L.F.; Salerno, M.S.; Ikenami, R.K. Unpacking the innovation ecosystem construct: Evolution, gaps and trends. *Technol. Forecast. Soc. Chang.* **2018**, *136*, 30–48. [CrossRef]
7. Peltoniemi, M.; Vuori, E. Business ecosystem as the new approach to complex adaptive business environments. In Proceedings of the eBusiness Research Forum, Tampere, Finland, 20–24 September 2004; Volume 2, pp. 267–281.
8. Huang, J.; Wang, H.; Wu, J.; Yang, Z.; Hu, X.; Bao, M. Exploring the key driving forces of the sustainable intergenerational evolution of the industrial alliance innovation ecosystem: Evidence from a case study of China's TDIA. *Sustainability* **2020**, *12*, 1320. [CrossRef]
9. Ritala, P.; Almpanopoulou, A. In defense of 'eco'in innovation ecosystem. *Technovation* **2017**, *60*, 39–42. [CrossRef]
10. Frosch, R.A.; Gallopoulos, N.E. Strategies for manufacturing. *Sci. Am.* **1989**, *261*, 144–152. [CrossRef]
11. Marshall, A. *Principles of Economics*, 8th ed.; Cengage Learning: London, UK, 2009.
12. Becattini, G.; Bellandi, M.; De Propris, L. *A Handbook of Industrial Districts*; Edward Elgar Publishing: Cheltenham, UK, 2014.

13. Porter, M.E. *Clusters and the New Economics of Competition*; Harvard Business School Press: Boston, MA, USA, 1998; Volume 76, pp. 77–90.
14. Freeman, C. *Technology Policy and Economic Performance*; Pinter Publishers: London, UK, 1989; p. 34.
15. Lundvall, B.Å. *National Systems of Innovation: Toward a Theory of Innovation and Interactive Learning*; Anthem press: London, UK, 2010; Volume 2.
16. Malerba, F. Sectoral systems of innovation and production. *Res. Policy* **2002**, *31*, 247–264. [CrossRef]
17. Cooke, P. Regional innovation systems: Competitive regulation in the new Europe. *Geoforum* **1992**, *23*, 365–382. [CrossRef]
18. Oh, D.S.; Phillips, F.; Park, S.; Lee, E. Innovation ecosystems: A critical examination. *Technovation* **2016**, *54*, 1–6. [CrossRef]
19. Gunderson, L.H.; Holling, C.S. *Panarchy: Understanding Transformations in Human and Natural Systems*; Island press: Washington, DC, USA, 2001.
20. Russell, M.G.; Smorodinskaya, N.V. Leveraging complexity for ecosystemic innovation. *Technol. Forecast. Soc. Chang.* **2018**, *136*, 114–131. [CrossRef]
21. Chesbrough, H.W.; Vanhaverbeke, W.; West, J. *Open Innovation: Researching a New Paradigm*; Oxford University Press: Oxford, UK, 2006.
22. Chesbrough, H.W.; Kim, S.; Agogino, A. Building an open innovation ecosystem. *Calif. Manag. Rev.* **2014**, *56*, 144–171. [CrossRef]
23. Rohrbeck, R.; Hölzle, K.; Gemünden, H.G. Opening up for competitive advantage–How Deutsche Telekom creates an open innovation ecosystem. *R&D Manag.* **2009**, *39*, 420–430.
24. Krishnamurthy, S.; Tripathi, A.K. Monetary donations to an open source software platform. *Res. Policy* **2009**, *38*, 404–414. [CrossRef]
25. Barney, J.B.; Hesterly, W.S. *Strategic Management and Competitive Advantage: Concepts and Cases*, 5th ed.; Prentice Hall: Upper Saddle River, NJ, USA, 2010.
26. Porter, M.E. *Competitive Advantage: Creating and Sustaining Superior Performance*; Free Press: New York, NY, USA, 1985.
27. Teece, D.J. Profiting from technological innovation: Implications for integration, collaboration, licensing and public policy. *Res. Policy* **1986**, *15*, 285–305. [CrossRef]
28. Moore, J.F. *The Death of Competition: Leadership and Strategy in the Age of Business Ecosystems*; Harper Business: New York, NY, USA, 1996.
29. Teece, D.J. Explicating dynamic capabilities: The nature and microfoundations of (sustainable) enterprise performance. *Strateg. Manag. J.* **2007**, *28*, 1319–1350. [CrossRef]
30. Jones, G.R. *Organizational Theory, Design, and Change*; Pearson: Upper Saddle River, NJ, USA, 2013.
31. Powell, W.W.; Koput, K.W.; Smith-Doerr, L. Interorganizational collaboration and the locus of innovation: Networks of learning in biotechnology. *Adm. Sci. Q.* **1996**, *41*, 116–145. [CrossRef]
32. Santos, F.M.; Eisenhardt, K.M. Organizational boundaries and theories of organization. *Organ. Sci.* **2005**, *16*, 491–508. [CrossRef]
33. Nelson, R.; Winter, S.G. *An Evolutionary Theory of Economic Change*; Belknap Press/Harvard University Press: Cambridge, MA, USA, 1982.
34. Ayres, R.U.; Ayres, L. *A Handbook of Industrial Ecology*; Edward Elgar Publishing: Cheltenham, UK, 2002.
35. Tiwana, A. *Platform Ecosystems: Aligning Architecture, Governance, and Strategy*; Murgan Kaufman Publishers: Burlington, MA USA, 2013.
36. Isckia, T. Amazon's evolving ecosystem: A cyber-bookstore and application service provider. *Can. J. Adm. Sci.* **2009**, *26*, 332–343. [CrossRef]
37. Gawer, A.; Cusumano, M.A. Industry platforms and ecosystem innovation. *J. Prod. Innov. Manag.* **2014**, *31*, 417–433. [CrossRef]
38. Iansiti, M.; Levien, R. *The Keystone Advantage: What the New Dynamics of Business Ecosystems Mean for Strategy, Innovation, and Sustainability*; Harvard Business Press: Brighton, MA, USA, 2004.
39. Gawer, A.; Cusumano, M.A. How companies become platform leaders. *MIT Sloan Manag.* **2008**, *49*, 68–75.
40. Isckia, T. Ecosystèmes d'affaires, stratégies de plateforme et innovation ouverte: Vers une approche intégrée de la dynamique d'innovation. *Manag. Avenir* **2011**, *6*, 157–176. [CrossRef]
41. Saxenian, A. *Regional Networks: Industrial Adaptation in Silicon Valley and Route 128*; Harvard University Press: Cambridge, MA, USA, 1994.

42. Radziwon, A.; Bogers, M.; Bilberg, A. Creating and capturing value in a regional innovation ecosystem: A study of how manufacturing SMEs develop collaborative solutions. *Int. J. Technol. Manag.* **2017**, *75*, 1–4.

43. Viitanen, J. Profiling regional innovation ecosystems as functional collaborative systems: The case of Cambridge. *Technol. Innov. Manag. Rev.* **2016**, *6*, 12. [CrossRef]

44. Grandadam, D.; Cohendet, P.; Simon, L. Places, spaces and the dynamics of creativity: The video game industry in Montreal. *Reg. Stud.* **2013**, *47*, 1701–1714. [CrossRef]

45. Cohendet, P.; Grandadam, D.; Simon, L. The anatomy of the creative city. *Ind. Innov.* **2010**, *17*, 91–111. [CrossRef]

46. Lange, B.; Schüßler, E. Unpacking the middleground of creative cities: Spatiotemporal dynamics in the configuration of the Berlin design field. *Reg. Stud.* **2018**, *52*, 1548–1558. [CrossRef]

47. Sarazin, B.; Simon, L.; Cohendet, P. *Les communautés D'innovation: De la Liberté Créatrice à L'innovation Organisée*; Éditions EMS: Caen, France, 2017.

48. Korhonen, J. Four ecosystem principles for an industrial ecosystem. *J. Clean. Prod.* **2001**, *9*, 253–259. [CrossRef]

49. Torre, A.; Zimmermann, J.B. Des clusters aux écosystèmes industriels locaux. *Rev. D'économie Ind.* **2015**, *4*, 13–38. [CrossRef]

50. Schumpeter, J. *Capitalism, Socialism and Democracy*; Routledge: Abingdon-on-Thames, UK, 1942.

51. Isckia, T.; De Reuver, M.; Lescop, D. Toward an Evolutionary Framework for Platform-Based Ecosystem Analysis. 2018. Available online: https://hal.archives-ouvertes.fr/hal-01992440 (accessed on 14 January 2020).

52. Fautrero, V.; Gueguen, G. Quand la domination du leader contribue au déclin. *Rev. Fr. Gest.* **2012**, *3*, 107–121. [CrossRef]

53. Pinkse, J.; Vernay, A.L.; D'Ippolito, B. An organisational perspective on the cluster paradox: Exploring how members of a cluster manage the tension between continuity and renewal. *Res. Policy* **2018**, *47*, 674–685. [CrossRef]

54. Franklin, J.F.; Spies, T.A.; Van Pelt, R.; Carey, A.B.; Thornburgh, D.A.; Berg, D.R.; Bible, K. Disturbances and structural development of natural forest ecosystems with silvicultural implications, using Douglas-fir forests as an example. *Forest Ecol. Manag.* **2002**, *155*, 399–423. [CrossRef]

55. Florida, R. *Cities and the Creative Class*; Routledge: Abingdon-on-Thames, UK, 2005.

56. Elzen, B.; Geels, F.W.; Green, K. *System Innovation And the Transition to Sustainability: Theory, Evidence and Policy*; Edward Elgar Publishing: Cheltenham, UK, 2004.

57. Bode, S.; Hübl, L.; Schaffner, J.; Twelemann, S. Discrimination Against Newcomers: Impacts of the German Emission Trading Regime on the Electricity Sector. Diskussionsbeitrag, Leibniz Universität Hannover, Hannover, Germany, 2005.

58. Lau, A.K.; Lo, W. Regional innovation system, absorptive capacity and innovation performance: An empirical study. *Technol. Forecast. Soc. Chang.* **2015**, *92*, 99–114. [CrossRef]

59. Porter, M.E. Competitive advantage, agglomeration economies, and regional policy. *Int. Reg. Sci. Rev.* **1996**, *19*, 85–90. [CrossRef]

60. Porter, M.E. Locations, clusters, and company strategy. In *The Oxford Handbook of Economic Geography*; Clark, G.L., Gertler, M.S., Feldman, M.P., Williams, K., Eds.; Oxford University Press: Oxford, UK; New York, NY, USA; pp. 253, 274.

61. Edquist, C. *Systems of Innovation: Technologies, Institutions and Organizations*; Routledge: Abingdon-on-Thames, UK, 2013.

62. Barnes, W.; Gartland, M.; Stack, M. Old habits die hard: Path dependency and behavioral lock-in. *J. Econ. Issues* **2004**, *38*, 371–377. [CrossRef]

63. Liebowitz, S.; Margolis, S.E. *Path Dependence and Lock-In*; Edward Elgar Publishing: Cheltenham, UK, 2014.

64. Puffert, D.J. *Path Dependence, Network Form and Technological Change*; Stanford University Press: Stanford, CA, USA, 2000.

65. Zucchella, A. Local cluster dynamics: Trajectories of mature industrial districts between decline and multiple embeddedness. *J. Inst. Econ.* **2006**, *2*, 21–44. [CrossRef]

66. Amable, B. Les systèmes d'innovation. In *Encyclopédie de L'innovation*; Mustar, P., Penan, H., Eds.; Economica: Paris, France, 2003; pp. 367–382.

67. Luhmann, N. The autopoiesis of social systems. In *Sociocybernetic Paradoxes: Observation, Control and Evolution of Self-Steering Systems*; Geyer, F., Van d. Zeuwen, J., Eds.; Sage: London, UK, 1986; pp. 172–192.

68. Oldenburg, R. *The Great Good Place: Café, Coffee Shops, Community Centers, Beauty Parlors, General Stores, Bars, Hangouts, and How They Get You Through the Day*; Da Capo Press: Cambridge, MA, USA, 1989.

69. Cohen, W.M.; Levinthal, D.A. Absorptive capacity: A new perspective on learning and innovation. *Adm. Sci. Q.* **1990**, *35*, 128–152. [CrossRef]

70. Pigford, A.A.E.; Hickey, G.M.; Klerkx, L. Beyond agricultural innovation systems? Exploring an agricultural innovation ecosystems approach for niche design and development in sustainability transitions. *Agric. Syst.* **2018**, *164*, 116–121. [CrossRef]

71. Granstrand, O.; Holgersson, M. Innovation ecosystems: A conceptual review and a new definition. *Technovation* **2019**, *90–91*, 102098. [CrossRef]

72. Xu, G.; Wu, Y.; Minshall, T.; Zhou, Y. Exploring innovation ecosystems across science, technology, and business: A case of 3D printing in China. *Technol. Forecast. Soc. Chang.* **2018**, *136*, 208–221. [CrossRef]

73. Andriopoulos, C.; Lewis, M.W. Exploitation-exploration tensions and organizational ambidexterity: Managing paradoxes of innovation. *Organ. Sci.* **2009**, *20*, 696–717. [CrossRef]

74. Eaton, B.; Elaluf-Calderwood, S.; Sørensen, C.; Yoo, Y. Dynamic structures of control and generativity in digital ecosystem service innovation: The cases of the Apple and Google mobile app stores. *Lond. Sch. Econ. Polit. Sci.* **2011**, *44*, 1–25.

Article

Proposal of a Holistic Framework to Support Sustainability of New Product Innovation Processes

Ana S. M. E. Dias [1,2,*], António Abreu [1,3], Helena V. G. Navas [2] and Ricardo Santos [1,4]

[1] ADEM-ISEL, Departmental Area of Mechanical Engineering of Superior Institute of Engineering of Lisbon, Polytechnic Institute of Lisbon, 1959-007 Lisbon, Portugal; ajfa@dem.isel.ipl.pt (A.A.); ricardosimoessantos84@ua.pt (R.S.)

[2] UNIDEMI, Department of Mechanical and Industrial Engineering, NOVA School of Science and Technology, Universidade NOVA de Lisboa, 2829-516 Caparica, Portugal; hvgn@fct.unl.pt

[3] CTS, UNINOVA, NOVA School of Science and Technology, Universidade NOVA de Lisboa, 2829-516 Caparica, Portugal

[4] GOVCOPP—University of Aveiro, 3810-193 Aveiro, Portugal

* Correspondence: ana.dias@isel.pt

Received: 27 March 2020; Accepted: 18 April 2020; Published: 23 April 2020

Abstract: The survival of companies in globalized and highly competitive markets, heavily depends on their ability to innovate through the creation of new products and/or services, supported by sustainable processes to prevent business failure. There are many factors regarding the interface company/stakeholders/market at all hierarchical levels, which have a major contribution to sustain innovation in processes regarding the creation of new products and services. A holistic approach of all these factors, as a whole, has not been a subject of scientific research conducting to the necessity of creating a proposal of a framework that can be integrated and systemic. Thus, this paper aims to propose a functional holistic model, which integrates the strategic, organizational and operational levels regarding market business and company interaction, as well as the set of factors to take into account to guarantee assurance that innovative processes are sustained, when new products and/or services are created or improved. Conducted through an investigation of the state of the art, by literature review, a comprehensive and integrated conceptual model was built in a deductive-inductive way. Then, the conceptual model was validated through four case studies. Finally, it was found that the conceptual framework became functional, because its applicability has been successfully tested in a business environment. As a result, the tool developed here, can be useful to measure and evaluate projects dedicated to companies that innovate in a sustainable way.

Keywords: sustainability; innovation; new products; functional framework; SIFSNPIP; case studies

---

1. Introduction

Competitive new products and services are the output of sustainable innovative processes that companies manage in a systematic way, challenged by demanding and dynamic business markets [1]. According to [2], sustainable innovative processes are crucial for the survival of competitive companies, being a major factor for business success [3]. Also, firms should undertake their ideas about sustainable innovative products and services, and bring it to market as quickly that they can, to be competitive in nowadays global markets [4]. But the sustainability of innovative processes that support new product creation is not an easy process, and therefore a project can fail even when it was initially estimated to succeed. So, the innovative processes developed to create new sustainable products, involves considerable and various risks due to the uncertainty associated to business markets, according to [5]. Thus, risk is a strong obstacle to be transposed in business market characterizes by uncertainty, complexity and turbulence, according to [2,6]. So, the shortening of the available time to manage projects with both efficiently and effectively, is a very important issue to be taken in account by managers, especially when they concern relating to products and services of radical innovative nature [7]. According to [6], companies should manage their projects in a proactive, structured and sustainable way, to survive and succeed in such competitive markets, and for that, sharing knowledge through collaborative networks is crucial [8], taking in account all possible variables and parameters that have influence in the strategic, organizational and operational hierarchical levels of companies [9]. This finding is extremely important regarding new products and services that emerge of innovative projects and their sustainable implementation, which requires an increasing rational and holistic approach [10]. It is extremely challenging for companies the development of successful sustainable innovative processes to create new products or services, and for that the path that involves generation of new ideas is crucial. New idea generation occurs normally in the beginning of sustainable innovation processes to generate new products and services. So, this point is especially important, since it determines companies potential to undertake promising new product and service ideas at reasonable costs. In contexts where resources are constrained, creativity seems to be extremely contributive to problem-solving processes [11].

During literature review, was found a lack of holistic approaches or frameworks that could encompass the strategic, organizational and operational hierarchical levels of companies, in order to create new products and services through sustainable innovative projects, aiming the minimization of the risks of business failure inherent to its implementation. To serve these needs, it's proposed in this paper a "Systemic and Integrated Framework for Sustainability of New Product Innovation Processes"—SIFSNPIP. Thus, an extensive literature review supporting the construction of conceptual version of SIFSNPIP was carried out and it's presented on Section 2 of this paper. In Section 3, the key phases regarding the research methodology approached in this paper are presented. Section 4 presents and describes the case studies that were carried out to validate conceptual SIFSNPIP model and shows the aspects that were found which allowed to transform the model from conceptual to functional. In Section 5 the full framework of SIFSNPIP it's presented. Finally, the main of this investigation conclusions are presented in Section 6.

2. Literature Review

In order to design a theoretical framework, that determines the sustainability of innovative processes in the creation of new products or services, it was needed to find the most relevant set of variables and parameters that comprise strategic, corporative and operational business levels, as well as the way they interact with each other.

Firstly the approach of the strategic level was needed in order to understand which aspects embrace the company/market interactions and its ways of articulation. Secondly was found that the organizational level can be decomposed into two sub-levels with the same importance to a company: the corporate culture with a structural nature and the management principles that normally respond with market situations. Thirdly, the operational level was approached, as well as its inherent processual variables. So, the developed literature review was organized in this sequence, as Figure 1 illustrates. In there, the arrows between all levels, show the way of the relationship that the three levels have among themselves.

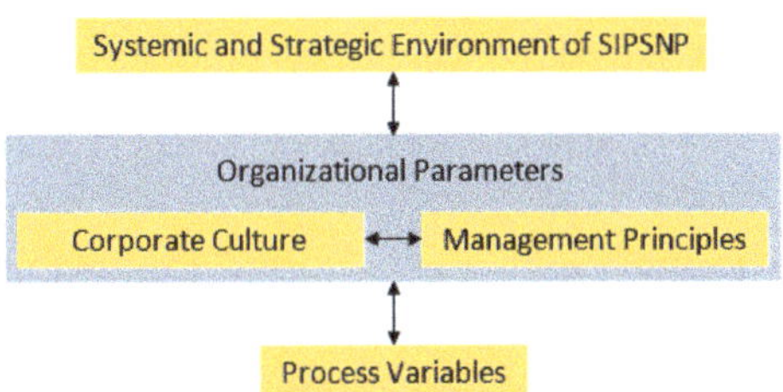

Figure 1. General framework of the approach of the Systemic and Integrated Framework for Sustainability of New Product Innovation Processes (SIFSNPIP) (authors' own elaboration).

2.1. Systemic and Strategic Environment in Sustainability of Innovative Processes that Support New Products (SIPSNP)

In the strategic level of the SIPSNP, is crucial for managers to take in account the relationship between: sustainable strategic innovation; source of new product or service creation; and the concept of business strategy. Sustainable strategic innovation in a highly competitive environment, embodies industries regarding both disruptive and incremental nature. The first one exists in markets without competitors, designated by "blue ocean strategy" (BOS), and the second exists whenever there is competition, designated by "red ocean strategy" (ROS) [12].

BOS is especially important when companies and businesses need to grow fast, and rather than compete with existing rivals as happens in ROS, BOS allows the creation of unique offerings for the emerging new markets [12].

Despite the above mentioned, is normal that hybrid strategies to be very common in business markets in which firms develop sustainable new products and services that emanate from innovation that embraces both radical and gradual processes, and in this context, hybrid strategies are called "purple ocean strategy" (POS) [13]. POS strategy corresponds to disruptive products and services with no competition at the beginning and while the possibility to remain like that exists. At the same time, other products regarding incremental innovation faces the existent competition. Beyond these strategies, companies have more challenges involving other factors of strategic nature that must be also considered, thus they require accurate analysis of correspondent trade-offs involved, namely facts concerning various risks and their interaction [14]. In a competitive environment along with high complexity of production processes risks of business failure must be analyzed in a systematic way, which means, with a continuous analysis of trade-offs involving the various risk factors of SIPSNP, especially those regarding quality, time and costs [15].

Independently of the SIPSNP strategies being radical, incremental or mixed, companies can't ignore the risks which they are exposed and must to be aware at the dynamics of competition through the implementation of systematic benchmarking practices [16]. For companies achieve the best performances of SIPSNP business, [17] concluded that teamwork, multidisciplinary and collaborative attitude have a huge impact in benchmarking practices efficiency and effectiveness, which should integrate the corporate culture of firms. According to [18], the globalization of markets and businesses is a trend that will remain strong in long term. Therefore an unavoidable aspect of globalization has been outsourcing practices, especially the knowledge-based services, such as the development of SIPSNP. Companies all over the world need to reduce costs, but the question is not the need of a particular practice of outsourcing or working abroad, but when and how it will be done to achieve greater competitive advantage in the market [18]. One special fact that arises in a particular process of outsourcing and/or offshoring is the "intellectual property" (IP) jointly developed. According to [18], exploitation and defense of IP, when generating both incremental and radical innovation, have impact on the strategic management of the focal company. Globalization and internationalization of business regarding SIPSNP projects, often correspond to engineering and management complex systems involving R&D and information highly reserved [19]. The strategic options pointed out are associated to risks of opportunistic expropriation of knowledge and related monitoring costs of the subcontracted partners, which sometimes are not only distant in geography but also in culture [19]. The focal company must have enough responsibility, knowledge and skills to ensure control over the processes regarding, onshore/offshore, third-party option (third-party logistics 3pl) decisions to guarantee that the final product or service fulfils the customer's needs. Therefore, all stakeholders involved in a business must integrate a network of synergies, to ensure the articulation of all processes regarding SIPSNP projects, but this fact increases complexity to the whole system and the concomitant risk that can emerge from the failure of each element. That's why the risks must be predicted so they can be avoided when trade-offs are considered [20].

A way of prevent the risks of business failure is the systematic relationship between companies and the market, meeting and even anticipating the customers' needs [21]. Therefore, the marketing performs an important point of articulation between a company and its customers, through the establishment of a systematic interaction between them promoting to companies the perception and understanding of the "voice of the customer" [22]. To this purpose, [22–24] point out some three important aspects: (1) a permanent interaction between the company and the customer to provide exchange of information, namely suggestions from the customer and a validation-built step by step between the designer and the customer, concerning the phases of products or services design; (2) in cases of products or services customization, it is crucial to considerate that the solution was obtained after experiments validated by the customer; (3) the existence of suppliers involvement with the company to ensure the sustainability the design of new products or services from the beginning.

In SIPSNP projects it is very important to consider, in strategic terms, if innovation requires the marketing function, because the role played by it in the whole new products or services design also depends on the level of innovation required, in order to obtain a positive trade-off regarding marketing/quality/cost/time [25].

From a general perspective, obtained from the literature review on the most relevant factors that comprise a strategic vision in SIPSNP environment described above, is presented in Figure 2 with a thereof summary of SIPSNP.

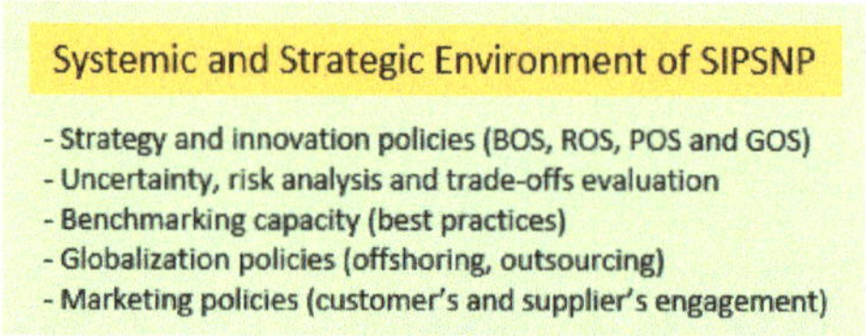

Figure 2. Systemic and strategic environment. Framework Approach (authors' own elaboration).

2.2. Organizational Parameters

From literature review it was found that several parameters have influence in firms that develop new products with sustainable innovative processes, in a structural way, and that they reported to the corporate culture; and the ones of a conjunctural order were derivate from management principles. It was also enlighten by the existing literature that each type of parameters is associated to specific factors, which will be covered in the following two sections.

2.2.1. Corporate Culture

In strategic management, the organizational parameters are considered part of the corporate culture. And one of the most relevant is the ability to function in the development of sustainable innovative projects with cross-functional teams perfectly connected in a systematic way. Therefore, many authors advise multidisciplinary, multifunctional and/or cross-functional organization type [1,26–29]. Such connections pass through the formation of collaborative teams, which should include at least employees of the organization, suppliers and customers. So, it is crucial to have a reliable information flow that ensure visibility and transparency in connecting people, processes and technologies.

The organizational strategy of working in multidisciplinary integrated teams (cross-functional) is increasingly suitable to companies that develop new products and services, due to markets globalization, existing together in partnerships and collaborative alliances with inter-organizational information sharing skills and sustainable innovation [30]. That is, a whole innovation capacity, radical and incremental, in organizations that work on network and that encompass collaboration with customers and suppliers [31].

Another way to characterize the sustainable innovative processes concerns to open innovation, in which resources move easily at the border or interface company/market [32]. Whenever open innovation must be shared as a partnership or strategic alliance, it assumes the designation of co-innovation [33]. This shared innovation, benefits the value chain to the customer, called a win-win relationship, and is of a major importance for companies to create value in the market. According to [34], with co-innovation different internal and external sources are integrated into a platform in order to turn the company more competitive and able to satisfy the customer's voice that means the existence of co-creation, co-design and joint development [35].

It follows that the corporate culture should incorporate another common inter-organizational factor: the competitiveness. That is, to incorporate in the company a competitive spirit associated with the effectiveness and success of sustainable new products and services available to the market [36].

2.2.2. Management Principles

Regarding the principles of management, the main organizational parameters that allow companies to respond to the market situations, are the following: compliance with legislation of the product inherent to each of the specific markets regarding new products or services development, manufacturing and commercialization [37]; product standardization that permits conformity with international rules and internal flexibility, facilitating modularization processes [38,39]; certification [39]; the agility of performance [40], connected with lean thinking [41,42] in the search for maximum efficiency, effectiveness and productivity [43].

In order to embody the paradigm of optimal productivity is needed to combine lean practices with flexibility and agility, especially when companies need to manufacture various types of products simultaneously [44,45]. Concluded about the importance to associate lean and agile concepts, so they proposed the term "leagility" aiming to integrate them in the paradigm of Supply Chain Management (SCM) in response to market's needs. Likewise, the terms flexibility and proactive flexibility were combined into the term "adaptability" according to [46]. On Figure 3, it's summarized the relevant factors of SIPSNP integrated at the organizational level.

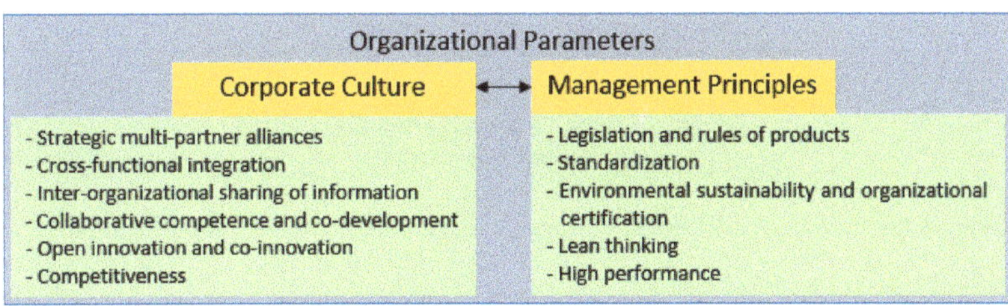

Figure 3. Organizational parameters. Framework Approach (authors' own elaboration).

2.3. Process Variables

According to literature review, are considered as relevant the following process variables: undertaking an idea of the product or service through a process of innovation management [47];

the organization and management of the project [19]; the quality and control of a project [22] and the engineering and technological capabilities [48,49].

Sustainable innovation management is a methodical process of new ideas generation that allows companies to create value in a proactive way [47], and that can be done in so many ways, therefore that's why each innovation process is unique [50]. From product and service perspective, innovation in a sustainable way to prevent and minimize risks of failure, is therefore a process of creation something new (a product, a service, a process, etc.) yet unknown by the market, and that is due mainly to creative capacity and the technologies available to undertake it. So, it is not a casual situation, but an overall process extending over time [51]. The generation of ideas is a fundamental part of an innovation process that can be convergent or divergent: it is convergent when the idea is the result of a systematic collective process based on trial and error; and it is divergent when a "flash of genius" of some bright and creative collaborator occurs. According to [52], an innovative value chain consists of three main phases: generation of ideas; sorting of ideas and their development; and its dissemination through organization to the market. If it is decided to undertake an innovative idea into a new product or service, the next step will be the project management of SIPSNP.

Many authors as [29,53], present classic models that are examples of "organizational architectures" that group, compose and arrange sub-teams, their inter-relationships and hierarchies. Information flows and "architecture of processes" allow to properly delegate the work to be carried out through the hierarchical levels of companies, as well and the respective flow-related information those levels, aiming to achieve projects goals [20]. Present a model regarding iterative project management, or spiral model, in which flexible changing of work specifications are possible, avoiding the need of restart the whole project from the beginning, modifying only the necessary steps when market changes occur. There are other kind of proposed managing models for SIPSNP projects that are more flexible and agile, widely used by companies worldwide and scientifically known, like simultaneous or concurrent engineering [54–57], and Stage–Gate® [58–60].

Another process variable to have in account in SIPSNP projects is the quality of products and services which are inherent to them [60]. And Taguchi (1986) method is a very important tool used by managers, because of its importance to achieve sufficiently robust outputs with high quality levels [25].

SIPSNP implementation also implies to know and to control engineering and technological capabilities [48,49,61]. Conclude about the importance of prototyping whenever it is needed, and since engineering, technology, quality and reliability are very important issues in those kind of processes, teamwork and collaboration across all hierarchical levels, are crucial for its success. Figure 4 illustrates the most relevant process variables approached by literature review.

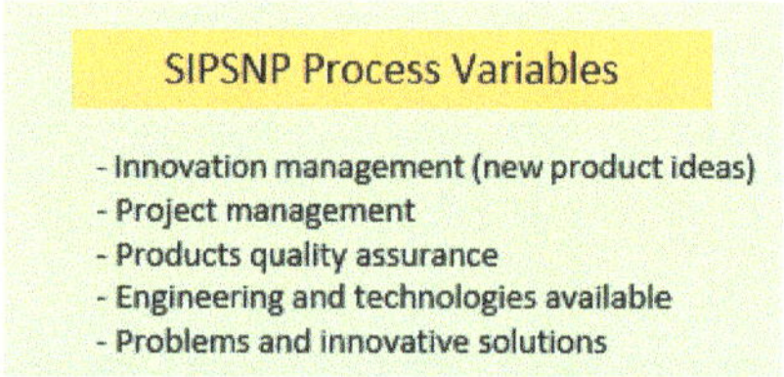

Figure 4. Operational level variables. Framework Approach (authors' own elaboration).

2.4. Problems and Innovative Solutions

Markets are increasingly demanding for sustainable and innovative products and services, additionally more information is required by customers, about the environmental impact of products and services provided by companies. Modern management must use sophisticated tools to meet such expectations, so it can be possible to improve monitoring processes of products and services impact, in order to understand how they can be made more sustainable. Regarding products and services lifecycle, which impact is not caused only by the industrial processes or even the usage of products and services, but also by natural methods of extraction and exploitation of raw materials and others, the transport and storage processes, etc. Therefore, a key factor for a successful sustainability management concerns to the availability and sharing of relevant data and knowledge that must be wisely shared and used along all logistic chain regarding to developed projects [62].

The tools and methodologies available to serve problem solving regarding SIPSNP, are one of the most important issues of a project [63,64]. Conducted a survey of about three dozen tools and techniques obtained through an extensive literature review and realization of several case studies on Taiwanese companies, as well as interviews with experts in the field. Based on [63,64] work, a sample of the most important tools for SIPSNP projects is presented in Table 1.

Table 1. Systemic and Strategic Environment in Sustainability of Innovative Processes that Support New Products (SIPSNP) tools and methodologies (authors' own elaboration).

Survey of Tools to Support SIPSNP	
Grouping	**Tools and Methodologies**
Creative and Innovative Solutions	TRIZ; DOE; DFX; Pugh analysis; Creative Design; Axiomatic Design
Focus on Quality Function	QFD (e.g.: Kano Model; Ishikawa diagram; DFMEA; Pareto law)
Focus on Precision Manufacturing	DFSS (DMAIC cycle and it's variants)
Focus on Involvement of Suppliers	SDI
Design Support	Robust Design; **Modular** Design; CE
Decision Support	AHP; CBR; DEA; Delphi Panel; Fuzzy logics; Neuronal Networks

Acronyms: TRIZ (Theory of Inventive Problem Solving); DOE (Design of Experiments); DFX (Design for Excellence); QFD (Quality Function Development); DFMEA (Design Failure Model and Effect Analysis); DFSS (Design for Six Sigma); DMAIC (Define-Measure-Analyze-Improve-Control); CE (Concurrent Engineering); AHP (Analytical Hierarchy Process); CBR (Case Based Reasoning); DEA (Data Envelopment Analysis).

Due to the fact that in the literature was found a general confusion about the application of the terms "tools" and "methodologies", for this paper they were grouped putting its focus on their use. Therefore, they were grated as follows: "Creative and Innovative Solutions"; "Quality Function"; "Precision Manufacturing"; "Involvement of Suppliers"; "Design Support" and "Decision Support". Also there was considerate that if similar problems occur, is not guaranteed that they will have similar solutions, because markets are dynamic, and along with that uncertainties can emerge [65]. Uncertainly can be reduced by if the company has an adequate portfolio of problems and respective solutions, and for that the methodology "Case-Based Reasoning" (CBR) can be extremely useful [66]. When a problem is new, there won't be any solutions to solve it obtained from the above mentioned portfolio, so it is necessary to use any of the available methodologies and presented in Table 1. But, if there are several solutions available in such portfolio, it is necessary to determine a ranking of solutions in order to adopt the more suitable one to the existent problem. And for that, one of the most commonly used method is the Analytical Hierarchy Process (AHP). According to [67,68], AHP is very useful to rank the various possible alternatives to support the decision making process. The different ways to achieve a solution to an existing problem, regarding SIPSNP projects, is illustrated in Figure 5.

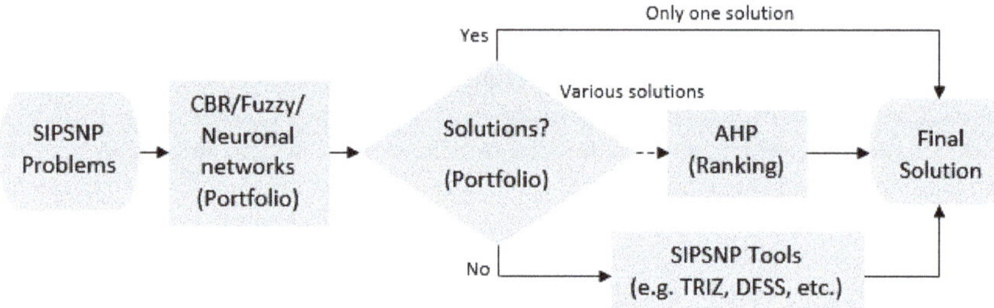

Figure 5. Problems and innovative solutions. Framework Approach (authors' own elaboration).

3. Research Methodology

As already mentioned, the objective of this investigation is to build a holistic model, which can become a support tool for the sustainable development of innovative processes for creating products and/or services (both incremental and radical), that is, a model that can serve as a roadmap for companies working in this area.

The literature review resulted in a survey, as exhaustive as possible, in a deductive way of the current state of the art and of the set of existing or proposed models for relevant but partial issues of this theme, as long as its methods and tools. The theoretical investigation was conducted in an exploratory way through a deductive-inductive strategy, given that the topic in question adapts to strategic management issues, whose problems and their solutions always involve so many aspects and perspectives, which vary with the changes that constantly occur in the markets that are currently globalized, hampering decision-making processes. It is intended that the main added value of this investigation is evidenced by the difference in the level of science in question between the situation of a possible pulverization of partial models and the construction of the holistic, comprehensive and integrated model, which is intended to be achieved with the work developed.

The proposed model was initially conceptual, as it was obtained thanks to the inductive jump performed after deduction in a qualitative type investigation, having been empirically tested at a later stage, with its external empirical validation through a set of case studies on sustainability innovative processes for creating products and/or services in a national and international industrial environment (in case studies involving projects with partners worldwide). The purpose of this validation was not to generate theory, but only to test, validate and improve it. And so, the conceptual model obtained became functional, because it was shown that it actually works in an industrial environment.

Therefore, it was decided from the beginning on a deductive-inductive structure research, through literature review that it was possible to achieve the conceptual model SIFSNPIP, and there was the need for its empirical validation. Then, for the validation of the model case studies in industrial environment were performed, since most issues to validate were questions of "how" and "why" type, in their qualitative and explanatory variant, as recommended by [69]. In accordance with this work goals, the research was generally regarded as descriptive, due to the fact that it aims to accurately describe the phenomena of reality studied and hence did not require the use of techniques and statistical methods. The methodologies used to validate models regarding exact sciences are often quantitative, while approaching social sciences they are often qualitative, given its high flexibility [69,70].

In the real meaning of the case studies, the case's target to be analyzed is called "unit of analysis" [71]. According to this definition, the units of analysis regarding this investigation are composed of innovative products and services, whose purpose was to test the proposed conceptual model. Since the case studies were performed though interviews complemented with guided tours of companies' manufacturing facilities (that were only possible to perform in Portugal), an interview script (protocol) was elaborated, according to the guidelines pointed out by [71]: interviews (recorded or not) at the place of analysis; telephone conversations; mail contacts; collection of written documents or computer data; collection of information from "key informants" (only one or a panel), that should be trustworthy people with the right technical and scientific knowledge, from the inside of the organization. Still according to [71], an important aspect to be agreed by both parts in each case study, is the confidentiality of data or information collected and the hypothesis of firm choose to remain anonymous. A final aspect pointed out by this author, refers to the importance of obtaining a formal authorization from the organization boards and provide to their representatives, as well as the "key informants", to review the material provided.

The criterion used in choosing the case studies presented in this article was based on the fact that its scope covers the objectives stipulated in the investigation. In other words, verify and validate all aspects inherent to the proposed conceptual model for the four product/service/incremental/radical combinations, as much as possible, in the perspectives of the national and international industry. The validation of the proposed model through case studies aims to assess its functionality in a business environment and to verify the need of its improvement, so that it can constitute, in the best possible way, a roadmap to be followed by of companies wishing to create new products and/or services in a sustainable manner in the business markets in which they operate. Figure 6 illustrates the key stages involving the research methodology followed in this paper.

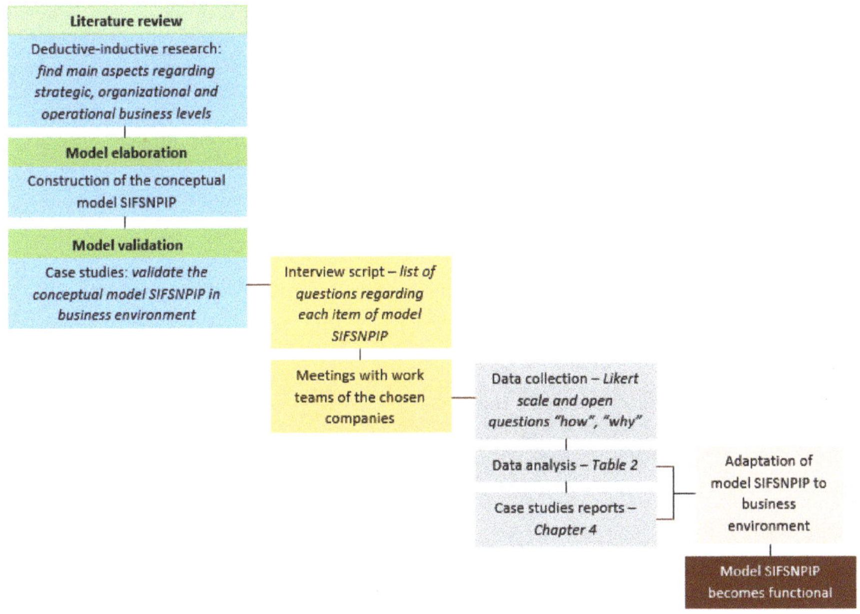

Figure 6. Key stages involving the research methodology (authors' own elaboration).

4. Case Studies

Four case studies were preformed, regarding products and services, which proved to be sufficient to validate the various parts of the developed model developed, based on literature review, and according to a specific protocol adopted, described by [72]. First, it was validated in the business field the proposed conceptual SIFSNPIP and, secondly, its usefulness was evidenced by demonstrating that it can successfully applied in the assessment of companies that design and develop sustainable new products, allowing to punctuate the evolutionary state of all their strategic, organizational and operational aspects and also its range of innovative products to market.

The ccompanies that collaborated to carry out the presented case studies, were the international business group Instituto de Soldadura e Qualidade (ISQ) and a Portuguese company of metalworking industry, which preferred to remain anonymous.

Of the four explanatory cases studied in industrial environment, two were related to products and were called by "HVAC" and "WJ-LASER" while the other two, were related to services and called by "NaturalHy" and "Brazing".

The case studies were carried out through meetings with work teams of the companies analyzed, belonging to the three levels of decision making: strategic; organizational and operational. For this, three meetings were carried out for each one of the four cases, with a total contribution of 12 work teams. To pursue this end, an interview script was elaborated, composed by questions inherent to each item covered in the model, which was made in accordance with [71,72].

The constitutive meetings of the case studies were aimed at obtaining answers to the questions in the interview script, in order to verify the extent to which all aspects evidenced in the model were performed by the company. The cases are described in Sections 4.1–4.4. Results regarding scores related to the degree of accomplishment by firms of each item of the model, using the Likert scale presented below, are presented in Table 2. Then, complementary open questions of type "how" and "why", made to better understand the extent to which the answers fit this scale to explain the scores obtained, also helped to present the description of the cases and to perform de discussion of the results presented in Section 4.5.

In order to define the scores attributed to each item of the model, the following Likert scale was used:

0. Nothing necessary is accomplished
1. It is performed below the necessary minimum
2. The minimum necessary is accomplished
3. The essentials are performed above the necessary minimum
4. Everything necessary is accomplished

The interview script open questions were designed to address the following aspects in a systemic way:

- Level of achievement of each item in the model;
- Interpretation of the results obtained;
- Maintenance of the model in the short and medium term;
- Elaboration of the conclusions that allowed to confirm (or even improve) the model.

Figure 7 shows that the four case studies were chosen to cover the four possibilities, regarding their application in products and services vs. incremental and radical innovation.

	Procuct	Service
Incremental	HVAC	Brazing
Radical	WJ-LASER	NaturalHy

Figure 7. Selection criteria for the case studies presented: Application vs. innovation type (authors' own elaboration).

4.1. HVAC Case

"HVAC" (Heating, Ventilating and Air Conditioning) case regards to the metalworking industry in which a Small-Medium Enterprise (SME) manufactures a high range of HVAC equipment to be commercialized in the market. This business activity requires a considerable capacity of innovation relating to the manufacture processes of its products, which are mainly: SPIRO® system; heat exchangers; silencers; air handling units (AHUs); rectangular, circular and oval ducts - those ones with

Ethylene-Propylene-Diene-Methylene (EPDM) sealing gasket; chilled beams; fan units; chilled water and storage heat tanks grilles and diffusers and CADvent software (for calculus and dimensioning of air duct installations). To guarantee the sustainability of its products in th metalworking industry e market the firm performs continuous improvement of the products, but because mostly all of them contain a huge number of components, as the AHUs. Therefore, the company can use its resources both to manufacturing and improvement processes, working in ROS to promote its customers satisfaction, and trough marketing, expanding itself in the market.

4.2. WJ-LASER Case

"WJ-LASER" case regards the use of both cutting processes water jet (WJ) and (light amplification by stimulated emission of radiation (LASER) on several kinds of materials to promote the manufacturing of customized products, that means, products defined by the customers. The use of this cutting technologies requires the tool Creative Design. The customers that requires this kind of technologies from the firm, are mainly from rehabilitation of ancient artifacts (oil paintings, pottery, papyri, etc.), art and advertising industries. Through the adjustment of cutting parameters in non-metal materials, this two technologies can be used on waste removal of high precision, without damaging the object material. When the firm uses this technologies in the cutting sector, due to the extreme flexibility of its equipment, the innovation level of the products obtained is radical, because all kind of geometries can be designed with high accuracy. Water jet and laser cutting technologies are suitable for small batches or even single parts, but not for mass production.

4.3. NaturalHy Case

NaturalHy case regards to a radical service provided by an international business group that focuses strongly on Research & Development (R&D).

The denomination "NaturalHy" defines a project that was recently concluded on which the group participated as part of the executive/steering committee. The project goal is the distribution and use of natural gas with hydrogen addition, so the chemical mix can be used across all Europe, with high level of safety and environmental sustainability through infrastructures designed and built for its distribution. The radical innovation of this case is inherent to the addition of hydrogen to natural gas with a combustion reaction, generating gaseous chemical reaction products with very low amount of carbon dioxide. The project also involved the building of pipelines and storage tanks to distribute and store this new gaseous product, and the distribution network can be used by both domestic and industrial fields, and beyond that, the project also covers permanent monitoring and tracing processes, to ensure its sustainability. The development of this project occurred between years 2004 to 2009, with the collaboration of over 38 business partners, involving a huge investment dimension. The amount invested in the project was about EUR 11 million (granted by the European Commission), having exceeded a profit of about EUR 17 million. Still during the period above mentioned, the project expanded to the Middle East, having participation in the building and operation of the Research Centre of the Petroleum Institute in Abu Dhabi laboratories. This case study was unique to the model purposed in this paper, because allowed it to pass to a functional level with the introduction of exportation policies as a factor not pointed out by literature review on strategic management, as one of the strategic inherent classical issues normally approached.

4.4. Brazing Case

Brazing case regards to the use of brazing technology in polymeric materials without lead alloys, as a service with incremental innovation, since this technology was already used worldwide in metal materials. Working with alternative chemical alloys rather than lead, in brazing of polymeric materials (that are the material basis of electronic circuit boards) allows to guarantee health and environmental sustainability. The company offers a testing service using this technology to serve projects regarding the manufacture of electronic and electrical components for several kinds of industries, namely audio visual, aerospatiale, appliances and electronics firms and business groups, in partnership with: Research & Development (R&D) institutions; airlines; governmental agencies and armed forces. From these business partners are highlighted: Crane-NSWC; American Air Force; Boeing; BAESystems; ITB Inc.; NASA; Texas Instruments; Northrop Grummam and Portuguese Association of Electrical and Electronics Industries (PAEEI). A disadvantage of using this technology is the difficulty on welding in polymeric materials using elements with high melting points, like tin to work with this type of circuits, and for this reason, it is crucial to conduct a high number of tests. When the risk of failure of an electronic board leads to catastrophic results, in the case of aircraft and military armament, lead alloys can be used again, so experiments are extremely important to make a decision about which alloy material can be used with a minimum risk of failure. Since testing is one the most important phase of high risk projects, DFSS, DOE, DFX, among others, are the support methodologies and tools to test performing with the highest accuracy possible.

4.5. Discussion of Results

In "HVAC" case it was found that the "systemic and strategic environment" level almost factor scores were of 1 and 2. This is an acceptable fact, because the firm works in ROS. In the factor levels of both "corporate culture" and "management principles", scores incidence occurred were of 1 and 3.

This was an expectable fact, because firm works with compliance by the rules and obligations to the market. Regarding "process variables" level, the scores were almost all between 1 and 2, due to the fact that the firm works with the same range of products aiming to make the best use of its resources and with a specific technological and engineering know-how. And from the panel of tools available in Table 1, the use of modular and tolerance design reached the score 4, because all production is composed of modular products, in which assembling processes need to obey to specific tolerances.

In "WJ-LASER" case, when rating the "systemic and strategic environment" level, almost factor scores were in 4, but not the one regarded to ROS vs. BOS that correspond to score 2, and the one regarded outsourcing with score 0. In the factor levels of both "corporate culture" and "management principles", scores incidence occurred were on 4. The factors regarding involvement of suppliers and customers scored in 2 or 3 respectively, because concerns to a radical innovation. In level "process variables", all scores obtained were equal to 4, because it is a project that must highly obey to all that issues to be extremely profitable. But relating Stage-gate® projects the score was 0, because this management tool is not applicable on water jet and laser technologies. It was found a large application of creative project methodology combined with modular design. And almost all other methodologies and tools referred in Table 1 were pointy used, except the TRIZ and DFSS, because there were no contradictions to be solved through TRIZ methodology, and DFSS has no application in individual parts or small batches of products concerning this case.

In "NaturalHy" case it was found that the "systemic and strategic environment" level, factors scores were of 4, except the one concerning the ROS, because it is the innovation is radical. This case showed up that exportation issue was not contemplated in the conceptual model so, it was a gap not found in the literature review, that when included on the "systemic and strategic environment" level of the conceptual model, turned it into functional. In the literature review about strategic management was found a single paper on this subject—the work of [73] - regarding business expansion and perspectives. Regarding levels of both "corporate culture" and "management principles" most of its factors were scored with 4, but not the ones concerning offshoring, outsourcing, and the need for low cost solutions. Regarding the level "process variables", its factors were scored between 1 and 4, because it is about a project regarding a huge amount of issues to take in account, and ones are more demanding than others. Regarding methodologies and tools presented on Table 1, the business group it was found that almost all of them were applicate in the project.

In "Brazing" case it was found that the "systemic and strategic environment" level, factors scores were practically all on 4. Regarding levels of both "corporate culture" and "management principles" almost all of them were scored with 4, and the same happened to level "process variables". Regarding methodologies and tools presented on Table 1, it was found that the company uses almost all of them, sometimes separately and other times combining the ones that are complementary. For example, the business group uses DFSS many times, because the level of accuracy required on the manufacturing of electronic circuit boards is a very accurate process.

All scores that were obtained using the already mentioned Likert scale on the application of all SIFSNPIP items, with the four case study carried out, are presented in Table 2.

Table 2. Summary of the scores of the factors measured by SIFSNPIP (authors' own elaboration).

Levels	Parameters and Variables	Cases			
		HVAC	WJ-LASER	NaturalHy	Brazing
Systemic and Strategic Environment	Strategy and innovation policies	2	2	2	4
	Risk analysis and trade-off evaluation	1	4	4	4
	Marketing policies; customers' and suppliers' engagement	1	2	4	4
	Benchmarking capacity	1	4	4	4
	Globalization policies	2	4	4	4
	Exploitation policies	——	——	4	——
Organizational (Culture)	Strategic multi-partner alliances	3	4	4	4
	Cross-functional integration	2	4	4	4
	Inter-organizational sharing of information	2	3	4	4
	Collaborative competence and co-development	3	4	4	4
	Open innovation and co-innovation	2	4	4	4
	Competitiveness	2	4	4	4
Organizational (Management Principles)	Legislation and rules of product	3	4	4	4
	Standardization	3	4	4	4
	Environmental sustainability	3	4	4	4
	Organizational certification	3	4	4	4
	Lean thinking	1	4	4	4
	High performance	2	4	4	4
Process Variables	New product ideas (conception and development)	2	4	4	4
	Project management	2	3	3	4
	Products quality assurance	3	4	4	4
	Engineering and technologies available	2	4	4	4
	Problems and innovative solutions	1	4	4	4
Problems and Innovative Solutions (Methodologies and Tools)	TRIZ; DOE; DFX; QFD; DFMEA; DFSS; DMAIC; CE; AHP; CBR; DEA; etc.	1	3	3	3

The four cases studies regarding both incremental and radical products and services, carried out on firms and business group's facilities, were found to be enough regarding the measurement of all issues that integrate its levels, because the results obtained fit into the analyzed realities. And conceptual model SIFSNPIP, built based on the literature review, became conceptual after its validation in industrial field through the described cases studies.

5. Proposal of a Conceptual/Functional Model

SIFSNPIP functional model was finally obtained, by putting Figures 1–5 altogether along with its interactions, as illustrated on Figure 8.

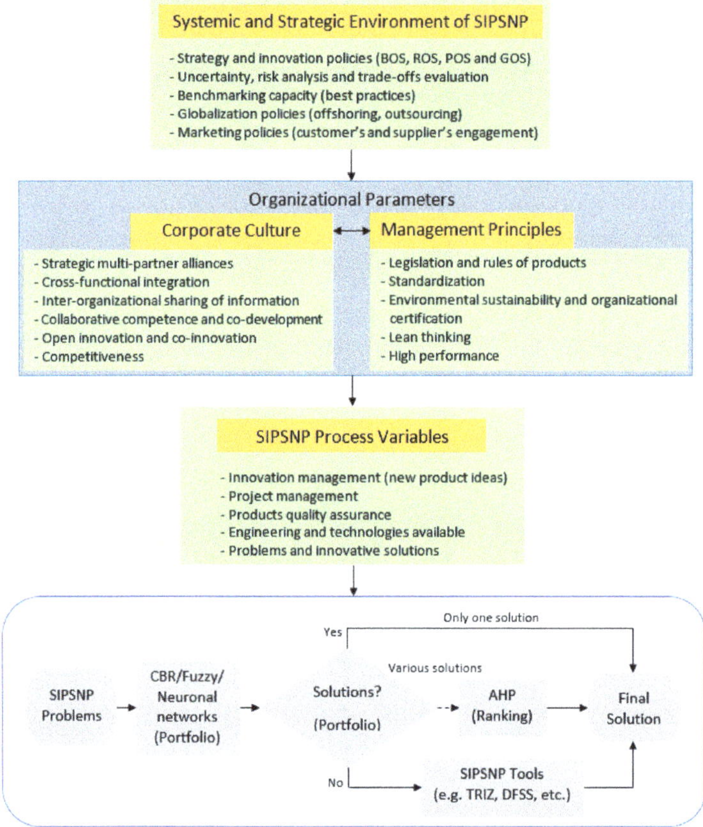

Figure 8. Functional model of SIFSNPIP (authors' own elaboration).

It is important to highlight that the item "export policy" integrates the "systemic and strategic environment" level of SIPSNP model, and also that this fact was the differentiator factor between the conceptual and functional model.

6. Conclusions

The research on literature review, was based on a deductive-inductive pathway, in order to construct a comprehensive and integrated conceptual model to support Sustainability of New Product Innovation Processes—the SIFSNPIP model. It was empirically validated, in the industrial environment through four explanatory case studies, referring to the implementation of sustainable new products and services, both incremental and disruptive. From the literature review it wasn't detect, until now, any holistic models to support sustainability of new product innovation processes, regarding this phenomenon as a whole appropriate for cases of enterprises or industries models, but only partial approaches. This fact justified the purpose of this investigation, and for all that was exposed in this paper, it was concluded that this goal was successfully achieved. The SIFSNPIP model that was initially conceptual, became functional after its external validation in industrial field through the performance of four case studies, regarding both incremental and radical products and services. So, functional SIFSNPIP model can be used as a diagnostic tool or roadmap for measurement of projects carried out by firms that innovate, design and develop new products with a sustainable way of management.

According to the initial objectives, model SIFSNPIP should allow at least two different uses:

- The first, with a purely scientific nature as an organized menu of solutions to problems that occurred in innovative processes to design new products, using known methodological and instrumental tools;
- The second, with an operational nature, in which the model will work as a diagnostic roadmap for measuring processes, projects, and products, with the purpose of reducing the risks of business failure as much as possible.

It seems clear that would be desirable that model SIFSNPIP could be better tested with a dozen or more case studies, with several business organizations that design new products, both incremental and radical, sustainably. Such applications could have, for organizations, a measurement and improvement of their own processes, in addition to any specific adaptations of the model, as well as data collection that would allowed a statistical treatment of the incidence and influence of the model factors on competitiveness of the national and international innovative industry itself.

Author Contributions: Conceptualization, A.S.M.E.D. and A.A.; methodology, A.S.M.E.D., A.A., H.V.G.N. and R.S.; validation, A.S.M.E.D., H.V.G.N. and R.S.; formal analysis, A.S.M.E.D. and A.A.; investigation, A.S.M.E.D., A.A., H.V.G.N. and R.S.; resources, A.S.M.E.D. and A.A.; data curation, A.S.M.E.D. and A.A.; writing—original draft preparation, A.S.M.E.D. and A.A.; writing—review and editing, A.S.M.E.D. and R.S.; visualization, A.S.M.E.D. and R.S.; supervision, A.S.M.E.D. and R.S.; project administration, A.S.M.E.D. and A.A. All authors have read and agreed to the published version of the manuscript.

Funding: This research received no external funding.

Acknowledgments: The authors of this paper wish to thank the Sustainability Journal for the opportunity to submit this paper to its Special Issue "Innovation Ecosystems: A Sustainability Perspective", a fact that gave inspired motivation to carry out this work. It is also intended to express gratitude to all of the members of the scientific community who, in the past, have written R&D works that served as references for the literature review done in this research work in particular. A "thank you" is sent to the companies that collaborated to carry out the presented case studies (international business group Instituto de Soldadura e Qualidade—ISQ and a Portuguese company of the metalworking industry that preferred to remain anonymous), which contributed to increase the knowledge about the subject in question. Lastly, the authors also wish to acknowledge the Fundação para a Ciência e a Tecnologia (FCT—MCTES) for its contact support that helped to carry out the case studies presented in this paper via the project UIDB/00667/2020 (UNIDEMI).

Conflicts of Interest: The authors declare no conflict of interest.

References

1. Akgün, A.E.; Byrne, J.C.; Lynn, G.S.; Keskin, H. New product development in turbulent environments: Impact of improvisation and unlearning on new product performance. *J. Eng. Technol. Manag.* **2007**, *24*, 203–230. [CrossRef]
2. Karniel, A.; Reich, Y. From DSM-Based Planning to Design Process Simulation: A Review of Process Scheme Logic Verification Issues. *IEEE Trans. Eng. Manag.* **2009**, *56*, 636–649. [CrossRef]
3. Kim, C.; Yang, K.H.; Kim, J. A strategy for third-party logistics: A case analysis using the blue ocean strategy. *Omega* **2008**, *36*, 522–534. [CrossRef]
4. Raehse, W. Compressed Development and Implementation of Innovations. *Chem. Ing. Tech.* **2012**, *84*, 588–596.
5. Mu, J.; Peng, G.; MacLachlan, D.L. Effect of risk management strategy on NPD performance. *Technovation* **2009**, *29*, 170–180. [CrossRef]
6. Liu, Y. Sustainable competitive advantage in turbulent business environments. *Int. J. Prod. Res.* **2013**, *51*, 2821–2841. [CrossRef]
7. Heising, W. The integration of ideation and project portfolio management—A key factor for sustainable success. *Int. J. Proj. Manag.* **2012**, *30*, 582–595. [CrossRef]
8. Karlsson, C.; Warda, P. Entrepreneurship and innovation networks. *Small Bus. Econ.* **2014**, *43*, 393–398. [CrossRef]
9. Patanakul, P.; Milosevic, D. The effectiveness in managing a group of multiple projects: Factors of influence and measurement criteria. *Int. J. Proj. Manag.* **2009**, *27*, 216–233. [CrossRef]
10. Bonabeau, E.; Bodick, N.; Armstrong, R.W. A more rational Approach to New-Product Development. *Harv. Bus. Rev.* **2013**, *86*, 96–102.

11. Liu, Z.; Feng, J.; Wang, J. Resource-Constrained Innovation Method for Sustainability: Application of Morphological Analysis and TRIZ Inventive Principles. *Sustainability* **2020**, *12*, 917–940. [CrossRef]

12. Markides, C. Strategic Innovation. *Sloan Manag. Rev.* **1997**, *38*, 9–23.

13. Gandellini, G.; Venanzi, D. Purple Ocean Strategy: How to Support SMEs' Recovery. *Procedia Soc. Behav. Sci.* **2011**, *24*, 1–15. [CrossRef]

14. Pollack, J.; Liberatore, M.J. Incorporating Quality Considerations into Project Time/Cost Tradeoff Analysis and Decision Making. *IEEE Trans. Eng. Manag.* **2006**, *53*, 534–542. [CrossRef]

15. Kim, J.; Kang, C.; Hwang, I. A practical approach to project scheduling: Considering the potential quality loss cost in the time–cost tradeoff problem. *Int. J. Proj. Manag.* **2012**, *30*, 264–272. [CrossRef]

16. Barczak, G.; Kahn, K.B. Identifying new product development best practice. *Bus. Horiz.* **2012**, *55*, 293–305. [CrossRef]

17. Cooper, R.G.; Edgett, S.J.; Kleinschmidt, E.J. Benchmarking Best NPD Practices—1. *Res. Technol. Manag.* **2004**, *47*, 31–43. [CrossRef]

18. Roy, S.; Sivakumar, K. Managing Intellectual Property in Global Outsourcing for Innovation Generation. *J. Prod. Innov. Manag.* **2011**, *28*, 48–62. [CrossRef]

19. Harmancioglu, N. Portfolio of controls in outsourcing relationships for global new product development. *Ind. Mark. Manag.* **2009**, *38*, 394–403. [CrossRef]

20. Choi, T.Y.; Dooley, K.J.; Rungtusanatham, M. Supply Networks and Complex Adaptive Systems; Control versus Emergence. *J. Oper. Manag.* **2001**, *19*, 351–366. [CrossRef]

21. Wang, C.-H.; Chen, J.-N. Using quality function deployment for collaborative product design and optimal selection of module mix. *Comput. Ind. Eng.* **2012**, *63*, 1030–1037. [CrossRef]

22. Campbell, R.I.; De Beer, D.J.; Barnard, L.J.; Booysen, G.J.; Truscott†, M.; Cain, R.; Burton, M.; Gy, D.; Hague, R. Design evolution through customer interaction with functional prototypes. *J. Eng. Des.* **2007**, *18*, 617–635. [CrossRef]

23. Schaarschmidt, M. Impediments to costumer integration into de innovation process: A case study in the telecommunications industry. *Eur. Manag. J.* **2014**, *32*, 350–361. [CrossRef]

24. Jayaram, J. Supplier involvement in new product development projects: Dimensionality and contingency effects. *Int. J. Prod. Res.* **2008**, *46*, 3717–3735. [CrossRef]

25. Kang, N.; Kim, J.; Park, Y. Integration of marketing domain and R&D domain in NPD design process. *Ind. Manag. Data Syst.* **2007**, *107*, 780–801.

26. Rihar, L.; Kušar, J.; Duhovnik Starbek, M. Teamwork as a Precondition for Simultaneous Product Realization. *Concurr. Eng. Res. Appl.* **2010**, *19*, 261–273. [CrossRef]

27. Kahn, B.K.; Barczak, G.; Nicholas, J.; Ledwith, A.; Perks, H. An Examination of New Product Development Best Practice. *J. Prod. Innov. Manag.* **2012**, *29*, 180–192. [CrossRef]

28. Brettel, M.F.; Engelen, H.A.; Neubauer, S. Cross-Functional Integration of R&D, Marketing, and Manufacturing in Radical and Incremental Product Innovations and Its Effects on Project Effectiveness and Efficiency. *J. Prod. Innov.* **2011**, *28*, 251–269.

29. Sharma, A. Computer-Aided Design Collaborative product innovation: Integrating elements of CPI via PLM framework. *Comput. Aided Des.* **2005**, *37*, 1425–1434. [CrossRef]

30. Athaíde, G.A.; Zhang, J.Q. The Determinants of Seller-Buyer Interactions during New Product Development in Technology-Based Industrial Markets. *J. Prod. Innov. Manag.* **2011**, *28*, 146–158. [CrossRef]

31. Al-Zu'bi Tsinopoulos, C. Suppliers versus Lead Users: Examining Their Relative Impact on Product Variety. *J. Prod. Innov. Manag.* **2012**, *29*, 667–680.

32. Robertson, P.L.; Casali, G.L.; Jacobson, D. Managing open incremental process innovation: Absorptive Capacity and distributed learning. *Res. Policy* **2012**, *41*, 822–832. [CrossRef]

33. Traitler, H.; Watzke, H.J.; Saguy, S.I. Reinventing R&D in an Open Innovation Ecosystem. *J. Food Sci.* **2011**, *76*, 62–68.

34. Lee, S.M.; Olson, S.M.; Trimi, S. Co-innovation: Convergenomics, collaboration, and co-creation for organizational values. *Manag. Decis.* **2012**, *50*, 817–831. [CrossRef]

35. Durugbo, C. Strategic framework for industrial product-service co-design: Findings from the microsystems industry. *Int. J. Prod. Res.* **2013**, *52*, 2881–2900. [CrossRef]

36. Shih, W.Y.C.; Agrafiotesb, K.; Sinhac, P. New product development by a textile and apparel manufacturer: A case study from Taiwan. *J. Text. Inst.* **2014**, *105*, 905–919. [CrossRef]

37. López-Mielgo, N.; Montes-Peón, J.M.; Vázquez-Ordás, C.J. Are quality and innovation management conflicting activities? *Technovation* **2009**, *29*, 537–545. [CrossRef]
38. Wang, Y.-M. A fuzzy-normalization-based group decision-making approach for prioritizing engineering design requirements in QFD under uncertainty. *Int. J. Prod. Res.* **2012**, *50*, 6963–6977. [CrossRef]
39. Salvador, F.; Villena, V.H. Supplier integration and NPD outcomes: Conditional moderation effects of modular design competence. *J. Supply Chain Manag.* **2013**, *49*, 87–113. [CrossRef]
40. Javier, T.-R.; Gutierrez-Gutierrez, L.; Ruiz-Moreno, A. The relationship between exploration and exploitation strategies, manufacturing flexibility and organizational learning: An empirical comparison between Non-ISO and ISO certified firms. *Eur. J. Oper. Res.* **2014**, *232*, 72–86.
41. Narasimham, R.; Swink, M.; Kim, S.W. Disentangling Leaness and Agility: An Empirical Investigation. *J. Oper. Manag.* **2006**, *24*, 440–457. [CrossRef]
42. Chen, J.C.; Li, Y.; Shady, B.D. From value stream mapping toward a lean/sigma continuous improvement process: An industrial case study. *Int. J. Prod. Res.* **2010**, *48*, 1069–1086. [CrossRef]
43. Yang, C.-H.; Lin, H.-L.; Li, H.-Y. Influences of production and R&D agglomeration on productivity: Evidence from Chinese electronics firms. *China Econ. Rev.* **2013**, *27*, 162–178.
44. Chang, A.-Y.; Hub, K.-J.; Hong, Y.-L. An ISM-ANP approach to identifying key agile factors in launching a new product into mass production. *Int. J. Prod. Res.* **2013**, *2*, 582–597. [CrossRef]
45. Naylor, J.B.; Naim, M.M.; Berry, D. Leagility: Integration the lean and agile manufacturing paradigms in the total supply chain. *Int. J. Prod. Econ.* **1999**, *62*, 107–118. [CrossRef]
46. Sawhney, R. Interplay between uncertainty and flexibility across the value-chain: Towards a transformation model of manufacturing flexibility. *J. Oper. Manag.* **2006**, *24*, 476–493. [CrossRef]
47. Junior, A.V.; Salerno, M.S.; Miguel, P.A.C. Analysis of innovation value chain management in a company from the steel industry. *Gestão Da Produção* **2014**, *21*, 1–18.
48. Pei, E.; Campbell, I.R.; Evans, M.A. A Taxonomic Classification of Visual Design Representations Used by Induatrial Designers and Engineering Designers. *Des. J.* **2011**, *14*, 64–91.
49. Wang, Y.; Li-Ying, J. When does inward technology licensing facilitate firms' NPD performance? A contingency perspective. *Technovation* **2014**, *34*, 44–53. [CrossRef]
50. Molenaar, N.; Keskin, D.; Diehl, J.C.; Lauche, K. Innovation Process and Needs of Sustainability Driven Small Firms. In Proceedings of the Knowledge Collaboration & Learning for Sustainable Innovation, ERSCP-EMSU Conference, Delft, The Netherlands, 25–29 October 2010.
51. Shéu, D.; Lee, H.-H. A proposed process for systematic innovation. *Int. J. Prod. Res.* **2011**, *49*, 847–868. [CrossRef]
52. Hansen, M.T.; Birkinshaw, J. The Innovation Value Chain. *Harv. Bus. Rev.* **2007**, *85*, 121–130. [PubMed]
53. Ulrich, K.T.; Eppinger, S.D. *Product, Design and Development*, 2nd ed.; Irwin McGraw-Hill: New York, NY, USA, 2000.
54. Bullinger, H.-J.; Warschat, J.; Fischer, D. Rapid product development-an overview. *Comput. Ind.* **2000**, *42*, 99–108. [CrossRef]
55. Yang, C.-C.; Yang, J.-K. An integrated model of value creation based on the refined Kano's model and the blue ocean strategy. *Total Qual. Manag.* **2011**, *22*, 925–940. [CrossRef]
56. Ko, Y.-T.; Yang, C.-C.; Kuo, P.-H. A Problem-Oriented Design Method for Product Innovation. *Concurr. Eng. Res. Appl.* **2011**, *19*, 335–344.
57. Yang, Q.; Lu, T.; Yao, T.; Zhang, B. The impact of uncertainty and ambiguity related to iteration and overlapping on schedule of product development projects. *Int. J. Proj. Manag.* **2014**, *32*, 827–837. [CrossRef]
58. Cooper, R.G. Perspective: The Stage-Gates Idea-to-Launch Process—Update, What's New, and NexGen Systems. *J. Prod. Innov. Manag.* **2008**, *25*, 213–232. [CrossRef]
59. Yang, K.; El-Haik, S.B. *Design for Six Sigma—A Roadmap for Product Development*, 2nd ed.; McGraw-Hill: New York, NY, USA, 2009.
60. Lenfle, S. Toward a genealogy of project management: Sidewinder and the management of exploratory projects. *Int. J. Proj. Manag.* **2014**, *32*, 931–932. [CrossRef]
61. Bogers, M.; Horst, W. Collaborative Prototyping: Cross-Fertilization of Knowledge in Prototype-Driven Problem Solving. *J. Prod. Innov. Manag.* **2013**, *31*, 744–764. [CrossRef]
62. Galati, F.; Bigliardi, B.; Petroni, A.; Pinna, C.; Rossi, M.; Terzi, S. Sustainable Product Lifecycle: The Role of ICT. *Sustainability* **2019**, *11*, 7003–7007. [CrossRef]

63. Yang, C.-C.; Jou, Y.-T.; Yeh, T.M. An Analysis of the Utilization and Effectiveness of NPD Tools and Techniques. In Proceedings of the 7th Asia Pacific Industrial Engineering and Management Systems Conference, Bangkok, Thailand, 17–20 December 2006; pp. 954–962.

64. Yeh, T.M.; Pai, F.Y.; Yang, C.C. Performance improvement in new product development with effective tools and techniques adoption for high-tech industries. *Qual. Quant.* **2010**, *44*, 131–152. [CrossRef]

65. Avramenko, Y.; Kraslawski, A. Similarity concept for case-based design in process engineering. *Comput. Chem. Eng.* **2006**, *30*, 548–557. [CrossRef]

66. Lee AH, I.; Lin, C.-Y. An integrated fuzzy QFD framework for new product development. *Flex Serv Manuf. J.* **2011**, *23*, 26–47. [CrossRef]

67. Chin, K.-S.; Xu, D.-L.; Yang, J.B.; Lam, J.P.-K. Group-based ER–AHP system for product project screening. *Expert Syst. Appl.* **2008**, *35*, 1909–1929. [CrossRef]

68. Chan, F.; Chan, H.; Lau, H.; Ip, R. An AHP approach in benchmarking logistics performance of the postal industry. *Benchmarking Int. J.* **2006**, *13*, 636–661. [CrossRef]

69. Yin, R.K. *Case Study Research, Design and Methods*, 2nd ed.; Applied Social Research Methods Series; Sage Publications Inc.: Thousand Oaks, CA, USA, 2009; Volume 5.

70. Stuart, I.; McCutcheon, D.; Handfield, R.; McLachlin, R.; Samson, D. Effective case research in operations, management: A process perspective. *J. Oper. Manag.* **2002**, *20*, 419–433. [CrossRef]

71. Yin, R.K. *Applications of Case Study Research*, 4th ed.; Sage Publications Inc.: Thousand Oaks, CA, USA, 2011; Volume 34.

72. Eisenhardt, K.M. Building Theories from Case Study Research. *Acad. Manag. Rev.* **1989**, *14*, 532–550. [CrossRef]

73. Erat, S.; Kavadias, S. Sequential testing of product designs: Implications or learning. *Manag. Sci.* **2008**, *54*, 956–968. [CrossRef]

 *sustainability*

Article

The Role of Territorially Embedded Innovation Ecosystems Accelerating Sustainability Transformations: A Case Study of the Transformation to Organic Wine Production in Tuscany (Italy)

Cristina Chaminade [1] and Filippo Randelli [2,*]

1   Department of Economic History, School of Economics and Management and Circle, Lund University, Box 7080, SE 220 07 Lund, Sweden; cristina.chaminade@ekh.lu.se
2   Department of Economics and Management, University of Florence, via delle Pandette, 9, 50127 Florence, Italy
*   Correspondence: filippo.randelli@unifi.it

Received: 21 April 2020; Accepted: 3 June 2020; Published: 5 June 2020

Abstract: Over the last few years, there has been a growing concern among academics and practitioners about the slow pace in which sustainability transformations unfold. While most socio-technical transformations tend to happen over extended periods, research shows that unless some dramatic changes are introduced, we are risking damaging the critical earth systems that sustain human life. In this context, understanding why and how transformations happen at a much faster pace in certain places than in others is of crucial importance. This paper investigates the rapid transformation of Panzano, from traditional wine production to organically produced wine. Using a combination of document analysis, participant observation, and face to face interviews in Panzano in 2019, this article examines the role of the territorially embedded innovation ecosystems facilitating this fast transformation. The study looks at place based-structural preconditions and different forms of agency at different stages in the transformation. Our findings illustrate that a place-based agency is paramount for accelerating sustainability transformations.

Keywords: innovation; ecosystem; organic wine; Tuscany

1. Introduction

In November 2018, the UN IPCC report indicated that profound transformations are needed before 2030 if we want to avert catastrophic environmental consequences, including the total loss of all coral reefs and a significant reduction of island communities [1]. A couple of months later, in April 2019, the UN launched the most comprehensive report on the state of world biodiversity and ecosystem services since the seminal Millennium Ecosystem Assessment [2]. The results of the three-year research project directly involving 150 scientists around the world were devastating. According to the report, two in five amphibian species are at risk of extinction, as one-third of reef-forming corals, and close to one-third of other marine species. In economic terms, the losses are jaw-dropping. Pollinator loss has put up to $577bn (£440bn) of crop output at risk, while land degradation has reduced the productivity of 23% of global land [2]. With slightly more than ten years to go to the IPCC deadline, the need for accelerated system transformations has come to the forefront of political and academic debates [3–8].

However, we still know very little about if and how system transformations might be accelerated [9]. In fact, in the real world, long-term incremental system changes are the norm, partly due to the systemic nature of innovation. On the one hand, systems are path-dependent. Prior investments in technologies and the related human capital, infrastructure, institutional frameworks, and other sunk costs can lock-in the system and prevent it from responding to radical changes [10]. On the other hand, science, technology, and innovation policies tend to focus on addressing failures in existing systems rather than changing them, thus perpetuating gradual change [8].

In contrast with the structuralist perspective, the literature on the governance of sustainability transformations has focused on the role of agency on the speed [11] and the directionality of the transformations [12]. According to this literature, the existence of protected spaces for experimentation or the role of intermediaries [13] can create new opportunities for transformation and affect the direction and speed of transformations. The literature has significantly contributed to our understanding of enabling factors supporting transformations and the role of agency. However, it has also received critiques, particularly from economic geographers, on the lack of attention to how innovative ecosystems embedded in particular territories, shape the speed and direction of transformations [14–16].

This paper aims to address this gap by looking at the role of territorially embedded innovation ecosystems (TEIE) in accelerated sustainability transformations. TEIEs belong to what Boyer (in this same special issue) calls the regional/local ecosystem approach, which highlights the territorial dynamics of the innovation process [17]. TEIE is an embracing concept that might include innovative clusters, industrial districts, or regional innovation systems. A more in-depth look into the role of structural preconditions and agency in the transformation of specific TEIE is essential for several reasons. On the one hand, structural preconditions differ significantly between different TEIEs. The economic structure of a region, its knowledge specialization, or its institutional frameworks are the results of historical processes of knowledge accumulation. The existence of shared institutional frameworks, strong networks, and a sense of place enable interactive learning and continuous innovation in TEIEs [18]. TEIEs are constructed relationally, through social capital and networks supporting the creation and exchange of knowledge [19]. It is in this respect that evolutionary economic geographers argue that history matters in regional transformations. In other words, the opportunities for transformation of particular TEIE are path-dependent [20].

On the other hand, the transformation capacity of particular territories can be shaped by particular forms of agency. Grillitch and Sotarauta suggest that different forms of agency can shape the opportunity space for transformation: entrepreneurial, institutional, and place-based. Regarding the latter, a strong sense of place [14] place frames [15] or place leadership [21] as a shared understanding of the identity of a place, and a shared vision of what that place might become, is crucial for transformations. Ecosystems, understood through this place-making lens, could potentially be a powerful mechanism for accelerating transformations.

While promising, this recent literature on agency and regional transformations remains at a rather general and theoretical level. It focuses on regional transformations in general, without paying specific attention to sustainability transformations as one specific transformation with a clear directionality [12]. Furthermore, for the current paper, there is a lack of studies analyzing how structural preconditions and agency in TEIEs affect the speed of transformation.

This paper aims at contributing to this gap in the literature. It investigates the role of the local innovation ecosystem in the rapid transformation of Panzano from traditional wine production to organically produced wine. Panzano is a county in Tuscany where the transition to organic wine has almost reached 100% of the territory in barely 25 years. This article investigates the role of place-based structural preconditions and different forms of agency in the transformation of the TEIE over time. The analysis allows us to identify which factors appear to be more significant at different stages in the transformation, from the incept to the acceleration and the consolidation. For instance, while structural preconditions and entrepreneurial agency are essential at the beginning of the transformation, it is place-based leadership that significantly contributes to the acceleration.

The paper is structured as follows. Section 2 discusses the concept of sustainability transformations and the determinants of its speed. We pay particular attention to the role of the specific territorial context in which these transformations emerge and deploy in shaping the speed and direction of transformations. Section 3 provides an overview of the methodology used for data gathering and analysis. Section 4 expands the study of Panzano, investigating the drivers behind its fast transformations, and Section 5 concludes.

2. Literature Review

2.1. Sustainability Transformations

Over the last few years, there has been a growing concern among academics and practitioners about the slow pace in which sustainability transformations unfold [22]. Research shows that unless some dramatic changes are introduced in the following decade, we are risking damaging the critical earth systems in which human life on this planet is sustained [23,24]. In this context, understanding why and how transformations happen at a much faster pace in certain places than in others is of crucial importance.

In the context of this paper, we follow Roggema et al. [25] definition of sustainability transformations. Transformation is seen as a change towards a future that is fundamentally different from the current situation [3]. Incremental change, on the other hand, is seen as a slow process, with imperceptible changes, and a transition is seen as a fluent change towards an improved version of the current status, but where the current system is not fundamentally changed [26]. In this respect, only the term "transformation" would capture the radical and non-linear nature of system change [27]. The latter is also referred to in the literature as deep transitions [8,27]. The current paper adopts this later understanding of sustainability transformations as embracing substantial change from the previous situation, including changes in practices, routines, beliefs, and policies.

A key characteristic of sustainability transformations of the radical and deep kind discussed above is that they take time. System transformations require structural changes in current economic, social, political, and technological regimes and meta-regimes [28], implying profound transformations of the dominant techno-economic paradigm [29] and the reconfigurations of actors, their relationships [30], and the formal and informal institutions that influence their behavior. The structuralist approaches to sustainability transformations see them as path-dependent. Prior investments in technologies and the related human capital, infrastructure, institutional frameworks, and other sunk costs prevent systems from responding to radical changes [10]. Economic actors set up routines to reduce the uncertainty driven by their bounded rationality. Due to their tacit and cumulative nature, routines are not easy to change, and very difficult to imitate for other firms [31]. In other words, firms are subject to cognitive constraints [31] that hinder the process of change. At the same time, science, technology, and innovation policies tend to focus on addressing failures in existing systems rather than changing them, thus perpetuating gradual change [8]. As a result of these structural constraints, it is argued that transformations will take decades and even centuries to complete.

It is only recently that the role of agency in sustainability transformations has come explicitly to the forefront of academic discussions [8,11,32]. Within this emerging stream of literature, researchers look at the capacity of individuals and organizations to act independently and to make their own free choices [32–34] and create opportunities for change [35]. The main finding is that understanding agency, which is how different actors might strategically join forces in networks to achieve particular goals, is key to overcoming the structural system inertia to incremental change and realizing transformations [11].

Both structural preconditions and agency vary significantly between different territories, and economic geographers have criticized the current literature on sustainability transformations for their lip service to the analysis on how innovative ecosystems embedded in particular territories shape the speed and direction of transformations [16–18]. TEIEs matter because regions have accumulated different skills and knowledge, relations, and institutional frameworks over time. But also, as place-based leadership is paramount for sustainability transformations [35]. In other words, both structural preconditions and agency are related to, and embedded, in particular territorial innovation ecosystems, creating a particular sense of place. The extent to which TEISs are important for understanding sustainability transformations will be discussed next.

2.2. Territorially Embedded Innovation Ecosystems and Sustainability Transformations

Economic geographers have long acknowledged the impact of territorially embedded ecosystems for innovation [36,37]. Firms and organizations closely located share network relations of (mainly) tacit and informal nature that are crucial for knowledge exchange. Geographical proximity also facilitates interaction between diverse and complementary capabilities [38,39] and a spatial neighbor effect conveyed through social interaction and visibility [40]. Firms can derive localized competitive advantages due to the joint and collective cumulative path of learning and coordination and close face to face interaction [41,42]. At the same time, informal control mechanisms, such as those exerted by social communities, are also powerful mechanisms to avoid free-riding [38].

In other words, innovation ecosystems are constructed relationally [19]. Moreover, those relations tend to happen with other organizations nearby. Place-making relations influence, and are influenced by, formal and informal institutions that are embedded in the territory [43]. Together, actors, relations, and institutions are the basis of a territorially embedded ecosystem.

It follows that the development of a particular ecosystem is path-dependent. The same relations, institutions, and actor configurations in the territory can act both as a promoter or deterrent of change. Similarly, the variety of knowledge present in one particular region affects the possibilities for path creation in that same region. Knowledge accumulates over time, and regions portray particular knowledge specializations. Thus, history matters in the transformative capacity of a particular TEIE, since it shapes the actors, network, and institutions of particular territories, as well as the knowledge base in which the innovation capacity of that particular territory is based.

While acknowledging the structural preconditions for the change of particular TEIEs, Grillitsch and Sotarauta [35] suggest that different forms of place-based agency can shape the capacity of a TEIE to transform. The authors propose to distinguish between three types of agency: Schumpeterian innovative entrepreneurship, institutional entrepreneurship, and place-leadership. Schumpeterian innovative entrepreneurship refers to attempts to break with existing growth paths through processes of Schumpeterian creative destruction, and it is observable through new ventures and new processes. Institutional entrepreneurship refers to individual and organizational attempts to change existing institutions, molding the rules of the game so that the Schumpeterian entrepreneurs can surface and succeed. Institutional entrepreneurship is observable through changes in rules and regulations. Finally, place leadership refers to "social processes involved in making things happen" [35]. Place leadership is the most difficult to observe due to its embeddedness in informal institutions. In the words of Sotarauta et al. [21] (p. 128), investigating place-based leadership is about revealing the types of social processes involved in "making things happen" and in "getting things done, more often than not, in an indirect manner". How different forms of agency and structural preconditions shape the capacity of a TEIE to start, speed up, and consolidate sustainability transformations remains to be studied.

Identifying small-scale accelerated transformations and analyzing them could be a first step in understanding how sustainability transformations can be accelerated. We do so by looking at the fast transformation of a region in Tuscany–Panzano, from conventional to organic wine production at a much faster pace than the neighboring regions. In particular, we look at how the combination of structural characteristics and agency influences the speed of change.

3. Research Design

This paper uses case study research, which is suitable for the holistic, in-depth exploration and understanding of complex issues, such as sustainability transformations [44]. This approach will help to explain both the process and outcome of human-institution-forest interactions, through the observation, reconstruction, and analysis of the case study outlined below.

3.1. Selection of Case Study: Organic Wine Production in Panzano

The case was selected following a purposive sample procedure. Purposive sampling is recommended when the aim is to focus on particular characteristics that enable one to explain the research questions [45], in this case, how the innovation ecosystem enabled the fast transition of Panzano to organic wine production. The case was selected because of its extreme characteristics; that is, the fact that the sustainability transformation has taken place at a much more accelerated pace than in the overall region.

Panzano is located in the municipality of Greve in Chianti within the Tuscany region (Italy), between the two cities of Firenze and Siena (see Figure 1). The municipality of Greve in Chianti is included in the Chianti Classico region, which covers 71,800 hectares (177,500 acres) of territory, of which 10,000 hectares are devoted to wine production. The Chianti Classico region also includes the entire territories of the municipalities of Castellina in Chianti, Gaiole in Chianti and Radda in Chianti and parts of those of Barberino Tavarnelle, Castelnuovo Berardenga, Poggibonsi, and San Casciano Val di Pesa.

Figure 1. The location of Panzano in Chianti (the borders are the municipality of Greve in Chianti), within Tuscany and Italy (small frame with Tuscan municipality borders). (Source: self-elaboration with the use of Quantum GIS and Open Street Map).

Rows of vines alternating with olive orchards are a characteristic feature of the Chianti Classico landscape. About 7200 hectares (17,290 acres) of vineyards are part of the DOCG (Denominazione di Origine Controllata e Garantita), for the production of Chianti Classico, one of the most famous red wine in Italy and the world, as 80% of the production is exported worldwide. The Chianti Classico region has multiple sub-zones, some of which have formed unions or associations to promote their wines. Panzano is a sub-zone in the municipality of Greve in Chianti, and it represents 10% of the entire Chianti Classico territory.

The Chianti Classico organic production has since the year 2000 and now represents approximately 35% of the total wine production in the region. What differentiates Panzano from the regional trends towards organic production is twofold: the percentage of organic production vis à vis conventional production and the speed of the transformation to organic production. The latter is the particular focus of this paper.

3.2. Data Collection and Analysis

For the data collection and analysis, we use innovation biographies (IBs). IBs are a "valuable methodology to reflect the evolutionary character of the dynamics of the social initiatives' and innovation processes in deepening the understanding of development paths, knowledge trajectories, and stakeholder interactions at the micro-level" [46] (p. 15). They are particularly useful for analyzing complex, emergent and non-linear events involving many actors across different levels [47].

In IBs, it is of paramount importance to use a combination of data collection techniques and sources to triangulate information, combining data from the individual and contextual level. Interviews constitute the individual level, as they reflect personal perspectives. Desk research using document analysis and participant observation enriches the biographic picture at the contextual level.

Primary data was collected through face to face semi-structured interviews conducted in 2018 to some wine producers in Panzano by both authors of this paper. The interviews were with the owners of the wineries and the local agronomist. The interviewees included all the key actors mentioned in all documents as paramount for the transformation of Panzano into organic wine production (five in total), which had been engaged in the transformation from the start. This provided depth of data rather than breadth, which was considered to be more suitable for the development of the IB. Moreover, an additional two interviews and a small survey were conducted with wineries in the Chianti Classico area for triangulation. The interviews were recorded. The interview guide was divided into four main blocks that enquired about the drivers and process of transformation, the role of networks, the sources of knowledge and the impact of the transformation on the economic, social and environmental sustainability of the firm to capture different aspects of the role of the innovation ecosystem in the transformation of Panzano.

The interviews took place either on the production field, where the interviewees could explain how much organic wine productions differed from conventional production or in their office. The notes taken during these field visits are part of the participant observation.

Writing and analyzing the IB "is a process of telling a real, detailed, and "thick" story covering all relevant aspects" [46] (p. 38). The use of IBs implies (i) the development of a biographical time-space path with the major milestones or events identified (sequence of events). (ii) For each event, information is collected on the actors involved, their relation (actors), and their location; (iii) the knowledge that they provided and if there was any conflict (barriers in the innovation process) [48]. The result is a chronological observation of the transformation of Panzano and the actors, networks, and institutions that enabled the transformation.

4. The Role of the Territorially Embedded Innovation Ecosystem in the Accelerated Transition of Panzano to Organic Wine Production

The transition towards organic agriculture in Panzano was not a linear process. Several key events can allow us to trace the historical process that brought Panzano to full organic wine production. The entire process is explained through three different phases.

4.1. Phase 1. Emergence—1992–2000

Organic production in Panzano can be traced back to the early nineties when a traditional wine producer in the area started experimenting with organic wine production in a small plot of his land (Interview 4). At that time, there was no sense of community in the region. Instead, each of the producers was fencing for itself (Interview 4).

The territory was just emerging from a profound crisis, which had forced producers to reduce the yield significantly to prevent the prices from sinking [49]. According to one of the interviewees, the region was making too much wine and of inferior quality. The wines were very light and not capable of age for a long time and, thus, not competitive in the international market.

Because of the crisis, many producers sold their farms. The new entrants did not come exclusively from the territory, but from other parts of Italy and from abroad. In other places in Tuscany and Chianti, foreign people bought farms as an investment or for exploitation (Interview 2). Instead, those that acquired the farms in Panzano moved with their families to Panzano, reflecting a life-choice.

The latter might be one of the reasons why the newcomers were committed almost from the start to the production of organic wine. They had a strong sense of responsibility to the environment and their family (Interviews 1, 3 and 4).

To revert to methods of production that were not chemical-intensive, the producers turned to the older employees of their farm that had worked in the vineyards before the fifties (Interviews 1 and 4), their agronomists (Interview 3) and old books (Interview 4), in search of the required knowledge. They also started to share their experience and visit other neighboring farms, to get to know how others were experimenting with organic productions, thus creating the first seeds of social capital in the region. Cognitive proximity and social proximity facilitated knowledge sharing. Regarding cognitive proximity, all newcomers had higher education, although not necessarily as oenologist or agronomists. Concerning social proximity, most of the newcomers were of the same age, share a similar philosophy of life (Interviews 1, 3–5), and were open-minded (Interview 3).

Furthermore, newcomers almost immediately developed a strong sense of place. Place identity reflects how the producers talk about the wine—they wanted a wine that could "reflect the sense of place" (Interview 1)—something more "subtle and reflective of the place".

Most of the wine producers were also part of a formal association of the wine producers of Panzano (Unione Viticoltori di Panzano in Chianti), which had been created in 1995 and agglutinated 20 out of the 35 Panzano producers. However, the association focused on the promotion of the local wine through a wine festival, rather than on creating a shared vision of the territory. It was not until the beginning of the year 2000 that the association started to play a different role, as discussed next.

4.2. Phase 2. Acceleration—2000–2016

A trigger event in the acceleration of the transformation to organic wines happened in 2000 when the Italian Ministry for Agriculture and Forestry forced all wine producers to use chemicals to fight an insect vector of the Golden Flavescence ("Scaphoideus titanus") a disease that causes enormous damages to the vineyards. The disease can be transmitted from plant to plant if geographically close together. So, the Ministry made mandatory two specific strategies to fight it: a regular spray of pesticide on all the vineyards, attacked or not, and uprooting of all the infected plants.

The Tuscany region had historically few attacks of Scaphoideus titanus since it prefers humid habitats, and therefore the insect was not well known by the winemakers. To prepare a strategy that would allow limiting the use of pesticides, the Winemakers Association of Panzano in Chianti reacted as a group and asked for the support of a local agronomist (Interviewee 5). Together, they decided to propose a monitor program to understand the real danger of the pest in the area. The region reacted positively and changed the decree from the obligation to spree to the obligation to monitor and control. The monitoring was carried out with the scientific support of the CRA (Consiglio per la Ricerca in Agricoltura e l'Analisil'Analisi dell "Economia Agraria) of Florence and the University of Pisa, and the Tuscany region approved it. The monitoring exercise showed that there was no presence of the insect in the region, and the use of pesticides was avoided.

The wine producers understood very clearly that, as an association, they had a strong influence and very tangible gains (Interview 4). The monitoring system based on the use of traps to identify the presence of insects was offered for free for the Panzano wineries. Therefore, the plague was a significant event in the transformation of Panzano, as they saw the value of being associated and acting together (Interviews 1–5). All the producers resisted using insecticides.

The success of the monitoring system strengthened the links among producers. It also allowed local producers to put their trust in the local agronomist (Interview 5), which is one of the pioneers of organic wines in Italy and on his approach to grape growing, which is rooted in two different levels of defense against pathogens: direct and indirect. The direct defense acts on the harmful agent, trying to neutralize it only in case of its presence in the vineyards. The direct approach makes use of copper and sulfur-based compounds to stop the pathogen agent. Under the present legislation, organic agriculture can also use a direct approach of defense: synthesized chemicals are limited but not banned, and any winemaker can decide to spray copper and sulfur-based compounds once the risk is high. It follows that monitoring is crucial to reduce the use of them.

Indirect defense views the outbreak of pathology as a result of a two parts interaction: the pathogen agent and the plant. Therefore, it is reductive to focus only on the chemical elimination of the harmful agent. Instead, it is more efficient (and less impactful on the environment) to prevent detrimental outbreaks by focusing on the natural defense that every plant has. This approach is based on prevention, with the use of chemical compounds just for extraordinary attacks of pathogens.

The implementation of the monitoring system enabled the establishment of connections outside Panzano, with the CRA of Florence and with the universities of Florence and Pisa. Since 2000, the monitoring project never stopped, and the Ministry for Agriculture and Forestry revoked the obligation of regular treatments with pesticides, and the monitoring system was officially included in the allowed strategies to fight the "Scaphoideus titanus".

In other words, the need for prevention against the "Scaphoideus titanus" was a challenge transformed into an opportunity. It is only due to the monitoring strategy success that the agronomist could start supporting some of the local winemakers in the transition to organic production (Interview 5).

A further step in the acceleration towards organic wines in Panzano happened in 2005, with the opening of SPEVIS (Stazione Sperimentale per la Viticoltura Sostenibile). SPEVIS is an experimental agronomical center for the development of sustainable wine production and the "Sustainable Panzano" strategy (Panzano Sostenibile) (Interviews 1, 3–5). SPEVIS was founded by the local agronomist in collaboration with local producers and research institutes, and it could be considered as a natural outcome of already existing informal collaborations. A unique characteristic of SPEVIS was that, although it was funded by only a handful of wine producers in the area, it provided advice and support to any local producer interested in shifting to organic wine production practices for free. The collective action taken around SPEVIS is another example of new place leadership involved in making things happen.

Nevertheless, the disruptive innovation of SPEVIS was not only the leadership per se but rather the way information was shared. In the case of SPEVIS, the goal was not only supporting their clients in the transition to organic wine production but rather to widespread as much and soon as possible the organic agriculture in the area. As a matter of fact, over time, SPEVIS published various types of booklets and researches studies online and for free. In 2008, SPEVIS presented its vision of sustainable viticulture, which became a "manual on organic wine production" the following year. Then, in 2011, they published a research-based manual, a yearbook for winemakers, a complete book with the description of Panzano's methodology and philosophy (updated in 2014 in a new edition), and a combination of books and cd-roms on grape diseases.

Therefore, it is clear how any case study on Panzano wine cluster cannot leave behind this free flow of knowledge, available for not only the Winemakers Association's members or the SPEVIS' clients. The place leadership of SPEVIS was functioning as a meta-actor, and the free flow of knowledge on organic viticulture was nurturing the Panzano innovation ecosystem.

After the introduction of SPEVIS, the conditions for the acceleration of the transition toward organic winemaking were set. Panzano reached 50% of vineyards under organic production exceptionally soon. Part of this excellent result was due to the initial excitement. Some producers were already interested in organic agriculture, and the interaction between SPEVIS and the local Winemakers association opened up the way to a radical change. Since 2000, it has taken 12 years to reach 75% of organic vineyards and 16 to reach 95%, which is an incredible result and represents a strong example in the Italian scenario (see Table 1). The types of actors can partly explain the dramatic acceleration of the last years. If initially, it was only the small and medium-size producers that changed to organic; in the past few years, the large producers have finally adopted 100% organic agriculture (Interview 4).

Table 1. The role of the territorially embedded innovation ecosystem in the three stages.

	Emergence 1992–2000	Acceleration 2000–2016	Stabilization 2016 Onwards
Estimated number of organic producers	5 (14%)	26 (75%)	33 (95%)
Triggering event	Overproduction crisis; 2/3 drop of prices; obligation to reduce yield	Top-down decree to combat the potential damage of an insect on the vineyards	Creation of the Bio-district in Greve in Chianti
Place-based structural preconditions	Variety of wine producers moving into the area; knowledge variety	Creation of SPEVIS	Accumulation of knowledge and skills in the area with regards to organic wine production
Place-based agency	Informal contacts among organic wine producers	Collective action to prevent the use of chemicals; change of the law; free advice and fast knowledge transfer among producers; increase awareness; creation of a joint vision of the place "Panzano sostenibile."	The association of wine producers in Panzano as well as SPEVIS has triggered the establishment of bio-districts, first in the county of Greve in Chianti and later on expanding to all Chianti Classico region. Structuration through the formalization of national laws on organic production and bio-districts
Role of the territorially embedded innovation ecosystem	Limited. Most of the producers acting individually. Formal association focused on the organization of the local wine fair. No shared vision	Fundamental for the rapid spread of information; consolidation of social capital and the structuration of actions around a collective vision of the place	Oriented towards scaling up and spreading the skills and experiences to other localities

Source: Own elaboration based on the interviews.

4.3. Phase 3. Stabilization and Scaling Up—2016 Onwards

Thanks to the propulsive thrust of SPEVIS, to the commitment of the AIAB (Associazione Italiana Agricoltura Biologica), and the will of the mayor of Greve in Chianti, on September 27, 2016, the Bio-district of Greve in Chianti was officially established [50]. According to one of the interviewees (Interview 3), the success of the joint response to the "Scaphoideus titanus" crisis triggered the idea of the bio-district. The bio-district aims to bridge the gap between farmers and citizens and enable spaces for the dialogue between policymakers, producers, consumers, and citizens to build shared visions about what the region could be (Interview 4). The latter is entirely in line with the sense-of-place discussed earlier [14].

Similar bio-districts soon followed the Bio-District of Greve in Chianti in Gaiole in Chianti, and more recently, the entire Chianti Classico became a bio-district (Interview 4). The latter refers to the entire DOCG Chianti Classico wine production area, and it has the goal to support the transition to organic agriculture of the entire area of Chianti Classico. Further structuration of the change is taking place in the forms of new laws. According to one of the interviewees (Interview 4), the Italian government is discussing a new law regarding organic agriculture, and one of the chapters in the law will be about bio-districts.

The success of Panzano is now widespread in the entire wine region, and today, 35% of agricultural land in the Chianti Classico area is organic [50]. Interestingly, not all the wine producers of Panzano identify themselves with the bio-district as strongly as they do with the Association of Panzano wine producers (Interview 1). In this sense, the regional bio-district does not trigger as strong a sense of place as the Panzano area does.

The transition to organic was not only influenced by place, but it also influenced place. Physically, the transformation to organic was also visible from the outside. In the words of one of the interviewees (Interview 1), it changed the landscape. For example, in particularly rocky places, the stones were not visible after a while, because the soil was so fertile that the rocks would not separate from it. Going back to more traditional production techniques also reinforced the linkages with the place.

5. Discussion

This paper analyzes a wine region in which the sustainability transformation towards organic wine production took place much faster than in the neighborhood regions. Our point of departure was trying to understand why this was the case. We used innovation biographies to capture the trajectory within the innovation ecosystem: we build a timeline, identifying the stages in the transformation (emergence, acceleration, and stabilization) and the triggering events that supported the transformation from one stage to the other. We focus on how the whole transformation could be achieved at a much faster pace than in surrounding localities. We look at actors and formal and informal networks and their embeddedness in the territory and how they influence institutional change at the regional and national levels.

The main contribution of this paper is to bring the role of the territorially embedded ecosystem to the discussion on the speed of sustainability transformations. In doing so, we look at both place-based structural preconditions and well as the place-based agency.

Our findings suggest that structural preconditions and place-based agency are essential for sustainability transformations, but not at the same stages in the transformation. Structural preconditions are paramount in the early stages of the transformation, together with the entrepreneurial agency. However, the place-based agency, and to a lesser extent, institutional agency, and becomes paramount in the acceleration phase.

In the case of Panzano, sharing knowledge about the transition to organic agriculture was crucial for the speed of the transformation. Both pioneers' farmers and SPEVIS, in collaboration with universities, were working as a source of knowledge for the conventional farmers that found it easy and convenient to shift to organic agriculture. The social and geographical proximity fostered the share of knowledge, and the result confirms that innovation ecosystems are constructed relationally [17] on a strong sense of place [14].

The role of a local leader and intermediary–SPEVIS- was crucial. The local leader reduced the gap between the farm and the market, supporting the farmers in the transition to organic viticulture. A unique characteristic of SPEVIS was that it provided free advice and support to any local producer interested in shifting to organic wine production practices. The success of SPEVIS may suggest that supporting the access to knowledge through the action of local meta-actors cooperating with the local firms is crucial for sustainability transformations. Furthermore, conventional farmers consider the first step towards organic as the most "expensive" due to the need for new skills, materials, and suppliers. In this perspective, the free advice and support received from SPEVIS was paramount to convince conventional wine producers to move towards organic production. In other words, the institutional agency is essential in the acceleration phase, but it should be intertwined with the place-based agency that only local leaders can guarantee.

Interestingly, the regional bio-district in Chianti is not transiting to organic at the same fast pace, because it does not trigger the strong sense of place as the Panzano area did. This insight of our research opens up a relevant field for future research: how to reproduce at a larger territorial scale the sense of place and the social capital, due to the density of relations that might accelerate the transition?

Is it better to refer to a larger region for a transformative change or to a mosaic of small territorially embedded innovation ecosystems?

These findings may have important policy implications. Policies tend to look at sustainability transformations, either at the macro-level of the market or the micro-level of firms, neglecting the role of the territory (meso level). A territorial approach to policy-making could illuminate the discussion on how to accelerate sustainability transformations. Our findings show that a strong sense of place facilitates a dense network of relations, which in turn fosters knowledge development and diffusion, a vital function of an innovation ecosystem. The pace of the transition can be accelerated by an effect of neighboring, which fosters social interaction and visibility [40].

Our study also reveals that the transition towards sustainability is not a linear process. Structural preconditions and different forms of agency play a different role at different stages in the transformation and thus require different policies over time, but also over space. For instance, while in some regions, it is still crucial to enhance better structural preconditions, in some others, fostering the cooperation among actors to widen the existing knowledge is paramount.

The case of Panzano highlights the importance of place-based approaches to the study of sustainability transformations [14,15]. Innovation ecosystems, understood through this place-making lens, could potentially be a powerful mechanism for accelerating transformations. While promising, our research is based only on a single case study. The observed mechanisms that facilitated an accelerated transformation to organic agriculture in Panzano may also operate in other regions or sectors but remain to be studied. For that, a systematic comparative analysis of different accelerated regional transformations is needed.

Author Contributions: Conceptualization, C.C. and F.R.; methodology, C.C and F.R.; validation, C.C. and F.R.; formal analysis, C.C. and F.R.; investigation, C.C. and F.R.; resources, C.C. and F.R.; data curation, C.C. and F.R.; writing—original draft preparation, C.C. and F.R.; writing—review and editing, C.C. and F.R.; visualization, C.C. and F.R.; supervision, C.C. and F.R.; funding acquisition, C.C.. All authors have read and agreed to the published version of the manuscript.

Funding: This work was partly supported by the EU-RISE project Makers (Smart Manufacturing for EU Growth and Prosperity). H2020 Marie Skłodowska-Curie Actions (Grant number 691192).

Acknowledgments: Cristina Chaminade would like to thank Marco Bellandi at the University of Florence and Fondazione per la Ricerca e L'Innovazione (Florence), for hosting her in Florence for half a year and supporting the research on regional sustainability transformations. Both authors would like to thank the interviewees for the richness of their insights and their generosity with their time. We thank the three anonymous reviewers for their careful reading of our manuscript and their many insightful comments and suggestions.

References

1. International Panel on Climate Change (IPCC) Summary for Policymakers. In *Global Warming of 1.5 °C. An IPCC Special Report on the Impacts of Global Warming of 1.5 °C Above Pre-Industrial Levels and Related Global Greenhouse Gas Emission Pathways, in the Context of Strengthening the Global Response to the Threat of Climate Change, Sustainable Development, and Efforts to Eradicate Poverty*; Masson-Delmotte, V.; Zhai, P.; Pörtner, H.-O.; Roberts, D.; Skea, J.; Shukla, P.R.; Pirani, A.; Moufouma-Okia, W.; Péan, C.; Pidcock, R.; et al. (Eds.) World Meteorological Organization: Geneva, Switzerland, 2018; p. 32.

2. Leemans, R.; De Groot, R.S. *Millennium Ecosystem Assessment, Ecosystems and Human-Wellbeing*; Island Press: Washington, DC, USA, 2005.

3. Scoones, I.; Stirling, A.; Abrol, D.; Atela, J.; Charli-Joseph, L.; Eakin, H.; Ely, A.; Olsson, P.; Pereira, L.; Mehrotra, R.P.; et al. Transformations to Sustainability. Working Papers from the STEPS Centre. 2018. Available online: https://steps-centre.org/publication/transformations-to-sustainability-wp104/ (accessed on 13 May 2020).

4. Leininger, J.; Dombrowsky, I.; Breuer, A.; Ruhe, C.; Janetschek, H.; Lotze-Campen, H. Governing the transformations towards sustainability. In *Transformations to Achieve the Sustainable Development Goals*; Kriegler, E., Messner, D., Nakicenovic, N., Riahi, K., Rockström, J., Sachs, J., van der Leeuw, S.,

van Vuuren, D., Eds.; Report prepared by The World in 2050 Initiative; International Institute for Applied Systems Analysis (IIASA): Laxenburg, Austria, 2018.

5. TWI2050. *Transformations to Achieve the Sustainable Development Goals. Report Prepared by e World in 2050 initiative*; International Institute for Applied Systems Analysis (IIASA): Laxenburg, Austria, 2018. Available online: www.twi2050.org (accessed on 5 June 2020).

6. Hölscher, K.; Wittmayer, J.M.; Loorbach, D. Transition versus transformation: What's the difference? *Environ. Innov. Soc. Transit.* **2018**, *27*, 1–3. [CrossRef]

7. Fagerberg, J. *Mission (im) Possible? The Role of Innovation (and Innovation Policy) in Supporting Structural Change & Sustainability Transitions*; Working Papers on Innovation Studies 20180216; Centre for Technology, Innovation and Culture, University of Oslo: Oslo, Norway, 2018.

8. Schot, J.; Kanger, L. Deep transitions: Emergence, acceleration, stabilization and directionality. *Res. Policy* **2018**, *47*, 1045–1059. [CrossRef]

9. Chaminade, C. Innovation for What? Unpacking the Role of Innovation for Weak and Strong Sustainability. *J. Sustain. Res.* **2020**, *2*, Online.

10. Dodgson, M.; Hughes, A.; Foster, J.; Metcalfe, S. Systems thinking, market failure, and the development of innovation policy: The case of Australia. *Res. Policy* **2011**, *40*, 1145–1156. [CrossRef]

11. Köhler, J.; Geels, F.W.; Kern, F.; Markard, J.; Onsongo, E.; Wieczorek, A.; Alkemade, F.; Avelino, F.; Bergek, A.; Fünfschilling, L.; et al. An agenda for sustainability transitions research: State of the art and future directions. *Environ. Innov. Soc. Transit.* **2019**, *31*, 1–32. [CrossRef]

12. Addressing Directionality: Orientation Failure and the Systems of Innovation Heuristic. In Towards Reflexive Governance (2016). Available online: https://ideas.repec.org/p/zbw/fisidp/52.html (accessed on 5 June 2020).

13. Kivimaa, P.; Boon, W.; Hyysalo, S.; Klerkx, L. Towards a typology of intermediaries in sustainability transitions: A systematic review and a research agenda. *Res. Policy* **2019**, *48*, 1062–1075. [CrossRef]

14. Frantzeskaki, N.; van Steenbergen, F.; Stedman, R.C. Sense of place and experimentation in urban sustainability transitions: The Resilience Lab in Carnisse, Rotterdam, The Netherlands. *Sustain. Sci.* **2018**, *13*, 1045–1059. [CrossRef]

15. Binz, C.; Coenen, L.; Murphy, J.T. Geographies of transition—From topical concerns to theoretical engagement: A commentary on the transitions research agenda. *Environ. Innov. Soc. Transit.* **2020**, *34*, 1–3. [CrossRef]

16. Hansen, T.; Coenen, L. The geography of sustainability transitions: Review, synthesis and reflections on an emergent research field. *Environ. Innov. Soc. Transit.* **2015**, *17*, 92–109. [CrossRef]

17. Boyer, J. Toward an Evolutionary and Sustainability Perspective of the Innovation Ecosystem: Revisiting the Panarchy Model. *Sustainability* **2020**, *12*, 3232. [CrossRef]

18. Asheim, B.T.; Isaksen, A. Regional innovation systems: The integration of local "sticky" and global "ubiquitous" knowledge. *J. Technol. Transf.* **2002**, *27*, 77–86. [CrossRef]

19. Murphy, J.T. Human geography and socio-technical transition studies: Promising intersections. *Environ. Innov. Soc. Transit.* **2015**, *17*, 73–91. [CrossRef]

20. Boschma, R.; Martin, R. *Handbook on Evolutionary Economic Geography*; Edward Elgar: Cheltenham, UK, 2010.

21. Sotarauta, M.; Beer, A.; Gibney, J. *Making Sense of Leadership in Urban and Regional Development*; Routledge: Abingdon-on-Thames, UK, 2017.

22. Geels, F.W. *Technological Transitions and System Innovations: A Co-Evolutionary and Socio-Technical Analysis*; Edward Elgar: Cheltenham, UK, 2005.

23. Steffen, W.; Persson, Å.; Deutsch, L.; Zalasiewicz, J.; Williams, M.; Richardson, K.; Crumley, C.; Crutzen, P.; Folke, C.; Molina, M.; et al. The Anthropocene: From global change to planetary stewardship. *AMBIO J. Hum. Environ.* **2011**, *40*, 739–761. [CrossRef]

24. Steffen, W.; Richardson, K.; Rockström, J.; Cornell, S.E.; Fetzer, I.; Bennett, E.M.; Biggs, R.; Carpenter, S.; de Vries, W.; Folke, C.; et al. Planetary boundaries: Guiding human development on a changing planet. *Science* **2015**, *347*, 1259855. [CrossRef] [PubMed]

25. Roggema, R.; Vermeend, T.; Dobbelsteen, A. Incremental change, transition or transformation? Optimising change pathways for climate adaptation in spatial planning. *Sustainability* **2012**, *4*, 2525–2549. [CrossRef]

26. Folke, C.; Carpenter, S.R.; Walker, B.; Scheffer, M.; Chapin, T.; Rockström, J. Resilience thinking: Integrating resilience, adaptability and transformability. *Ecol. Soc.* **2010**, *15*, 20. [CrossRef]

27. Kanger, L.; Schot, J. Deep transitions: Theorizing the long-term patterns of socio-technical change. *Environ. Innov. Soc. Transit.* **2019**, *32*, 7–21. [CrossRef]

28. Geels, F.W. Technological transitions as evolutionary reconfiguration processes: A multi-level perspective and a case-study. *Res. Policy* **2002**, *31*, 1257–1274. [CrossRef]

29. Perez, C. Technological revolutions and techno-economic paradigms. *Camb. J. Econ.* **2009**, *34*, 185–202. [CrossRef]

30. Randelli, F.; Rocchi, B. Analysing the role of consumers within technological innovation systems: The case of alternative food networks. *Environ. Innov. Soc. Transit.* **2016**, *25*, 94–106. [CrossRef]

31. Nelson, R.R.; Winter, S.G. *An Evolutionary Theory of Economic Change*; The Belknap Press: Cambridge, MA, USA, 1982.

32. Fischer, L.-B.; Newig, J. Importance of actors and agency in sustainability transitions: A systematic exploration of the literature. *Sustainability* **2016**, *8*, 476. [CrossRef]

33. Wittmayer, J.M.; Avelino, F.; van Steenbergen, F.; Loorbach, D. Actor roles in transition: Insights from sociological perspectives. *Environ. Innov. Soc. Transit.* **2017**, *24*, 45–56. [CrossRef]

34. Patterson, J.; Schulz, K.; Vervoort, J.; Van Der Hel, S.; Widerberg, O.; Adler, C.; Hurlbert, M.; Anderton, K.; Sethi, M.; Barau, A. Exploring the governance and politics of transformations towards sustainability. *Environ. Innov. Soc. Transit.* **2017**, *24*, 1–16. [CrossRef]

35. Grillitsch, M.; Sotarauta, M. Trinity of change agency, regional development paths and opportunity spaces. *Prog. Hum. Geogr.* **2019**. Online first. [CrossRef]

36. Storper, M.; Venables, A.J. Buzz: Face-to-face contact and the urban economy. *J. Econ. Geogr.* **2004**, *4*, 351–370. [CrossRef]

37. Sonn, J.W.; Storper, M. The increasing importance of geographical proximity in knowledge production: An analysis of US patent citations, 1975–1997. *Environ. Plan. A* **2008**, *40*, 1020–1039. [CrossRef]

38. Nooteboom, B. Learning by interaction: Absorptive capacity, cognitive distance and governance. *J. Manag. Gov.* **2000**, *4*, 69–92. [CrossRef]

39. Boschma, R. Proximity and innovation: A critical assessment. *Reg. Stud.* **2005**, *39*, 61–74. [CrossRef]

40. Graziano, M.; Gillingham, K. Spatial patterns of solar photovoltaic system adoption: The influence of neighbors and the built environment. *J. Econ. Geogr.* **2015**, *15*, 815–839. [CrossRef]

41. Asheim, B.; Gertler, M. The geography of innovation. In *The Oxford Handbook of Innovation*; Fagerberg, J., Mowery, D., Nelson, R., Eds.; Oxford University Press: New York, NY, USA, 2005; pp. 291–317.

42. Asheim, B.T.; Isaksen, A.; Trippl, M. *Advanced Introduction to Regional Innovation Systems*; Edward Elgar: Cheltenham, UK, 2019.

43. Zukauskaite, E.; Trippl, M.; Plechero, M. Institutional thickness revisited. *Econ. Geogr.* **2017**, *93*, 325–345. [CrossRef]

44. Yin, R. *Case Study Research: Design and Methods*; Sage Publishing: Thousands Oaks, CA, USA, 1994.

45. Ritchie, J.; Lewis, J.; Nicholls, C.M.; Ormston, R. *Qualitative Research Practice: A Guide for Social Science Students and Researchers*; Sage Publishing: Thousands Oaks, CA, USA, 2013.

46. Kleverbeck, M.; Terstriep, J. Analysing the Social Innovation Process–The Methodology of Social Innovation Biographies. *Eur. Public Soc. Innov. Rev.* **2018**, *2*, 15–29. [CrossRef]

47. Moore, M.-L.; Westley, F.R.; Tjornbo, O.; Holroyd, C. The loop, the lens, and the lesson: Using resilience theory to examine public policy and social innovation. In *Social Innovation. Blurring Boundaries to Reconfigure Markets*; Nicholls, A., Murdock, A., Eds.; Springer: Berlin/Heidelberg, Germany, 2012; pp. 89–113.

48. Butzin, A.; Widmaier, B. Exploring Territorial Knowledge Dynamics through Innovation Biographies. *Reg. Stud.* **2015**, *50*, 220–232. [CrossRef]

49. Randelli, F.; Romei, P.; Tortora, M. An evolutionary approach to the study of rural tourism. *Land Use Policy* **2014**, *38*, 276–281. [CrossRef]

50. Biodistretto del Chianti Home Page. Available online: https://www.biodistrettodelchianti.it/index-en.html (accessed on 4 June 2020).

 *sustainability*

Article

A Framework for Risk Assessment in Collaborative Networks to Promote Sustainable Systems in Innovation Ecosystems

Ricardo Santos [1,2,*], António Abreu [1,3], Ana Dias [1,4], João M.F. Calado [1,5], Vitor Anes [1,5] and José Soares [6]

1. Instituto Superior de Engenharia de Lisboa (ISEL), Instituto Politécnico de Lisboa, 1959-007 Lisboa, Portugal; ajfa@dem.isel.ipl.pt (A.A.); ana.dias@isel.pt (A.D.); jcalado@dem.isel.ipl.pt (J.M.F.C.); vitor.anes@isel.pt (V.A.)
2. GOVCOPP-Universidade de Aveiro, 3810-193 Aveiro, Portugal
3. CTS Uninova, Faculdade de Ciências e Tecnologia, Universidade Nova de Lisboa, 2829-516 Lisboa, Portugal
4. UNIDEMI, Department of Mechanical and Industrial Engineering, NOVA School of Science and Technology, Universidade NOVA de Lisboa, 2829-516 Caparica, Portugal
5. IDMEC-IST, Universidade de Lisboa, 1049-001 Lisboa, Portugal
6. ADVANCE-Instituto Superior de Economia e Gestão (ISEG), Universidade de Lisboa, 1200-109 Lisboa, Portugal; josesoares@iseg.ulisboa.pt
* Correspondence: ricardosimoessantos84@ua.pt

Received: 16 July 2020; Accepted: 31 July 2020; Published: 2 August 2020

Abstract: Nowadays—and due to an increasingly competitive world—organizations need to collaborate in an open innovation context to be efficient and effective by achieving high levels of innovation with their products and services. However, the existing resources—as well as the innovation achieved from the diversity of partners involved—brings challenges to the management; in particularly with risk management. To fulfill such needs, risk management frameworks have been created to support managers, on preventing threats with systems development, although without properly account the influence of each system component, on the entire system, as well as the subjectivity within human perception. To account for these issues, a framework supported by fuzzy logic is presented in this work, to evaluate the risk level on system development in open innovation environment. The approach robustness is assessed by using a case study, where the challenges and benefits found are discussed.

Keywords: collaborative networks; virtual enterprise; open innovation; risk management; fuzzy logic; systems engineering

---

1. Introduction

Nowadays, small and medium-sized enterprises (SMEs) plays a crucial role on most economies in the world, by creating jobs that reduces the unemployment, besides the contribution for the gross domestic product (GDP) of each country. In most countries, its presence, is expressed by more than 87 percent of all existed enterprises [1].

Furthermore, the unpredictability around the economic context and the highly competitiveness of the actual marketplace, have forced organizations into a position where its crucial to find forms of survivance in such context. From the literature, there is a consensus that innovation should being part on new product/system development to organizations, reach competitive advantage in highly competitive markets or even in new markets to be explored [2–4].

In this context, the innovation ecosystems, and in particular, virtual enterprises (VEs) allows to share the necessary competencies and resources, to develop products/systems to better respond to the market opportunities, by operating in a collaborative network context [5].

Furthermore—and based on the existed studies from the literature—one-way to promote sustainability on innovation ecosystems (IE), is by identifying and assessing the risks involved in such environments, namely on the VE, created, with the purpose of having a collaborative environment, where knowledge, competences and risk are shared between the partners involved.

However, the development and the innovation on new systems, it is not easy to be managed especially for VEs [6,7], since that the system or the product to be developed, is normally associated with several risks, regarding the different system components involved. Which are developed by the different partners of the collaborative network.

Furthermore, there is a certain subjectivity degree on risk assessment, related with human perception, which increases with the number of risk managers, as well as with the number of partners involved [8]. This becomes even more difficult, by knowing that the responsibility on each system component, is shared by more than one partner/actor of the same network [9].

Thus, the selection of the right partner, in order to minimize the risk on each system component and then, to minimize the global risk of the system development, assumes one of the main reasons to develop models that allows the risk management in such context.

Additionally, companies are relatively vulnerable to external events, such as political, economic, technical and financial, among others.

In this sense, risk management, could help VEs to mitigate unwanted risks that could avoid the success with its system development. Without managing the risks properly, VEs could face severe consequences, such as losing clients, negative environmental impact—and even financial bankruptcy [4].

To do this, risk management (RM), act as a process to identify, assess, monitor and report the risks involved, in terms of their impacts and probabilities of occurrence [5]. The identification and evaluation allow of such risks, allows the elaboration of a set of actions to minimize any negative effects that may occur [6].

There are several methods that can be found on literature (e.g., decision trees, Delphi analysis, failure modes and effects analysis (FMEA), risk matrix, among others), which can be used to identify and analyze the risks involved. However, there is a lack of approaches to manage risk on VEs that simultaneously analyses the influence between each system component and the final system or product to be developed, as well as to include human perception on classifying each risk according to its impact and probability of occurrence [7].

Furthermore, the subjectivity around risk assessment, referred before, it is not included in most existent models found on literature, which brings the necessity to be included into an integrated approach, in order to deal with the human perception on risk assessment.

Therefore, we have considered a set of risks adopted here, on behalf of VE, by considering the design of each system component as an innovative project itself, with a set of risks involved.

The innovation, reached on system component, contributes to the innovation of the final product or system, developed on behalf of the VE created for that purpose.

The VE as well as the system/product to be developed, can have multiple risks categories related to each system domain (SD), whose relationship with each risk from each system component, should be accounted.

Thus, in this study, it will be presented an approach to incorporate the issues referred above, namely, the inclusion of a possible influence from each system component risk on the system/product domains, as well as the subjectivity around human perception on risk assessment, by using fuzzy systems.

The model developed here, will also be tested, by using a case study based on a developed system to produce "green" energy, in order to identify the challenges to be accounted on future research, as well as the benefits achieved, such as the prioritization of each risk, regarding each system component, in order to define the actions to mitigate such risks in an open innovation context.

Therefore, Section 2 describes a literature review, followed by Section 3, in which the research method is described, while Section 4 describes the case study used to validate the proposed approach through the achieved results. Section 5 ends this study, with some concluding remarks and future work.

2. Literature Review

2.1. Open Innovation Ecosystems

The open innovation (OI) and its impacts on economic, social and even political markets, are widely valued among several stakeholders (which naturally includes the policymakers), since it is an important key factor to achieve competitiveness. Several authors have recently increased the literature production, regarding the importance of innovation into our societies [7]. The innovation concept has become an important issue to be considered in the political and public discourse, with bringing therefore, an important impact on several scientific fields. Several definitions of this concept can be found on literature. An important definition, refers innovation, has a phenomenon that occurs when an invention is introduced in the market [8,9], although a formally definition, can be found by Chesbrough [10], broking therefore, with the classical linear approach, related with the closed innovation paradigm and by introducing new challenges within the innovation process.

The same author [10] claims that the closed innovation model type is no longer sustainable, from the economic point of view and organizations should engage into an open innovation approaches to be more competitive, by being a new way to create value to a company or other organization [10]. In essence, these OI models, can be defined by having two different types of knowledge flow and resources [10]. The first one, regards to the outside-in knowledge flow, i.e., it occurs when a company brings knowledge from the exterior, as well as resources from its partners, customers, scientific centers, universities, among other stakeholders involved, in order to improve its performance in terms of innovation, which allows to reduce costs and even time through the acquisition or borrow of the resources that they need to achieve its goals.

The second one, regards, the inside-out knowledge flow, where the companies search for solutions to share knowledge, already available from each partner, and/or other resources within the exterior environment in order to add value to the organization. An example of this, is the transfers of rights and the out-licensing.

With regards to OI, there are on literature, some approaches that can be found, namely the works of [11,12], with two of them, being very used among the managers of OI projects.

An OI model, is the InnoCentive [12], which was created in 2001. This model, runs in a software platform, available on web and it is based on six steps, which starts to identify the ideas and problems, followed by the challenge formulation, the intellectual property agreement specification, the challenge publication, the solution assessment and finishing with a price regarding the intellectual property transfer. The other model, widely used by organizations, is the one designed by Procter & Gamble (P&G), named "Connect + Develop", which allows to work in an inwards and outwards approach, by including issues such as the marketing for engineering, as well as the commercial services for product design.

Although an open innovation approach could release positive achievements, it also brings risk to the organizations, which leads them, to design a proper strategy to protect innovation, in order to produce boundaries and to turn the same outcomes measurable [13,14]. Furthermore and in an OI context, organizations should also consider the risks involved, to be aware of the greater ones involved, to maintain their competitive performance.

Based on the OI concept, an OI ecosystem, can be referred as an innovation ecosystem (IE) with a number of supported activities, being classified as an open innovation actions [15].

Regarding the IE, some works has explored the types of stakeholders involved in OI ecosystems, as well as the relation between the focal firms and the partners involved, with their influences on innovation, by recurring to different IE perspectives.

An example of such perspective is the one from [15], which has stated that a IE, is an aligned framework, formed by a set of multilateral partners, with the aim of interacting with each other, to achieve a certain (and focal) value. Another perspective can be given by [16], where they consider IE, as a business ecosystem with a high level of interaction among the different key partners, namely customers, universities, suppliers and competitors and suppliers.

Additionally—and based on works from [17]—it seems to exist the evidence that the product/ system innovation of nowadays, has increasingly depended on the interaction between the organizations that participates in the IE.

Other studies seem to reinforce this interdependence in terms of collaboration, between the innovation actors involved (e.g., [18,19]), although there is a lack of research that explores the interdependences between the actors with regards to a particular innovation ecosystem ([17]).

Usually, most of the literature, only focus on situations with a single mode of interaction, such as the cooperation between the university and industry (e.g., [16–19]), while other works claims for more integration of IE modes to accelerate firm innovation, which brings the need to explore more the OI ecosystem modes in order to provide a more comprehensive analytical framework for understanding OI ecosystems. Authors such as [18] have done this work, by determining that an IE is formed by a multilateral set of partners, which needs to interact to get a focal value.

Other authors, such as [19], referred the fact that IEs are more than just a typical OI with an outside-in flow—or even an inside-out flow—between the partners of the network, being more an interrelated actors to achieve and sustain an IE, to create multidisciplinary knowledge and achieve high levels of innovation with a product/system.

2.2. Risk Assessment

Apart from the development of innovation management approaches, organizations also tries to predict the changes (and also responding to them) that may occur with the development of a systems/product before and after its launch in the market, in order to be more competitive, effective and efficient. Such attempts have been performed, despite the lack of practical approaches, to support their risk management attempts [20].

The uncertainty, related with the innovative processes, is bonded not only to the inherent failure risk, but also to inherent success possibility, which brings the need to adequately manage the risk management within an innovative process/activity [20].

Therefore, and since that the aim of a risk management model is to facilitate process innovation, instead of stifle the same innovation—organizations require a strategy that allows to perform early risk diagnosis and also the management [20] of the risks involved.

From the literature, we can identify several risk assessment approaches, widely used by organizations, such as the balanced scorecard (BSC) [21], the failure mode effects analysis (FMEA), analytic hierarchy process (AHP) [22], Fault Tree analysis (FTA) [23] and the risk diagnosing methodology (RDM) [24].

The analytic hierarchy process (AHP) could be considered as a multivariate analysis method with the purpose to reduce the randomness, related to the subjective evaluations, by taking into consideration several and different objectives, based on different criteria [23]. This method is mostly used on selection of scenarios [24]. Another risk evaluation approach referred here, is RDM approach, which has the main purpose of providing strategies to support (and even improve) the possibility of a given project to be succeeded, through the identification and management of their (potential) risks [23,24]. The RDM approach, is also implemented to assist the systematic diagnosis of organizations, by accounting issues like the consumer acceptance, the commercial viability, the organization competitive answers, the external responses, the product and manufacturing technology process, in which the evaluation of the project risk is defined not only by the risk probability of occurrence and its correspondent effects, but also defined by the organizations' capability to influence the destiny of the actions, regarding each risk.

3. Research Methodology

3.1. Proposed Approach

Based on what was stated before, a virtual enterprise (VE), commonly arises to develop specific products/systems, by joining different competences and resources from a set of partners, to achieve high levels of innovation at a low-cost level, when compared to a "traditional enterprise" that must acquire from the market, the same resources and competencies.

Therefore, and based on the importance of VE on having a framework to assess the risk in such context—in this work it will be presented an approach to evaluate the existence of a possible relationship between the individual risk from a given system component n (*Scn*) (which will be assessed according to a set of different domains or risk categories) and the risk, regarding each domain of the virtual enterprise, represented here by the product/system to be developed.

Thus—and based on [25,26]—the different risks involved can be categorized based on the taxonomy adopted in this work, which is organized according to the 2 hierarchical risk levels considered here, namely; the system risk (SR) and the system component risk (Scn) (Figures 1 and 2).

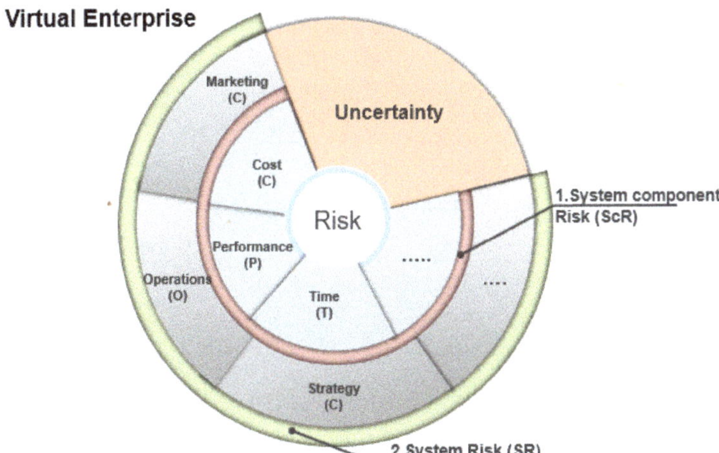

Figure 1. Risk taxonomy adopted (adapted from [21]).

By bearing in mind that a system component (Sc), can be considered (and managed) as a project, the completion of such project, even with or without success, could bring an impact on the various VE domains through its system to be developed, given the expected changes that may be occur with the development of each Sc itself.

Instead of VE domains, we have system domains (SD).

Therefore, each system component n (*Scn*) of the system to be develop, has a development project involved, which have a risk associated with it that can be managed on behalf of project risk management (PRM) techniques. Each risk, associated with the *Scn*, can be considered, as part of the systems' risk management (SRM), managed by the VE management board, which is responsible for the entire system to be developed (Figure 2).

Furthermore, the risk regarding to a given system component n (ScR_{Scn}), can be achieved by accounting the risk, obtained in several domains or risk categories, namely; performance (P), quality (Q), cost (C), time (T), among others [25,26].

Apart from the risk, associated with the system component (ScR_{Scn}), there is the risk associated with the system to be developed (SR), where and according to the risk taxonomy presented on Figure 1, it can be assessed on behalf of the following risk categories strategy (S), operations (O), finance (F), marketing (M), information systems (IS), environment (E), among others.

The possible existence of a relationship between the two risks referred before, namely ScR_{Scn} and SR, as well as its domains or risk categories involved, are described on Figure 2.

Thus, the risk regarding each system component (ScR_{Scn}), can influence the risk value of the system risk (SR), on its different domains, which in this study we have considered the following system domains (SD), namely; strategy (S), operations (O) and marketing (M).

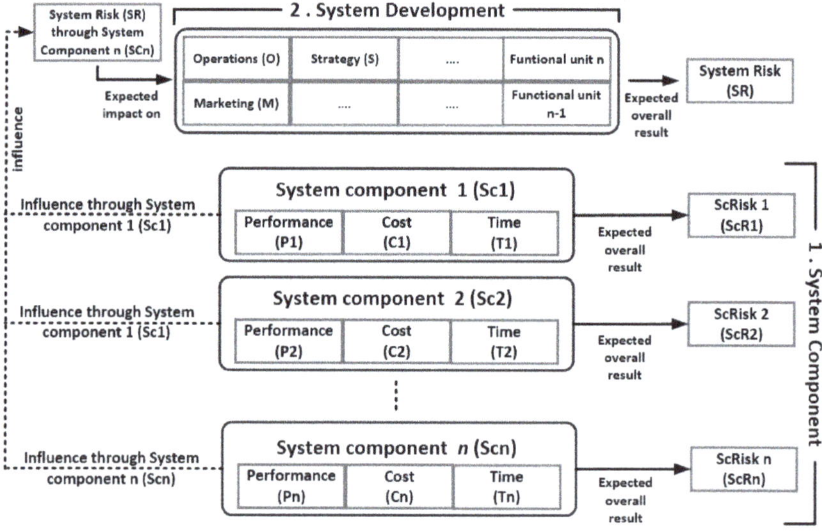

Figure 2. Relationship between the ScR_{Scn} and system risk (SR).

The need to define and quantify such influence, allows to predict the effect that a risk, regarding to a given system component n (ScR_{Scn}) can originate in the overall system risk (SR), with the transition from a current stage into another (and future) one, reached through its deployment ([2]).

The system component risk (ScR), regarding its domains, namely "time", "performance" and "cost", are presented and discussed on [26–28], as a set of risk categories considered on behalf of project risk management, according to some studies found on literature.

Therefore, and based on [25–31]—the ScR was categorized, as it follows:

- Time–accomplishment degree of the timeframe to complete the project within the planned;
- Cost—accomplishment degree of the allocated budget constrain, regarding the project completion;
- Performance–accomplishment degree of business and technical goals of the project, through the process outputs.

Regarding the system risk domains, it was considered the following ones:

- Strategy (S)—resulting from the errors in strategy (e.g., by developing a technology regarding a component that cannot work with other technologies from other product components or even a product technology that cannot meet the consumer needs) [27];

- Operational (O)—resulted from the risks regarding the production process implementation, the existence of problems around the procurement and distribution or even the delay (due to the production) with the product to be lunched [27];
- Marketing (M)—resulted from the value perceived by the costumers, which is related to the effectiveness of marketing actions (e.g., failure to generate demand for a product lunch and other risks related to demand, customer feels uncertain that the product do not meet the needs or expectation) [28,29].

Even the SR domains, as well as the ScR domains, could be expanded into additional and other risk categories apart from the ones presented before.

3.2. Model Architecture

The approach presented in this work, is based on the concept of system development (Figure 2) and it consists into an integrated approach that includes fuzzy logic to incorporate the uncertainty and ambiguity, regarding the human perception on risk evaluation with regards to system development.

Therefore, and based on the relationship shown on Figure 2—the approach presented in this work, resorts to fuzzy logic to make a qualitative and quantitative analysis of product risk evaluation in an open innovation context, by integrating two hierarchical risk levels, namely; system component and system development as a whole.

The approach, it is presented on Figure 3.

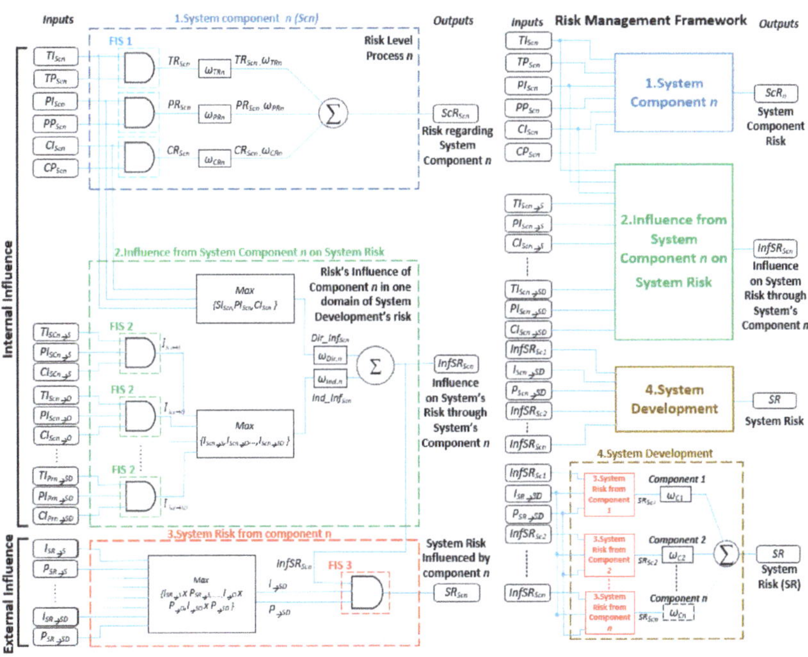

Figure 3. Proposed model.

The system component level, has the aim of evaluating the system component risk (ScR_{Scn}), by taking into consideration a given system component (Sc) n. On the other hand, the second one, evaluates the system risk level (SR), by including the existence of a possible influence ($InfSR_{Scn}$), regarding each system component n (Scn) on the entire system risk (SR) (Figure 3).

For each risk category, regarding each system component (Figure 1), namely performance (P), cost (C) and time (T), there is an individual risk ($^{-}R_{Scn}$), which results from the combination of each (expected) impact ($^-I_{Scn}$) and the correspondent probability of occurrence ($^-P_{Scn}$) considered here (Figure 3), i.e.,:

$$TR_{Scn} = TP_{Scn}.TI_{Scn} \tag{1}$$

$$PR_{Scn} = PP_{Scn}.PI_{Scn} \tag{2}$$

$$CR_{Scn} = CP_{Scn}.CI_{Scn} \tag{3}$$

Based on Figures 2 and 3, the risk category, regarding each system component n (ScR_{Scn}), can be achieved by combining all the individual risk category considered, i.e.,:

$$ScR_{Scn} = TR_{Scn}.\omega_{TRn} + PR_{Scn}.\omega_{PRn} + CR_{Scn}.\omega_{CRn} \tag{4}$$

With the weights (ω_{-Rn}), corresponding each one, to a risk category, regarding a system component n, namely cost (C), time (T) and performance (P). These weights, should satisfy the following condition:

$$1 = \omega_{TRn} + \omega_{PRn} + \omega_{CRn} \tag{5}$$

The existence of a (possible) influence, regarding the risk with a given System component n (ScR_{Scn}), over each system development domain (SD) considered here, is also included in this approach, by taking the correspondent (and expected) impact variable ($^-I_{Scn \to SD}$), which will affect one (or even more) domains of the system development (SD).

In this work, it was considered three domains, regarding to the system, developed on behalf of the VE created, namely, marketing (M), operations (O) and strategy (S).

Through the risk categories time (T), performance (P) and cost (C), related to a system component n (Sc_n), it was achieved a set of expected impact values ($^-IScn \to SD$), regarding to each system development domain considered. Then and for each system domain (SD) considered here, only one value of $^-IScn \to SD$ is selected (Figure 3), between the different values. Such a process is preformed based on the risk manager perception.

Furthermore and based on each SD considered here, fuzzy logic is then deployed, by using a set of linguistic variables and inference rules, to obtain the correspondent impact value of $^-IScn \to SD$.

Regarding the indirect influence, resulted from the system component n (Ind_Inf_{Scn}) on each SD, it is achieved, by accounting the maximum value of the three corresponding (and expected) impact values ($^-I_{Scn \to SD}$) related to each system domain, achieved before, i.e.,:

$$Ind\_Inf_{Scn} = \max\{SI_{Scn \to S}, PI_{Scn \to S}, CI_{Scn \to S}, \ldots, SI_{Scn \to O}, PI_{Scn \to O}, CI_{Scn \to O}\} \tag{6}$$

The existence of a (possible) direct influence, from each system component and through its different domains, is also considered here, since that the resources used on each system component, can affect resource availability in terms of the system development context.

Thus, the Sc domain with more impact, allows to estimate the direct influence (Dir_Inf_{Scn}) of a given Scn, which is obtained, by achieving the maximum value of the three $^-IScn \to SD$ values, mentioned before, i.e.,:

$$Dir\_Inf_{Scn} = \max\{PI_{Scn}, TI_{Scn}, CI_{Scn}\} \tag{7}$$

Therefore, the total influence ($InfSR_{Scn}$) from a given Scn on system development, will be resulted by adding the Dir_Inf_{Scn} with Ind_Inf_{Scn}, i.e.,:

$$InfPR_{Scn} = Ind\_Inf_{Scn}.\omega_{Ind.n} + Dir\_Inf_{Scn}.\omega_{Dir.n} \tag{8}$$

The existence of the weights $\omega_{Dir.n}$ and $\omega_{Ind.n}$, intends to define the relative importance, given by the risk manager to the direct and indirect influences achieved before, with both parameters, satisfying the following condition, i.e.,:

$$1 = \omega_{Ind.n} + \omega_{Dir.n} \tag{9}$$

Therefore, the system development risk (SR_{Scn}), resulted from each *Scn*, is achieved by combining the probability of occurrence of the event ($P_{SR \rightarrow SD}$), the external impact on each system domain (SD) ($I_{SR \rightarrow SD}$) and the influence from each *Scn* on system risk ($InfSR_{Scn}$), i.e.,:

$$SR_{Scn} = P_{SR \rightarrow SD}.I_{SR \rightarrow SD}.InfSR_{Scn} \tag{10}$$

The values of $I_{SR \rightarrow SD}$ and $P_{SR \rightarrow SD}$, are achieved, by selecting the maximum value of product, between $I_{SR \rightarrow SD}$ and $P_{SR \rightarrow SD}$, both considered for each system domain (SD), namely marketing (M), operations (O) and strategy (S), i.e.,:

$$\langle I_{SR \rightarrow SD}, P_{SR \rightarrow SD} \rangle = \max\{I_{SR \rightarrow M} \times P_{SR \rightarrow M}, I_{SR \rightarrow O} \times P_{SR \rightarrow O}, I_{SR \rightarrow S} \times P_{SR \rightarrow S}\} \tag{11}$$

The system risk (SR) is therefore resulted from the contribution of each SR_{Scn}, related to the correspondent *Scn* considered and based on its relative importance (ω_{Scn}), i.e.,:

$$SR = \omega_{Sc1}.SR_{Sc1} + \omega_{Sc2}.SR_{Sc2} + \omega_{Scn}.SR_{Scn} \tag{12}$$

where ω_{Sc1}, ω_{Sc2} and ω_{Scn}, are the weights, respectively regarded to the relative importance given to each system component n (*Scn*).

3.3. Fuzzy Implementation

The levels of risk, related to ScR_{Scn} and SR values, were obtained from each correspondent fuzzy inference system (FIS) (Figure 3) and they were based on a set of inference rules from the type of "If-And-Then" and according to the risk category.

According to Figure 3 and regarding the system component level, the FIS "F1", is based on the expressions (1–3), regarding the inputs ($^-P_{Scn}$) and ($^-I_{Scn}$), i.e.,:

$$^-R_{Scn} = {}^-P_{Scn} \cap {}^-I_{-Scn} \tag{13}$$

Thus, the inference rules, regarding F1 system (Figures 3 and 4), can be formulated as: "IF probability (-PPrn) is P AND impact (-IPrn) is I, THEN the process risk is R.

The same approach can be achieved for the impact $^-IScn \rightarrow SD$, which allows to obtain the indirect influence Ind_Inf_{Scn}, defined on (6). Therefore, and based on 13, the impact $IScn \rightarrow SD$, is achieved through the following expressions:

$$I_{Scn \rightarrow S} = TI_{Scn \rightarrow S} \cap PI_{Scn \rightarrow S} \cap CI_{Scn \rightarrow S} \tag{14}$$

$$I_{Scn \rightarrow M} = TI_{Scn \rightarrow M} \cap PI_{Scn \rightarrow M} \cap CI_{Scn \rightarrow M} \tag{15}$$

$$I_{Scn \rightarrow O} = TI_{Scn \rightarrow O} \cap PI_{Scn \rightarrow O} \cap CI_{Scn \rightarrow O} \tag{16}$$

The possible influence of the risk, regarding the system component n, related to each system domain (SD), can also be considered by using the FIS F2 (Figures 3 and 4), which is based on a set of linguistic rules and variables, which can be formulated as "If impact ($TIScn \rightarrow SD$) is it and the impact ($TIScn \rightarrow SD$) is Ip and the impact ($TIScn \rightarrow SD$) is Ic, Then the average impact of the system component n, on each SD is $^-IScn \rightarrow SD$.

Similar approach can be proposed, regarding the system risk level (SR_{Scn}), where, based on (10)—and by considering the influence from Scn on system risk development ($InfSR_{Scn}$)—SR_{Scn} is obtained by using the following expression:

$$SR_{Scn} = P_{SR \rightarrow SD} \cap I_{SR \rightarrow SD} \cap InfSR_{Scn} \tag{17}$$

Thus, the inference rules regarding the FIS F3 (Figures 3 and 4), is defined as "If probability $(P_{Scn \rightarrow SD})$ is P, impact $(I_{SR \rightarrow SD})$ is I and Scn influence on system risk $(InfSR_{Scn})$ is Inf. Then the system risk level (SR_{Scn}) is SR.

3.4. Definition of Linguistic Variables: Values and Pertinence Functions

Regarding the linguistic variables—and according to some works existing on the literature ([25,26])—it is normally recommended that the number of levels shouldn't be more than nine levels, since that a bigger value, surpass human perception limits, when it concerns to value discrimination.

Therefore, and based on these recommendations—five levels were adopted for each of the linguistic variables considered in this work. The definition of the five levels, as well as the correspondent pertinence functions, are described on Tables 1–3. Regarding each pertinence function, it was used the triangular type function, whose parameters a, b and c, were also described on the same tables.

Table 1. Values regarding the linguistic variable type "probability of occurrence (P)" and pertinence functions.

Pertinence Levels.	Description	Frequency	Fuzzy Parameters [a, b, c]
Rare	It is accounted that the event will happen only in certain circumstances.	Event has occurred or is expected to occur once in the next 48 months	(0, 0,0.25)
Unlikely	The event is not likely, although it can occur.	Event has occurred or is expected to occur once in the next 24 months	(0, 0.25,0.50)
Likely	Probable occurrence event	Event has occurred or is expected to occur once in the next 18 months	(0.25,0.50,0.75)
Very Likely	The event will likely occur	Event has occurred or is expected to occur once in the next 12 months	(0.5,0.75,1.0)
Expected	The event is expected to occur	Event has occurred or is expected to occur once in the next 6 months	(0.75, 1, 1)

Thus—and with regards to all off the domains of the Sc probability of occurrence $(^{-}P_{Scn})$, i.e., cost (C), time (T) and performance (P)—it was defined the pertinence functions and linguistic variables, which are presented on Table 1.

Same approach, was also deployed to define the probability of occurrence, regarding the system risk $(P_{SR \rightarrow SD})$, by also using Table 1, with SD, being the system domain considered here, i.e., strategy (S), operations (O) and marketing (M).

Regarding the linguistic variables, related to the expected impact, considered for each Scn $(-I_{Scn})$ and on behalf of the three Sc domains considered here (cost (C), time (T) and performance (P)), on Table 2 are described the correspondent linguistic values, as well as the pertinence functions, used to define the same variables.

Table 2. Values regarding the linguistic variable type "expected impact (I)" and pertinence functions.

Pertinence Levels	Process Domain			Fuzzy Parameters (a, b, c)
	Time (T)	Performance (P)	Cost (C)	
Neglectable	Insignificant impact on the processes required to obtain deliverables. No changes in established activities	Insignificant impact on the initial project budget (<2%)	Timing delay is easily recoverable.	(0,0,2.5)
Low	Prevents the fulfillment of one or more activities established for each project task. No task changes.	Low impact on project budget (2–5%)	Low schedule delay is not recoverable.	(0,2.5, 5.0)
Moderate	Prevents the fulfillment of one or more tasks. No requirement changes.	Moderate impact on the initial project budget (5–10%)	Moderate delay in the completion of the project. Without compromising the project requirements.	(2.5,5.0,7.5)
High	Prevents the fulfillment of one or more project requirements. Scope change required.	High impact on the initial project budget (10–30%)	Acceleration in the fulfillment of tasks with anticipation of the project calendar.	(5.0,7.5,10.0)
Severe	It prevents the fulfillment of the project objective(s) and it is not possible to achieve it even with changes in scope.	Impact on the initial heavy budget making the project unfeasible (>30%)	Project deadline exceeded making it impossible to complete the project since the project is no longer adequate to the organizational reality.	(7.5,10.0,10.0)

Similar approach was conducted for the variable—$I_{Scn \to SD}$, regarding the correspondent expected impact value of each Sc domain on each system domain (SD) considered here, namely, marketing (M), operations (O) and strategy (S), with Table 2, also being applied in this case, given the same criteria, established for -I_{Scn} variable.

With regards to the Sc risk ($^-R_{Scn}$)—and on behalf of its Sc domains (cost (C), time (T) and performance (P))—the correspondent linguistic variable, is defined on Table 3, through its pertinence functions and correspondent triangular parameters (a, b, c). A similar approach was also deployed to define the linguistic variable regarding the system risk, related to the SR_{Scn} variable.

Table 3. Values regarding the linguistic variable type "risk level (R)" and pertinence functions.

Pertinence Levels	Description	Fuzzy Parameters (a, b, c)
Very low	Risk can be accepted as it does not pose a threat to the project/organization, it must be monitored to ensure that its level does not change.	(0, 0, 0.25)
Low	Risk can be accepted. Risk control must be carried out based on a cost–benefit analysis	(0, 0.25, 0.50)
Moderate	Risk must be mitigated; the effectiveness of controls must be monitored.	(0.25, 0.50, 0.75)
High	Efforts should be made to mitigate risk as soon as possible.	(0.50, 0.75,1.0)
Very High	Immediate action must be taken to mitigate the risk.	(0.75, 1.0, 1.0)

In addition to the parameters, presented on Tables 1–3, it is possible to correspond the linguistic values into numeric ones, based on a set of intervals (Table 4), which will help to analyze the final results obtained, regarding the risks with system components, as well as the overall system development risk.

Table 4. Variable types used: linguistic values and the correspondent numeric ones.

Variable Type					
Impact of Occurrence		Probability of Occurrence		Risk of Occurrence	
Linguistic Levels	Numeric Correspondence	Linguistic Levels	Numeric Correspondence	Linguistic Levels	Numeric Correspondence
Insignificant	[0,2]	Very Low	[0,0.2]	Very low	[0,2]
Low	[2,4]	Low	[0.2,0.4]	Low	[2,4]
Moderate	[4,6]	Moderate	[0.4,0.6]	Moderate	[4,6]
High	[6,8]	High	[0.6,0.8]	High	[6,8]
Severe	[8,10]	Very High	[0.8,1.0]	Very high	[8,10]

3.5. Fuzzy Deployment

Based on the model, presented on Figure 3, each one of the fuzzy inference system (FIS), was implemented by recurring to Matlab fuzzy Logic Toolbox^{TM} (version R2017a) (Figure 4), which is integrated on MATLAB® software platform. Thus, the definition of the membership functions, as well as the inference rules, was based on Tables 1–3—and also made by recurring to the same toolbox—which was also used to analyze the behavior of each FIS considered.

As it referred before, in all of the FIS considered here, it was used triangular functions, whose parameters were taken from Tables 1–3, as well as the membership functions, whose inference rules were implemented by using Mamdani inference mechanism (Figure 4), due to its intuitive approach, well-suited to human input and widespread acceptance on literature [32,33].

With regards to the system component (Sc) level, the FIS F1, has on its inputs the linguistic variables "Probability of occurrence ($^{-}P_{Scn}$)", as well as the correspondent expected Impact value ($^{-}I_{Scn}$), while the respective output, is the Sc risk level ($^{-}R_{Scn}$), which is related to each one of the Sc domains considered in this work.

The implementation of FIS F2 (influence of a given *Scn* on each system domain), has on its inputs, the linguistic variables, $TIScn \rightarrow SD$, $PIScn \rightarrow SD$ and $CIScn \rightarrow SD$, with the output variable of F2 having the *Scn* average impact, on each system domain (SD), i.e., $^{-}IScn \rightarrow SD$.

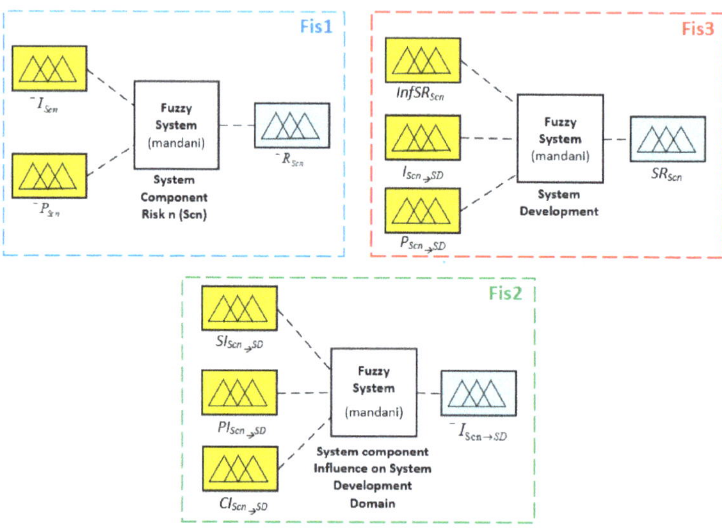

Figure 4. MATLAB implementation, regarding each fuzzy inference system (FIS) considered.

The FIS F3, has on its inputs the linguistic variables $P_{Scn \to SD}$, $IScn \to SD$ and $InfSR_{Scn}$, while the output is consistent with the linguistic variable SR_{Scn}.

With regards to the defuzzification method used in all FIS deployment, it was used the centroid approach, given its widely use in works from the literature ([26]).

In Figure 5, it is presented the surfaces, resulted from several simulations to assess the behavior of FIS 1, with concerns to each risk category (domain) considered here, regarding the Sc risk level.

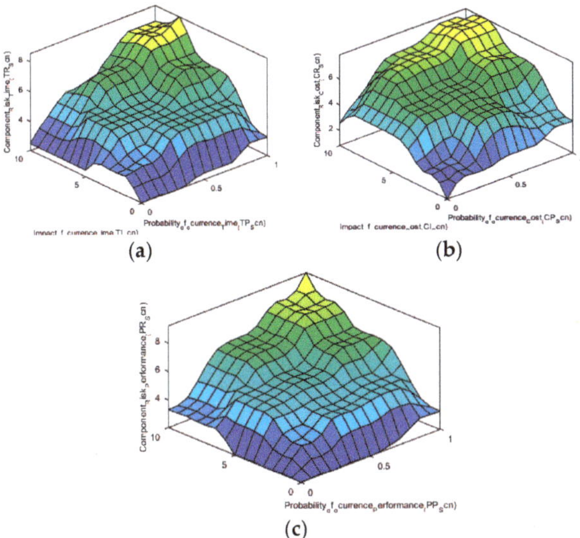

Figure 5. FIS surfaces, regarding FIS F1 and related to (**a**) time (T), (**b**) performance (P) and (**c**) cost (C).

Based on the three obtained surfaces, presented on Figure 5, it can be noticed some differences, regarding the three graphics, which are mainly associated with the difference regarding the inference rules adopted for each one of the Sc domains considered in this work, allowing therefore, to obtain different combinations of the output, for the same pair of inputs.

4. Case Study, Results & Discussion

The model robustness was evaluated by recurring to a case study, which was based on a virtual enterprise (VE). This VE, was established to develop an integrated and sustainable approach, to supply an industry, by producing its own energy through the integration of hydrogen with photovoltaic system.

The purpose of this system is to become a typical integrated product, to be further sold (and deployed) to the small industries, with a nominal electric power, ranging between 40–60 kW.

On these conditions, the system is designed so that the load to be fed, does not depend on the public electricity network, which makes the installation, self-sufficient in terms of electric energy.

The diagram presented on Figure 6, shows the developed system, with all of its components.

In order to ensure the total independence of the industry from the electric grid, the system produces electric energy during the day, by the photovoltaic and fuel cell systems, with the remained energy, being stored by using hydrogen as an energy carrier. This is preformed, by converting the same energy into the correspondent amount of hydrogen to be stored. The hydrogen stored, is then converted into electric energy by the fuel cell, during the night.

Based on Figure 6b, the system developed here, is composed by a series of components, which includes the photovoltaic system (PV panels and the converter), the electrolyzer to convert electric energy into hydrogen, the fuel cell to convert hydrogen into electric energy, the fuel tanks, the control system of the flow of air/oxygen (to control the purity of the O_2 consumed by the fuel cell), the remain power converters, the management system, as well as other auxiliar systems (e.g., electric valves, tubes, cables, electric bombs, among other components).

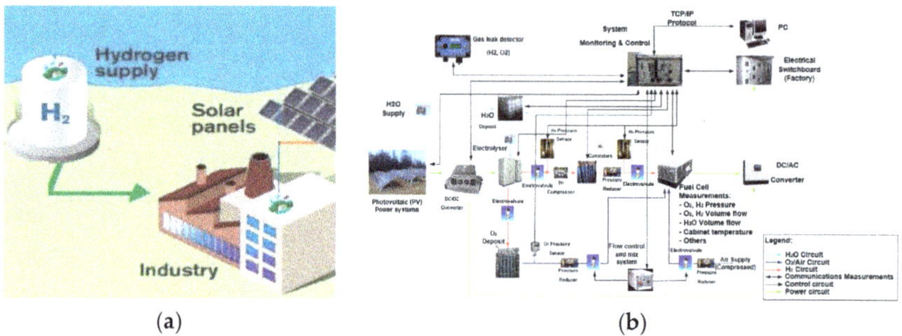

(a)   (b)

Figure 6. System developed (views). (**a**) Conceptual; (**b**) schematic.

Each system component is developed/manufactured by a set of partners (with indirect/direct involvement), through the open innovation network (OIN), created on behalf of the virtual enterprise (VE), established to develop the system presented on Figure 6.

This virtual enterprise, is formed by 13 partners (Figure 7), originated from different industries and sectors, including two Research & Development (R & D) centers from two universities.

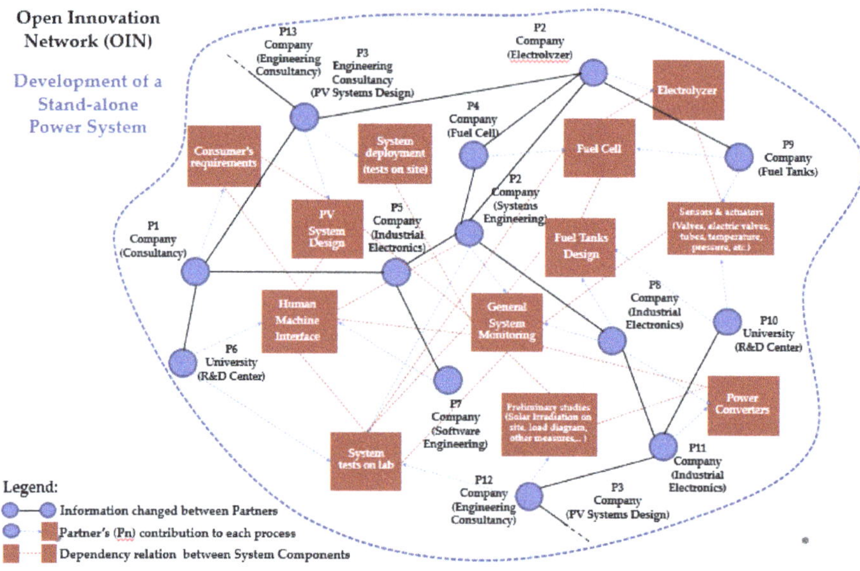

Figure 7. Virtual enterprise and diversity of processes involved [10].

In order to model the resources and the different competences involved here, shared among OIN partners, it was adopted the open innovation model, developed from the work of [10], which allows to manage the innovation (and the partner contribution) on behalf of the system/product do be developed (Figure 7).

The VE, shares different competences and resources between the partners involved, resulting therefore, into a set of innovations around the product development.

The interaction between the different partners from different areas (and therefore with different skills), allows to achieve innovations within this system, such as the reduction of available time to provide full power to specific loads/industries, where the load diagram suffers rapid changes in terms of power demand.

Based on the model, presented on Figure 7, it can even be mapped, the information changed between the different partners, as well as their contribution for each process, related to each system component to be developed—and also the interdependency relationship between each system component.

On Table 5, it can be seen some of the system components developed at the time, as well as the partners involved.

Table 5. Process description and partners involved (based on [10]).

Pr.	Ref.	Description	Partners Involved
1	K01Pr1	Consumer requirements	P8, P7
2	K02Pr3	PV system design	P3, P12
3	K87Pr4	System deployment (tests on site)	P3
4	K01Pr6	Human machine interface (HMI)	P6 e P7
5	K02Pr5	Systems tests on lab	P2, P6 e P12
6	K01Pr6	General system monitoring and control	P2, P8
7	K01Pr6	Fuel tanks design	P9, P10
8	K01Pr6	Fuel cell	P4, P9
9	K01Pr6	Preliminary studies (solar irradiation on site, load diagram, other measures)	P12
10	K01Pr6	electrolyzer	P2
11	K01Pr6	Sensors & actuators (valves, electric valves, tubes, temperature, pressure	P9, P10
12	K01Pr6	power converters	P8, P11

Each process/task is related to a specific component of the system, with the responsibility, being shared (in some cases) by more than one partner of the OIN.

Therefore, and to validate the approach developed here—it was considered a portfolio of 12 processes/system components, which are presented on Table 5.

For each process, there is a set of possible events, related to the schedule/time risk category (e.g., failure to accomplish the deadline on design the General System Monitoring and Control), cost (e.g., unexpectable additional costs during the execution of the electrolyzer) and performance (e.g., failure to meet the expected technical requirements of the developed fuel cell).

Thus, and, in order to apply the approach developed in this work, on behalf of the 17 inputs referred before (Figures 3 and 4), a group of risk managers, have used the model inputs, based on the three identified risks, which is presented on Table 6, namely: SI_{Scn}, SP_{Scn}, PI_{Scn}, PP_{Scn}, CI_{Scn}, CP_{Scn} (system component level), $SI_{Scn \to S}$, $PI_{Scn \to S}$, $CI_{Scn \to S}$, $SI_{Scn \to O}$, $PI_{Scn \to O}$, $CI_{Scn \to O}$, $SI_{Scn \to F}$, $PI_{Scn \to F}$, $CI_{Scn \to F}$ (process influence) and $I_{\to SD}$, $P_{\to SD}$, $InfSR_{Scn}$ (System level) related to each one of the 12 system components considered here. The inputs from the model presented on Figure 3, are shown on Table 6, regarding the different weights and related for each fuzzy inference system (FIS) considered here, namely FIS 1, FIS 2 and FIS 3.

Table 6. Fuzzy inference system (FIS) inputs used; FIS F1, FIS F2 and FIS F3.

	FIS F1					
N Sc.	TI_{Scn}	TP_{Scn}	PI_{Scn}	PP_{Scn}	CI_{Scn}	CP_{Scn}
1	Insignificant	Low	Low	Rare	Low	Unlikely
2	Low	Moderate	Low	Unlikely	Moderate	Unlikely
3	Insignificant	Low	Moderate	Likely	Low	Likely
4	Insignificant	High	Severe	Very Likely	High	Unlikely
5	Low	Moderate	Moderate	Expected	Moderate	Expected
6	Moderate	Expected	Insignificant	Rare	Insignificant	Very Likely
7	Moderate	Unlikely	Severe	Unlikely	Moderate	Very Likely
8	High	Likely	Moderate	Very Likely	Moderate	Rare
9	Severe	Very Likely	Low	Unlikely	Low	Unlikely
10	Low	Likely	Severe	Likely	Severe	Rare
11	Low	Rare	Low	Expected	Severe	Unlikely

	FIS F2								
Sc.	$TI_{Scn \rightarrow S}$	$PI_{Scn \rightarrow S}$	$CI_{Scn \rightarrow S}$	$TI_{Scn \rightarrow O}$	$PI_{Scn \rightarrow O}$	$CI_{Scn \rightarrow O}$	$TI_{Scn \rightarrow M}$	$PI_{Scn \rightarrow M}$	$CI_{Scn \rightarrow M}$
1	Insignificant	Insignificant	High	Insignificant	Insignificant	Insignificant	Low	Insignificant	High
2	Low	Moderate	Low	Severe	Moderate	Low	Insignificant	Severe	Low
3	Moderate	Moderate	Moderate	Moderate	Insignificant	Moderate	Moderate	Moderate	Moderate
4	Severe	Moderate	High	High	Moderate	High	Low	High	High
5	Moderate	Severe	Low	Moderate	Low	Severe	Low	Severe	Low
6	Insignificant	Moderate	Moderate	Moderate	Low	Insignificant	Low	Insignificant	Moderate
7	Severe	Moderate	Moderate	Severe	Severe	High	Moderate	Low	Insignificant
8	Moderate	Insignificant	Severe	Moderate	Insignificant	Moderate	Moderate	High	Moderate
9	Low	Low	Moderate	Insignificant	High	Low	Insignificant	Low	Moderate
10	Moderate	Moderate	High	Moderate	Low	Low	Insignificant	Moderate	High
11	Severe	Moderate	Low	Insignificant	Insignificant	Moderate	High	Moderate	Low
12	Moderate	Insignificant	Moderate	High	High	Moderate	Low	Low	Moderate

	FIS F3								
Sc.	$I_{\rightarrow S}$	$P_{\rightarrow S}$	$I_{\rightarrow O}$	$P_{\rightarrow O}$	$I_{\rightarrow M}$	$P_{\rightarrow M}$	$I_{\rightarrow SD}$	$P_{\rightarrow SD}$	$InfSR_{Scn}$
1	Insignificant	Rare	High	Rare	Insignificant	Rare	Moderate	Rare	Insignificant
2	Moderate	Unlikely	Low	Unlikely	Moderate	Unlikely	Severe	Unlikely	Low
3	Insignificant	Unlikely	Moderate	Unlikely	Moderate	Likely	Moderate	Unlikely	Moderate
4	Moderate	Very Likely	High	Very Likely	Moderate	Very Likely	High	Very Likely	High
5	Low	Very Likely	Low	Very Likely	Severe	Expected	Severe	Very Likely	Severe
6	Low	Rare	Moderate	Rare	Moderate	Rare	Severe	Rare	Insignificant
7	Severe	Unlikely	Insignificant	Unlikely	Moderate	Unlikely	Insignificant	Unlikely	Low
8	Insignificant	Very Likely	Moderate	Very Likely	Insignificant	Very Likely	Low	Very Likely	Low
9	High	Unlikely	Moderate	Unlikely	Low	Unlikely	Severe	Unlikely	Moderate
10	Low	Likely	High	Likely	Moderate	Likely	Moderate	Likely	Severe
11	Insignificant	Expected	Low	Expected	Moderate	Expected	Moderate	Expected	Moderate
12	High	Likely	Moderate	Likely	Insignificant	Expected	Insignificant	Likely	Insignificant

The outputs from the model presented on Figure 3, are shown on Table 7, considering the different weights and related for each fuzzy inference system (FIS) considered here, namely FIS 1, FIS 2 and FIS 3.

Table 7. Fuzzy inference system (FIS) outputs used.

	FIS F1			FIS F2			FIS F3
Sc.	TR_{Scn}	PR_{Scn}	CR_{Scn}	$I_{Scn \rightarrow S}$	$I_{Scn \rightarrow O}$	$I_{Scn \rightarrow M}$	SR_{Scn}
1	Very Low	Low	Low	Low	Very Low	Low	Very Low
2	Moderate	Low	Low	Moderate	High	Moderate	Low
3	Very Low	Moderate	Low	Moderate	Moderate	Moderate	Moderate
4	Moderate	Severe	High	High	Moderate	High	High
5	High	Low	Moderate	High	Moderate	Low	High
6	Low	Moderate	Very Low	Moderate	Low	Low	Moderate
7	Moderate	Moderate	Moderate	Moderate	Severe	Low	Low
8	High	High	Moderate	High	Moderate	Moderate	Low
9	High	Low	Low	Low	Moderate	Low	High
10	Moderate	Moderate	High	Moderate	Low	Moderate	Moderate
11	Low	Moderate	High	Moderate	Very Low	Moderate	Moderate
12	Low	Moderate	Moderate	Moderate	High	Low	Very Low

Regarding the model outputs presented on Table 7, it is possible to correspond the linguistic values into numeric ones, based on a set of intervals (Table 4), in order to support the risk manager, through the analyze of the final results obtained, with regards to the 12 Sc considered here, leading to the Sc overall risk (ScR), as well as the system development risk (SR) (Tables 8 and 9, respectively).

Table 8. Outputs, regarding the system component risk (ScR) level.

Sc.	TR_{Scn}	PR_{Scn}	CR_{Scn}	ω_{TRn}	ω_{PRn}	ω_{CRn}	*System Component Risk (ScR_{Scn})*
1	1.9	2.1	3.7	0.32	0.26	0.42	2.7
2	4.3	2.5	3.1	0.26	0.23	0.51	3.3
3	2.0	5.6	2.5	0.29	0.21	0.50	3.0
4	5.4	8.7	6.9	0.41	0.19	0.40	6.6
5	7.3	3.4	5.9	0.36	0.14	0.50	6.1
6	2.1	4.8	1.7	0.27	0.21	0.52	2.5
7	4.4	5.5	4.1	0.31	0.21	0.48	4.5
8	7.5	7.1	4.9	0.18	0.26	0.56	5.9
9	6.2	3.1	2.7	0.42	0.18	0.40	4.2
10	5.1	5.8	7.8	0.38	0.13	0.49	6.5
11	2.8	4.2	7.2	0.38	0.14	0.48	5.1
12	3.9	4.9	5.6	0.36	0.14	0.50	4.9

Therefore, on Table 8, it is presented the final values regarding the ScR level, as well as the values of the weights, namely ω_{TRn}, ω_{PRn} and ω_{CRn}, related to the system component (*Sc*) risk level (*ScR*).

Through Table 8, the Sc with more risk, is Sc4 (*Human Machine Interface (HMI)*), followed by Sc10 (*Electrolyzer*) and Sc5 (*Systems tests on lab*). However—and by applying the same relative importance (i.e., the same weights) used on Sc10, into Sc8 risk assessment, for instance—we have verified that Sc8 has more risk than Sc10, which illustrates the importance given by the risk manager to each one of the Sc risk component (i.e., TR_{Scn}, PR_{Scn} and CR_{Scn}) when preforming the risk assessment.

Furthermore and based on the results from Table 8, we verify that the Sc 10 (Electrolyzer), has more potential to surpass the initial planned budget constraint, bringing therefore more risk to the overall system project. However, and despite the risk values obtained, it has small influence into the final product/system, since that the value of the overall system risk from Sc10 (SR_{Scn}) is smaller, practically neglectable, due to the relative importance ω_{PRn}, given by the system risk manager.

On Table 9, are presented the values of the weights regarding the direct and indirect influence of each Scn (i.e., $\omega_{Dir.n}$ and $\omega_{Ind.n}$), into the overall system risk (SR_{Scn}). The values of the weights (ω_{Cn}), regarding each ScR contribution to the SR (SR_{Scn}), are also presented here, as well as the different outputs from FIS 2 (see Figure 3), obtained based on Tables 4 and 7.

Table 9. Outputs regarding the system risk (SR) level.

Sc.	$I_{Scn \rightarrow S}$	$I_{Scn \rightarrow O}$	$I_{Scn \rightarrow M}$	*Max* $\{I_{Scn \rightarrow SD}\}$	*Max* $\{SI_{Scn}, PI_{Scn}, CI_{Scn}\}$	$\omega_{Ind.n}$	$\omega_{Dir.n}$	$InfSR_{Scn}$	*Max* $\{I_{SR \rightarrow SD}$ x $P_{SR \rightarrow SD}\}$ PR_{Scn}	*Max* $\{I_{SR \rightarrow SD}$ x $P_{SR \rightarrow SD}\}$ TR_{Scn}	SR_{Scn}	ω_{Scn}	SR_{Scn} x ωScn
1	2.8	1.8	2.6	2.8	3.7	0.61	0.39	3.15	1.5	1.9	1.5	0.07	0.11
2	4.1	7.8	4.7	7.8	4.3	0.28	0.72	5.28	3.8	4.3	4.7	0.08	0.38
3	4.7	4.7	5.1	5.1	5.6	0.46	0.54	5.37	5.8	2	5.8	0.01	0.06
4	6.7	5.9	7.1	7.1	8.7	0.51	0.49	7.88	7.8	5.4	7.8	0.07	0.55
5	7.1	5.7	2.1	7.1	7.3	0.46	0.54	7.21	6.7	7.3	6.7	0.08	0.54
6	4.7	2.3	2.8	4.7	4.8	0.62	0.38	4.74	4.6	2.1	4.6	0.09	0.41
7	5.2	9.2	3.7	9.2	5.5	0.11	0.89	5.91	2.9	4.4	3.8	0.08	0.30
8	6.4	4.2	5.2	6.4	7.5	0.46	0.54	6.99	3.2	7.5	6.3	0.09	0.57
9	2.4	4.8	3.1	4.8	6.2	0.56	0.44	5.42	7.4	6.2	7.4	0.09	0.67
10	4.3	3.1	5.8	5.8	7.8	0.57	0.43	6.66	5.9	5.1	5.7	0.09	0.51
11	4.9	1.1	5.4	5.4	7.2	0.66	0.44	6.73	6.2	2.8	6.3	0.13	0.82
12	5.8	6.5	3.8	6.5	5.6	0.64	0.36	6.18	0.8	3.9	4.1	0.12	0.49
system risk (SR)													5.40

On the other hand—and according to Table 9—the Sc which will inspire more concerns to the risk manager is the Sc 5 (*Systems tests on lab*), given the high value of risk, achieved not only with the process itself (SR_{Sc5}), but also with its influence on system risk, expressed through ScR5.

This previous detection, have allowed to identify the Sc with more risk, the risk source and also the partners involved, in order to study new ways to reduce such risk, even before the Sc risk *Systems tests on lab* (Sc5) took place.

From the results presented on Table 9, we can also prioritize the 12 Scs, according to its risks (srscn). Thus, the Sc with the highest risk is the process number 4 (*human machine interface (hmi)*), followed by the Sc number 9 (*preliminary studies (solar irradiation on site, load diagram, other measures)*), Sc number 5 (*systems tests on lab*), Sc number 8 (fuel cell) and Sc number 11 (*sensors & actuators (valves, electric valves, tubes, temperature, pressure)*), Sc number 3 (*electrolyzer system deployment (tests on site)*), Sc number 10 (*electrolyzer*), Sc number 2 (*pv system design*), Sc number 6 (*general system monitoring and control*), Sc number 12 (*power converters*), Sc number 7 (*fuel tanks design*) and Sc number 1 (*consumer requirements*). by considering all of the systems components risks and according to the correspondent relative importance (ω_{Scn}) it is also possible to assess the overall risk regarding the system to developed, which in this case is 5.40, being therefore a system/product with a *moderated risk*, according to the model risk classification and given the conditions, associated with the case study, used in this work.

5. Conclusions

In this study, it was developed a methodology to assess the risk, regarding the development of a system (or a product), in an innovation ecosystem context, by considering a virtual enterprise (VE).

For this purpose, it was included the risks regarding each system component, from the partners involved on that component, as well as the risks regarding the VE system domain to be developed.

Apart from the identification and assessment of the risks involved, it was also considered the influence regarding the risk from each system component on each final system domain, related to the final system/product to be development by the VE.

The problem with subjectivity regarding the risk assessment, was also considered in this work, by integrated fuzzy logic on this approach, in order to try to avoid the possible biases, associated with human perception on risk assessment.

Therefore, the possible influence of each system component in the system risk domain, was also accounted and evaluated, allowing not only to act on system component level context and thus to estimate the correspondent risk, but also to assess its contribution for system-level risk, by considering each domain involved.

The method presented here, also allows to evaluate and prioritize each system component considered and according to its risk, by identifying at the same time, the sources of the risk (i.e., risk category) with more severity and therefore to act, in order to reduce or even mitigate such risk.

By reducing the risks involved with the correspondent system component, it is possible to reduce the system (or product) overall risk, contributing therefore to sustain the innovative ecosystem created to develop such system or product.

Additionally, the use of fuzzy logic to evaluate the different risks involved, have contributed to reduce the uncertainty and ambiguity with such analysis, which characterizes the risk evaluation, strongly dependent on the risk manager perception.

As a future work, the framework developed here, could also consider the individual risk with each activity, belonged to each system component involved on product/system development, by considering not only the risks associated with threats, but also the opportunities involved.

Author Contributions: Conceptualization, R.S. and A.A.; methodology, V.A., J.M.F.C. and A.A.; validation, R.S. and A.D.; formal analysis, A.D. and J.S.; investigation, J.S., A.A., and R.S.; resources, J.M.F.C. and A.A.; data curation, J.S. and A.A.; writing—original draft preparation, A.D. and V.A.; writing—review and editing, J.S. and R.S.; visualization, V.A. and J.S.; supervision, J.M.F.C. and A.A.; project administration, A.A. and R.S. All authors have read and agreed to the published version of the manuscript.

Funding: This research received no external funding.

Conflicts of Interest: The authors declare no conflict of interest.

References

1. Xie, X.; Wang, H. How can open innovation ecosystem modes push product innovation forward? An fsQCA analysis. *J. Bus. Res.* **2020**, *108*, 29–41. [CrossRef]
2. Urze, P.; Abreu, A. *System Thinking to Understand Networked Innovation*; Springer: Berlin/Heidelberg, Germany, 2014; Volume 434, pp. 327–335.
3. Januška, M. Communication as a key factor in Virtual Enterprise paradigm support. In *Innovation and Knowledge Management: A Global Competitive Advantage*; Int. Bus. Inf. Manag. Assoc. (IBIMA): Kuala Lumpur, Malaysia, 2011; pp. 1–9. ISBN 978-0-9821489-5-2.
4. Abreu, A.; Camarinha Matos, L.M. An Approach to Measure Social Capital in Collaborative Networks. In *IFIP International Federation for Information Processing*; Adaptation and Value Creating Collaborative Networks; Camarinha-Matos, L.M., Pereira-Klen, A., Afsarmanesh, H., Eds.; Springer: Berlin/Heidelberg, Germany, 2011; pp. 29–40.
5. Enkel, E.; Gassmann, D.O.; Chesbrough, H. Open R&D and open innovation: Exploring the phenomenon. *Rd Manag.* **2009**, *39*, 311–316. [CrossRef]
6. Granstrand, O.; Holgersson, M. Innovation ecosystems: A conceptual review and a new definition. *Technovation* **2020**. [CrossRef]
7. Arenal, A.; Armuña, C.; Feijoo, C.; Ramos, S.; Xu, Z.; Moreno, A. Innovation ecosystems theory revisited: The case of artificial intelligence in China. *Telecommun. Policy* **2020**, *44*, 101960. [CrossRef]
8. Pereira, L.; Tenera, A.; Wemans, J. Insights on Individual's Risk Perception for Risk Assessment in Web-based Risk Management Tools. *Procedia Technol.* **2013**, *9*, 886–892. [CrossRef]
9. Yin, D.; Ming, X.; Zhang, X. Sustainable and Smart Product Innovation Ecosystem: An integrative status review and future perspectives. *J. Clean. Prod.* **2020**, *274*, 123005. [CrossRef]
10. Santos, R.; Abreu, A.; Anes, V. Developing a Green Product-Based in an Open Innovation Environment. Case Study: Electrical Vehicle. In *Security Education and Critical Infrastructures*; Springer Science and Business Media LLC: Berlin/Heidelberg, Germany, 2019; Volume 568, pp. 115–127.
11. Innocentive. Available online: https://www.innocentive.com/ (accessed on 1 January 2019).
12. Rauter, R.; Globocnik, D.; Perl-Vorbach, E.; Baumgartner, R.J. Open innovation and its effects on economic and sustainability innovation performance. *J. Innov. Knowl.* **2019**, *4*, 226–233. [CrossRef]
13. Adner, R. Ecosystem as structure: An actionable construct for strategy. *J. Manag.* **2017**, *43*, 39–58. [CrossRef]
14. Reynolds, E.B.; Uygun, Y. Strengthening advanced manufacturing innovation ecosystems: The case of Massachusetts. *Technol. Forecast. Soc. Chang.* **2017**, *136*, 178–191. [CrossRef]
15. Gordon, T.J.; Greenspan, D. Chaos and fractals: New tools for technological and social forecasting. *Technol. Forecast. Soc. Chang.* **1988**, *34*, 178–191.
16. Wei, F.; Feng, N.; Yang, S.; Zhao, Q. A conceptual framework of two-stage partner selection in platform-based innovation ecosystems for servitization. *J. Clean. Prod.* **2020**, *262*, 121431. [CrossRef]
17. Robaczewska, J.; Vanhaverbeke, W.; Lorenz, A. Applying open innovation strategies in the context of a regional innovation ecosystem: The case of Janssen Pharmaceuticals. *Glob. Transit.* **2019**, *1*, 120–131. [CrossRef]
18. Xu, G.; Wu, Y.; Minshall, T.; Zhou, Y. Exploring innovation ecosystems across science, technology, and business: A case of 3D printing in China. *Technol. Forecast. Soc. Chang.* **2017**, *136*, 208–221. [CrossRef]
19. Sun, S.L.; Zhang, Y.; Cao, Y.; Dong, J.; Cantwell, J. Enriching innovation ecosystems: The role of government in a university science park. *Glob. Transit.* **2019**, *1*, 104–119. [CrossRef]
20. Madsen, H.L. Business model innovation and the global ecosystem for sustainable development. *J. Clean. Prod.* **2020**, *247*, 119102. [CrossRef]
21. Keizer, J.; Halman, J.; Song, M. From experience: Applying the Risk Diagnosing methodology. *J. Prod. Innov. Manag.* **2002**, *19*, 213–232. [CrossRef]
22. West, J.; Salter, A.; Vanhaverbeke, W.; Chesbrough, H. Open innovation: The next decade. *Res. Policy* **2014**, *43*, 805–811. [CrossRef]
23. Henschel, T. Risk Management Practices of SMEs: Evaluating and Implementing Effective Risk Management Systems. ESV—Erich Schmidt Verlag. (Edwards & Bowen, 2005). 2008. Available online: https://www.researchgate.net/publication/259812085_Risk_Management_Practices_of_SMEs_Evaluating_and_Implementing_Effective_Risk_Management_Systems (accessed on 4 April 2020).

24. Hülle, J.; Kaspar, R.; Möller, K. Analytic network process—An overview of applications in research and practice. *Int. J. Oper. Res.* **2013**, *16*, 172. [CrossRef]
25. Piet, G.J.; Culhane, F.E.; Jongbloed, R.; Robinson, L.; Rumes, B.; Tamis, J. An integrated risk-based assessment of the North Sea to guide ecosystem-based management. *Sci. Total. Environ.* **2019**, *654*, 694–704. [CrossRef]
26. Pereira, L.; Tenera, A.; Bispo, J.; Wemans, J. A Risk Diagnosing Methodology Web-Based Platform for Micro, Small and Medium Businesses: Remarks and Enhancements. In *Biomedical Engineering Systems and Technologies*; Springer Science and Business Media LLC: Berlin/Heidelberg, Germany, 2015; Volume 454, pp. 340–356.
27. PMBOK Guide. A Guide to the Project Management Body of Knowledge: PMBOK Guide. *Int. J. Prod. Res.* **2017**, *53*, 13–14.
28. Abreu, A.; Martins, J.D.M.; Calado, J.M. Fuzzy Logic Model to Support Risk Assessment in Innovation Ecosystems. In Proceedings of the 2018 13th APCA International Conference on Control and Soft Computing (CONTROLO), Azores, Portugal, 17 July 2018; Institute of Electrical and Electronics Engineers (IEEE): Piscataway, NJ, USA, 2018; pp. 104–109.
29. Meyer, W.G. Quantifying risk: Measuring the invisible. In Proceedings of the PMI® Global Congress 2015 EMEA, London, UK, 11–13 May 2015; PMI: Newtown Square, PA, USA, 2015.
30. Grace, M.F.; Leverty, J.T.; Phillips, R.D.; Shimpi, P. The Value of Investing in Enterprise Risk Management. *J. Risk Insur.* **2014**, *82*, 289–316. [CrossRef]
31. Mansor, N. Risk factors affecting new product development (NPD) Performance in small Medium Enterprises (SMES). *Int. J. Rev. Appl. Soc. Sci.* **2016**, *27*, 18–25.
32. Ricardo Ricardo Santos ISEL- Instituto Superior de Engenharia de Lisboa, Instituto Politécnico de Lisboa, University of Aveiro, Portugal GOVCOPP, University of Aveiro; Antonio Antonio Abreu ISEL- Instituto Superior de Engenharia de Lisboa, Instituto Politécnico de Lisboa, CTS Uninova, FCT, Universidade Nova de Lisboa, Portugal; Calado, J.M.; Anes, V. An Approach Based on Fuzzy Logic, to Improve Quality Management on Research and Development Centres. In Proceedings of the 3rd International Conference on Vision, Image and Signal Processing 2019, Vancouver, BC, Canada, 26–28 August 2019; Volume 36, pp. 1–6. [CrossRef]
33. Abreu, A.; Santos, R.; Calado, J.M.; Requeijo, J. A Fuzzy Logic Model to Enhance Quality Management on R&D Units. *Kne Eng.* **2020**, *5*, 285–298. [CrossRef]

 *sustainability*

Article

Technopreneurial Intentions among Bulgarian STEM Students: The Role of University

Desislava Yordanova [1], José António Filipe [2,*] and Manuel Pacheco Coelho [3]

[1]   Faculty of Economics and Business Administration, Sofia University "St. Kliment Ohridski",
     125 Tzarigradsko Shosse Blvd., blok 3, 1113 Sofia, Bulgaria; d_yordanova@feb.uni-sofia.bg
[2]   Departamento de Matemática, ISTA–Escola de Tecnologias e Arquitectura, Iscte-Instituto Universitário de
     Lisboa, Information Sciences, Technologies and Architecture Research Center (ISTAR-IUL),
     Business Research Unit-IUL (BRU-IUL), 1649-026 Lisbon, Portugal
[3]   SOCIUS, ISEG/University of Lisbon, 1649-004 Lisboa, Portugal; coelho@iseg.ulisboa.pt
*   Correspondence: jose.filipe@iscte-iul.pt

Received: 3 July 2020; Accepted: 5 August 2020; Published: 11 August 2020

Abstract: Entrepreneurship, innovation and technology are essential to the economic development of societies. Universities are increasingly involved in creating an internal favourable environment supporting entrepreneurship and innovation. In our work, we aimed to study the role of university for the development of technopreneurial intentions in a sample of Bulgarian STEM (STEM refers to any subjects that fall under the disciplines of science, technology, engineering or mathematics.) students exhibiting entrepreneurial intentions. The empirical findings of the study are in line with previous empirical evidence about the role of university support for entrepreneurial intentions among students; results also show that students in universities with better research in their scientific field of study are more likely to exhibit technopreneurial intention. Determinants of entrepreneurial intentions identified in the literature such as entrepreneurial role models, perceived support from social networks, willingness to take risks and gender may not be relevant specifically for technopreneurial intentions. The results of the study have important practical implications.

Keywords: entrepreneurship; innovation; ecosystem; technopreneurial intentions

---

1. Introduction

The bodies of literature on innovation ecosystems and entrepreneurship ecosystems acknowledge the important role of universities in the development of human capital, knowledge capital and entrepreneurship capital [1]. Markkula and Kune [2] argued that universities may cultivate the spirit of entrepreneurial discovery which drives innovation in regional innovation ecosystems. Researchers, students and companies may discover new opportunities and ideas, experiment, and take risks for the purpose of creating value [2]. Although the research on the role of university for entrepreneurship development has focused predominantly on patent-based activities, technology transfer, and scientific research by academic staff, empirical evidence suggests that students may be as twice as likely as faculty members to start a new venture after graduation and that the ventures founded by recent graduates are not of low quality [3]. The role of entrepreneurship education for fostering entrepreneurial intentions among students is a relatively well-researched topic. A large body of literature on entrepreneurship education demonstrates that entrepreneurial intentions and behaviour among students are positively associated with entrepreneurship education [4–6]. Entrepreneurship among students and recent alumni is also influenced by the university context and there is a need for greater understanding of the nature and determinants of student entrepreneurial systems [7].

Technology entrepreneurship emerges between two major scientific fields: entrepreneurship and technology-based innovation [8], and therefore may play a vital role in enhancing sustainable development. It was acknowledged that entrepreneurship may contribute in diverse ways to the economic, social and environmental dimensions of sustainable development [9–11]. Entrepreneurship and innovation can provide solutions to pressing economic, social and environmental problems [12,13]. Countries with strong performance in terms of innovative entrepreneurship are placed in the top of the rankings for sustainable development, while countries with weaker performance on this indicator have lower scores of sustainable development [13]. In the United Nations plan of actions "The 2030 Agenda for Sustainable Development," entrepreneurship and innovation are expected to contribute to achieving many of the established sustainable development goals and targets including promoting employment and economic growth, developing and diffusing environmentally sustainable technologies, increasing of capacity-building for developing countries, enhancing technology capabilities in different sectors, etc. [12].

Academic research in technology entrepreneurship has generated a complex and interdisciplinary literature which relies on diverse theoretical backgrounds and addresses a wide number of topics [14]. Several important research gaps in this literature have been identified. Shane and Venkataraman [15] highlighted the need for more research on the context for technology entrepreneurship, the process of new technology venture creation and the drivers to create new technology ventures. Mosey, Guerrero and Greenman [16] called for more research exploring the role of various university support measures including entrepreneurship education for the generation of talent and the experience of individuals in relation to technology entrepreneurship. Mosey [17] suggested that universities are an ideal research setting for investigating technology entrepreneurship.

Technology entrepreneurship is viewed as a specific type of entrepreneurship that is essentially different from other types of entrepreneurship such as social entrepreneurship, small business management and self-employment [18]. Lee and Wong [19] demonstrated that scientists and engineers exhibit different entrepreneurial intentions in relation to the type of business they intend to start. However, the available research on entrepreneurial intentions has regarded this construct as largely homogeneous [20]. Liñán and Fayolle [21] identified 409 papers researching entrepreneurial intentions between 2004 and 2013, but only 18 papers address specific types of intentions. Fayolle and Liñán [22] emphasized the need to investigate various entrepreneurial scenarios related to this construct. Given these gaps, the research objective of the present study was to identify university-level and individual-level factors related to the university, which are associated with the likelihood of technopreneurial intentions in a sample of 337 Bulgarian STEM (science, technology, engineering or mathematics) students exhibiting entrepreneurial intentions controlling for other individual-level factors.

The paper is structured as follows. Section 2 contains a review of the literature on technology entrepreneurship and entrepreneurial intentions and hypotheses about the role of university-related factors for the likelihood of technopreneurial intentions among STEM students. Section 3 describes the research methodology adopted in the study. Next, the empirical findings of the study are presented in Section 4. Lastly, Section 5 provides discussion, conclusions, limitations, practical implications and recommendations for future research.

2. Background, Conceptual Model and Hypotheses

2.1. Technology Entrepreneurship: At the Crossroad of Entrepreneurship and Technology-Based Innovation

Technology entrepreneurship is receiving increasing attention among academics, policy makers, entrepreneurs, managers and investors since the first symposium on technology entrepreneurship at Purdue University (USA) in 1970 [18]. During the past decades, academic research in the field of technology entrepreneurship has progressed rapidly in terms of volume, breadth and diversity [14,18], but there is still no consensus among academics about the definition of this concept. Burgelman, Christensen and Wheelwright [23] (p. 3) argued that technological entrepreneurship involves "activities that create new resource combinations to make innovation possible, bringing together the technical and commercial worlds in a profitable way." It includes "exploiting opportunities, and assembling resources around a technological solution" [14] (p. 2) and "the transformation of promising technologies into value" [24] (p. 9). Petti [25] adopted a systemic view of technology entrepreneurship, positing that in, addition to entrepreneurial and managerial components, technology entrepreneurship involves also an environmental component. Technological opportunities emerge and are exploited in a system of interactive actors engaged in various activities related to the development and identification of technologies, recognition of opportunities, product development and business development and creation [20,21]. The environmental component consists of external institutions, relational configurations and resources that condition the development of technology entrepreneurship [24,25]. Garud and Karnøe [26] conceptualized technology entrepreneurship as a distributed agency involving not only technology entrepreneurs themselves but also customers, actors who develop complementary assets and those in institutional forums. The authors emphasized that all these actors may actively participate in the entrepreneurial process to shape the emerging technology in different ways.

Hsu [27] emphasized that technology entrepreneurship, by its nature, is innovation-based. Technology entrepreneurship may be distinguished from mainstream entrepreneurship, because it is concerned with new opportunities stemming particularly from innovation in science, and engineering [8,25] (p. XIII) views technology entrepreneurship as a process incorporating four main sets of activities: the creation or identification of technologies; recognition of opportunities; technology development/application; and creation of a business that utilizes the technology/application developed to generate value. ([18], p. 9) emphasizes that technology entrepreneurship is not about small businesses or general management practices in such businesses owned by engineers or scientists. It is about creating and capturing value through producing and adopting technology ([18], p. 9).

Technology entrepreneurship involves various outcomes including value creation [18,24], value capture [18], creation of new resource combinations [23], creation of new technology-based firms [28–30], creation of (new/innovative) products, services or processes [31,32].

As a research field, technology entrepreneurship involves different levels of analysis including individual level, product/service, business/firm, and the system as a whole [23,31,33]. Burgelman, Christensen and Wheelwright [23] (p. 3) distinguished between individual technology entrepreneurship and corporate technology entrepreneurship. Phan and Foo [33] (p. 2) outlined several levels of analysis: individual level, (scientists/entrepreneurs, venture capitalists and other individuals with contribution to technology entrepreneurship); organizational level (technological teams, structures, processes and interorganizational linkages influencing value creation); and systems level (players in the ecology of value creation including governing factors, industry standards and the economics of geographical locations). Spiegel and Marxt [31] also identified three levels of analysis related to product/service, business/firm and the system as a whole and distinguished between new entrants and existing firms. They argued that both new technology-based firms and incumbent technology-based firms play a significant role for the commercialization of new technologies, but different issues in technology entrepreneurship may receive a different focus in new and existing technology-based firms [31].

A number of differentiating aspects of technology entrepreneurship relative to economics, entrepreneurship and management have been identified [18] (p. 10):

i.   the interdependence between scientific and technological change and the selection and development of new products, assets and their attributes;

ii. the application of technology entrepreneurship to both new and established firms as well as to both small and large firms;

iii. conceptualization of technology entrepreneurship as an investment in a project;

iv.   the interdependence between technology entrepreneurship and the resource-based view of sustainable competitive advantage;

v.   the interdependence between technology entrepreneurship and the theory of the firm.

As a relatively under-researched topic, technology entrepreneurship is seen as a promising area for entrepreneurship research and practice [34].

2.2. Entrepreneurial Intentions and Their Antecedents

It was recognized by researchers in entrepreneurship that new venture formation is a planned behaviour, an intentional act which requires planning how the perceived opportunity will be exploited [35–39]. Since entrepreneurship could be treated as planned behaviour, models of intentions and their antecedents are a useful framework for studying entrepreneurial behaviour [36]. The emphasis on intentions is especially valuable when investigating phenomena that are rare, hard to observe and involving unpredictable time lags such as entrepreneurship [40]. "Thus, a strong intention to start a business should result in an eventual attempt, even if immediate circumstances such as marriage, child bearing, finishing school, a lucrative or rewarding job or earthquakes may dictate a long delay" ([36], p. 414). In this case, entrepreneurial intentions provide understanding about entrepreneurial behaviour without witnessing that behaviour [37].

Intentions-based models allow us to better understand why individuals made certain decisions in the process of new venture creation and to predict entrepreneurial behaviour [36]. Entrepreneurial intentions play a mediating role between potential exogenous antecedents such as demographics, traits, perceived availability of critical resources and situational role beliefs and the act of new venture formation [36]. The dominant models in the research on entrepreneurial intentions are Bird's model [35] for implementing entrepreneurial ideas; Shapero and Sokol's model [41] of the entrepreneurial event; and Ajzen's [42] theory of planned behaviour, and only the last two models have received empirical support in the literature [22].

Shapero and Sokol's model [41] of the entrepreneurial event was developed specifically for the domain of entrepreneurship [36]. For [41] (p. 72), there is an association between the entrepreneurial event and "initiative-taking, consolidation of resources, management, relative autonomy, and risk-taking." These authors suggested that an entrepreneurial event is affected by the social and the cultural environment and several other social factors. Shapero and Sokol's model [41] of the entrepreneurial event reveals how cultural and social variables influence entrepreneurial behaviour [43]. In this model, entrepreneurial intentions are derived from three attitudinal variables: perceived desirability, perceived feasibility and a propensity to act upon opportunities [36]. Perceived desirability is defined as the perceived attractiveness of starting a business, while perceived feasibility is the degree to which a person feels capable of starting a business [36]. Krueger, Reilly and Carsrud [36] commented that the Theory of Planned Behaviour (TPB) and the Shapero and Sokol's model overlap considerably with regard to several elements. Moreover, the models explain similar proportions in the variation of intentions [32].

Attitudinal antecedents of entrepreneurial intentions posited by the TPB and other similar theoretical models have received empirical support. Attitudinal constructs have statistically significant influence on entrepreneurial intentions [36,44–47]. Individuals with close entrepreneurial role models are more likely to exhibit entrepreneurial intentions [43,48,49]. Preferences, perceptions, other cognitive factors and psychological traits tend to be linked to entrepreneurial intentions [44,48,49].

2.3. University as an Entrepreneurship and Innovation Ecosystem

The mission and academic tasks of the university have changed profoundly as a result of two academic revolutions [50]. The first academic revolution led to the adoption of research as another university function in addition to the traditional academic task of teaching, while the second academic revolution added a new academic task related to economic and social development. Graham [51] (p. 1) emphasized that governments "are looking to technological innovation as a driver for national grown and to universities as the incubators of this national capacity." The emergence of the entrepreneurial university is a response to the increasing importance of knowledge in national and regional innovation systems [52]. Universities are increasingly involved in creating an internal favourable environment supporting entrepreneurship and innovation. Entrepreneurial and innovation activities within universities are supported by various support mechanisms, activities, structures, and intermediaries such as technology transfer offices, accelerators, incubators or science parks, etc. [7,53,54]. An entrepreneurial ecosystem for student entrepreneurship within university involves a continuum of activities (early-stage support, accelerators and incubators), variety of entrepreneurs (alumni, students, faculty, post-docs), support actors (internal and external actors) and investors [7]. In addition to direct mechanisms for supporting the transfer of technology from academia to industry, there are also indirect mechanisms supporting entrepreneurial activities via entrepreneurship education [55].

Drawing upon the growing literature about the role of university for the development of entrepreneurship and innovation, the conceptual model adopted in this study (Figure 1) suggests that the development of technopreneurial intentions among STEM students is influenced by the participation in entrepreneurship education, university support with concept and business development and university research excellence. The construct technopreneurial intentions indicates if the respondent thinks s/he will start a technology business [38]. Kraaijenbrink, Bos and Groen [56] differentiated among various types of university support. University support with concept development refers to increasing students' awareness and motivation and providing students with ideas and knowledge needed to start a new business. University support with business development refers to support to start-ups such as provision of financial means. University support with concept development is provided to students in the early stages of the entrepreneurial process, while university support with business development is oriented to start-ups in the later stages of the entrepreneurial process. The construct participation in entrepreneurship education refers to participation in both compulsory and elective entrepreneurship courses. The construct university research excellence refers to the research productivity of university based on its publication and citation records.

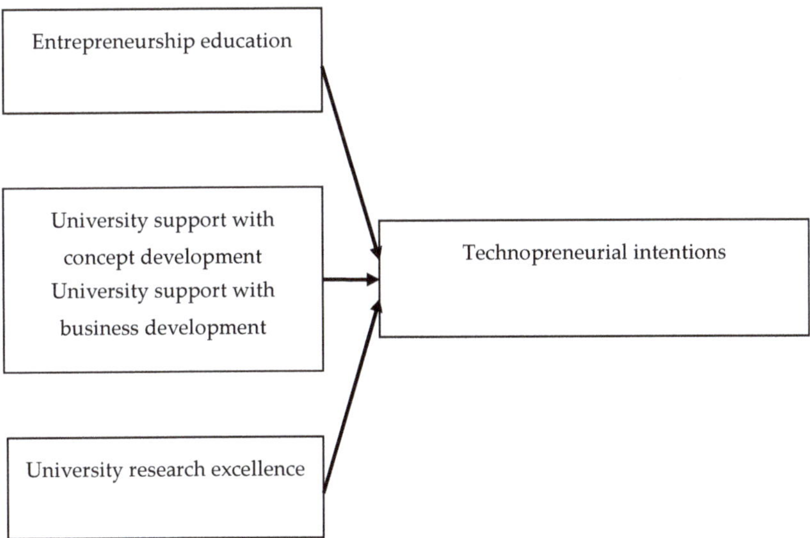

Figure 1. Conceptual model of the role of university for students' technopreneurial intentions.

The research on entrepreneurship education reveals that entrepreneurship education enhances entrepreneurship knowledge and skills, intentions and behaviour. Johannisson [57] suggested that learning from entrepreneurship education occurs at different levels: 1/know-why: values, attitudes, motivation; 2/know-what: knowledge; 3/know-how: abilities and skills; 4/know-when: experience and intuition; 5/know-who: social skills, networks. Falkäng and Alberti [58] identified several positive effects for participants in entrepreneurship education including self-employment and ability to act as an independent operator of a venture; personal and career satisfaction; knowledge and understanding acquisition; skills acquisition; identification of individual potential; changed attitudes. Knowledge is an important antecedent of opportunity identification related to technological change [59]. Martin et al. [4] undertook a quantitative review of the literature and found a significant positive relationship between entrepreneurship education and training and positive perceptions of entrepreneurship, entrepreneurship knowledge and skills, and entrepreneurial intentions. Hahn et al. [60] found that both elective and compulsory entrepreneurship education is conductive for the development of entrepreneurial skills. Rideout and Gray [61] reviewed the empirical research on the outcomes of university-based entrepreneurship education and confirmed the association between entrepreneurship education and psychological outcomes and significant effects of entrepreneurship education on entrepreneurial intentions and behaviour. Dickson et al.'s [5] review of the research on entrepreneurship education published between 1995 and 2006 demonstrated that entrepreneurship education positively influences entrepreneurial intentions. Bae et al. [62] showed that the correlation between entrepreneurship education and entrepreneurial intentions is greater than the correlation between business education and entrepreneurial intentions. Souitaris, Zerbinati and Al-Laham [47] illustrated that entrepreneurship education provides inspiration that raises entrepreneurial attitudes and intentions among STEM students. Technopreneurial learning has a positive impact on the technopreneurial intentions of students [63]. Therefore, we suggest that the participation in entrepreneurship education may lead to the development of technopreneurial intentions instead of other entrepreneurial intentions among STEM students:

Hypothesis 1 (H1). *Participation in entrepreneurship education increases the likelihood of technopreneurial intentions among STEM students.*

The university environment may stimulate and support entrepreneurial intentions and behaviour of students in various ways. The university environment and support for entrepreneurship may positively influence students' perceptions, attitudes and awareness about entrepreneurship as a career option and may encourage them to choose an entrepreneurial career [64]. It can also enhance students' motivation and self-confidence needed for an entrepreneurial career [65]. University may provide access to crucial start-up resources including experts, finance, contacts, know-how, counselling, etc., necessary for exploring entrepreneurial opportunities [64,65]. Concept development support may raise students' awareness of entrepreneurial opportunities, and thus may motivate them and increase their entrepreneurial self-efficacy [66]. Concept development support and business development support play significant roles for students' entrepreneurial self-efficacy and students' entrepreneurial intentions [56,65–67]. University entrepreneurial support is associated with students' entrepreneurship engagement [68].

STEM students exhibit certain characteristics such as aversion to financial and bureaucracy issues, lack of knowledge in the management field, lack of market focus and encounter certain obstacles related to lack of entrepreneurial skills, lack of skills for start-up development, and lack of financial resources for new product development and registering a patent [69]. Therefore, STEM students seek support from universities, entrepreneurship centres and incubators to overcome these specific obstacles [69]. University support factors positively affect the entrepreneurial intentions of technical students [44] and the creation of technology spin-offs versus nontechnology spin-offs [70]. Therefore, we suggest that the presence of concept development support and business development support within the university may help STEM students to develop technopreneurial intentions:

Hypothesis 2 (H2). *University support with concept development increases the likelihood of technopreneurial intentions among STEM students.*

Hypothesis 3 (H3). *University support with business development increases the likelihood of technopreneurial intentions among STEM students.*

Technological advances based on scientific research represent an important source of technological opportunities [71]. Intellectual eminence of the university is an important factor that increases new firm formation [72]. Universities with higher levels of research productivity tend to develop higher entrepreneurial activity [73]. Rasmussen et al. [74] stressed that the extent of entrepreneurial activity and the types of the established new ventures are affected by university-level factors. Shane [59] discovered that the recognition of business opportunities stemming from technological innovations is determined by prior education and experience. Research-intense universities are a source of knowledge spillovers, creating entrepreneurial opportunities for exploiting new knowledge by starting high-tech firms [75–79]. Research institutions are an important source of knowledge for exploring new business opportunities among younger and well-educated founders of high-tech firms [80]. Technological knowledge from universities has a significant role also for the growth of academic spin-offs [81]. By fulfilling their educational function, academic institutions are transferring new knowledge to their students [77], who may exploit it for starting new high-tech ventures. Bonaccorsi et al. [82] found that the creation of knowledge-intensive firms is positively affected by new knowledge generated by high-quality universities, while low-quality universities have little effect on knowledge-intensive entrepreneurship. Research excellence has a significant positive impact on the discovery of technological opportunities [71]. Beyhan and Findik [83] reported that research excellence of the university is associated with the odds of new technology venture creation among students and new graduates. They concluded that high-quality research-oriented universities provide a supportive environment for exploration and exploitation of new knowledge and development of entrepreneurial competences among students [83]. A university's research is seen as an important resource for aspiring entrepreneurs, which can result in new knowledge and technologies that can be eventually commercialized [64]. Singhry [84] reported that technological capabilities positively influence technopreneurial intentions. STEM students in universities characterized with research excellence may be more likely to have access to advanced knowledge, skills, the latest advancements and innovations, new technologies, and therefore may be more likely to perceive opportunities for technology entrepreneurship and to develop technopreneurial intentions. Such access to the latest scientific knowledge, technologies and inventions may also increase the perceptions of desirability and feasibility for technology entrepreneurship and eventually technopreneurial intentions versus other entrepreneurial intentions among STEM students. Therefore, we suggest that:

Hypothesis 4 (H4). *University research excellence increases the likelihood of technopreneurial intentions among STEM students.*

3. Research Methodology

This study utilized a database about technology entrepreneurship among Bulgarian science and engineering students. The database was collected using a cross-sectional survey among science and engineering students in Bulgarian universities to investigate the influence of entrepreneurship education on technopreneurial intentions and their antecedents. Science and engineering students were selected for the empirical analysis because they exhibit the potential to start technology ventures [47]. The survey was administrated to students in science or engineering majors in 15 Bulgarian universities in 2015 and 2016. The selected universities are located in Sofia and several other major Bulgarian cities. Rectors, deans and department heads in all Bulgarian universities providing bachelor and master programs in science and engineering study fields were contacted and invited to participate in the survey. Only 14 public universities and one private university expressed a consent to participate in the survey. A quota sampling technique was adopted for data collection based on the total number of science and engineering students enrolled in each university, which was obtained from the Ministry of Education and Science. The database has the same proportions of science and engineering students from the different universities as the entire population of science and engineering students enrolled in the selected 15 universities in the respective year in which the survey was conducted. The database includes 1061 students. Some of them are technology business owners or have started and manage a technology business for their employers. Others are in a process of starting a technology business for themselves or their employers. A third group of students exhibited technopreneurial intentions. The rest of the respondents did not report entrepreneurial intentions. The students in the database are enrolled in various study fields such as communication and computer equipment, informatics and computer sciences, biotechnologies, electrical engineering, electronics and automation, power engineering, transport, navigation and aviation, general engineering, biological sciences, chemical sciences, chemical technologies, architecture, construction and geodesy, earth sciences, minerals prospecting, extraction and processing, mechanics, energetics and food technologies. Students enrolled in the study fields of social sciences, humanities, medicine, national security and military science were excluded from the survey. The questionnaire used in the study includes questions which requested a broad array of information related to demographic characteristics of respondents, entrepreneurial intentions, attitudes toward entrepreneurship, entrepreneurial behaviour, and entrepreneurship education. In a short introduction about the aims of the survey in the questionnaire, technology entrepreneurship is defined as the creation of a new technology-based business, while technology-based business is described as a business whose products or services depend largely on the application of scientific or technological knowledge [85]. A pilot study was conducted among 15 students (8 males and 7 females) in order to pre-test the initial version of the questionnaire. Due to comments from some students, minor changes were introduced in some questions. With the approval and cooperation of rectors, deans, department heads and lecturers in 15 Bulgarian universities, a questionnaire was distributed during class sessions. Students were informed that participation in the survey was voluntary and that the questionnaires were only for research purposes. In the instructions to respondents with regard to the filling-in procedure, they were advised that the instrument should be completed anonymously and that it was important to answer all questions. In order to secure a high response rate, to monitor respondents while they were answering the questionnaire, and to be able to answer further questions from respondents, the author was present during the data collection in most occasions. If missing information was identified when the respondents were submitting the filled questionnaires, the respondents were politely asked to complete it. Questionnaires with missing answers were removed from the database and data collection from each university continued until the required quota fixed by the researcher was fulfilled. The sample for this study was extracted from the database and is composed of 337 STEM students who exhibit entrepreneurial intentions. They reported that they think they will start a business [38,86]. All respondents reporting entrepreneurial intentions were asked if they think they will start a technology business [38,86]. In the sample, 299 students reported that they think they will start a technology business [38,86]. They are neither business owners nor in a process of starting a business. More than 79% of the respondents in the sample are undergraduate students. Female students represented less than 41% of the sample. The great majority of the respondents are full-time students. Only 21.7% of the sample is composed of part-time students.

The dependent variable in this study is technopreneurial intentions (TECH_INT). It takes value 1 if the respondent thinks s/he will start a technology business [38,86], and value 0 otherwise. As all respondents reported that they think they will start a business, the respondent who do not think they will start a technology business tend to exhibit entrepreneurial intentions to start a nontechnology business. The study employed several independent university-level and individual-level variables. The university-level variables include university research excellence, university support with concept development, and university support with business development, while the individual-level variable is participation in entrepreneurship education. The variable participation in entrepreneurship education (ENTR_EDU) takes value 1 if the respondent was/is enrolled in an entrepreneurship course within the university, and value 0 if not. The variable university research excellence (RES_EXC) is measured with the H-index of the university in the scientific field of study of the respondent in SCOPUS.

Two perceptual measures of entrepreneurship development support provided by the university were adopted. It was suggested that, although universities can support entrepreneurship with objective measures, it is important to take into account the extent to which such objective measures can influence students by evaluating students' perceptions of entrepreneurship development support provided by the university [56]. The variable university support with concept development (CONCEPT_DEV) is measured with a four-item 7-point Likert scale developed by [56], which reveals students' perceptions of the university support for concept development beyond teaching. The scale exhibits high reliability (Cronbach's alpha = 0.910). The variable university support with business development (BUS_DEV) is measured with a three-item 7-point Likert scale developed by [56], which reveals students' perceptions of the university support for starting and developing a new business (Cronbach's alpha = 0.669). Both scales exhibit acceptable reliability which exceeds the minimum acceptable level of 0.6 [87].

The study controlled for individual differences related to the support from social networks, positive entrepreneurial role models, gender, willingness to take risks, perceived new technology venture feasibility and desirability, previous experience in a technology company. These variables have been identified as significant predictors of entrepreneurial intentions in the literature [21,36,44]. The variable perceived new technology venture feasibility (FEASIBILITY_TE) reveals new technology venture perceived feasibility and is measured with an index composed by 4 items measured on a 7-point Likert scale [36,38,88]. The Cronbach's alpha of the scale is 0.597, which is close to the minimum acceptable level of 0.6 [87]. The variable perceived new technology venture desirability (DESIRABILITY_TE) indicates how desirable technology entrepreneurship is for respondents. It is measured with an index composed by 3 items measured on a 7-point Likert scale [36,38,88]. The Cronbach's alpha of the scale is 0.719. The variable willingness to take risks (RISK) indicates students' willingness to take risks and is measured with 4 items adopted from [89] (Cronbach's alpha = 0.736). All scales exhibit acceptable reliability, which is roughly equal to or higher than the minimum acceptable level of 0.6 [87]. The variable positive entrepreneurial role models (ROLE_MODELS) takes value 1 if the respondent has at least one entrepreneur among parents, relatives, friends or acquaintances whose success gave her/him a positive impression of entrepreneurship [63], and value 0 otherwise. The variable support from social networks (SOC_NET_SUP) takes value 1 if the respondent can count on support from family, partner, friends and acquaintances if s/he becomes an entrepreneur after his/her studies [64], and 0 otherwise. The variable gender (GENDER) takes value 1 if the respondent is male and value 0 if she is female. The variable previous experience in a technology company (TECH_EXP) takes value 1 if the respondent has previous professional experience in a technology company, and value 0 otherwise. Table 1 contains the description of the variables used in the study.

Table 1. Description of the variables used in the study.

Variables	Description
Dependent variable	
TECH_EXP	1 = the respondent thinks s/he will start a technology business; 0 = otherwise [34,68]
Independent variables	
(ENTR_EDU)	1 = the respondent was/is enrolled in an entrepreneurship course within the university; 0 = otherwise
RES_EXC	the value of the H-index of the university in the scientific field of study of the respondent in SCOPUS
CONCEPT_DEV	a four-item 7-point Likert scale developed by [65]
BUS_DEV	a three-item 7-point Likert scale developed by [65]
Control variable	
FEASIBILITY_TE	a four-item 7-point Likert scale [36,38,88]

Table 1. *Cont.*

Variables	Description
DESIRABILITY_TE	a three-item 7-point Likert scale [36,38,88]
RISK	a four-item 7-point Likert scale [89]
ROLE_MODELS	1 = the respondent has at least one entrepreneur among parents, relatives, friends or acquaintances whose success gave her/him a positive impression of entrepreneurship; 0 = otherwise [63]
SOC_NET_SUP	1 = the respondent can count on support from family, partner, friends and acquaintances if s/he becomes an entrepreneur after his/her studies; 0 otherwise [63]
GENDER	1 = the respondent is male; 0 = the respondent is female
TECH_EXP	1 = the respondent has previous professional experience in a technology company; 0 = otherwise

Taking into account the objectives of this study and the properties of the data, we applied a binary logistic regression for data analysis [90]. A binary logistic regression was employed to deal explicitly with the dependent variable TECH_INT, which is a binary variable [90]. The logistic regression is a more robust method since according to [87,90,91]:

- the dependent variable needs not to be normally distributed;
- logistic regression does not assume a linear relationship between the dependent and the independent variables;
- the dependent variable needs not to be homoscedastic for each level of the independent variable(s);
- normally distributed error terms are not assumed;
- independent variables can be categorical;
- it does not require independent variables to be interval or unbounded.

The application of nonparametric techniques is adequate when the independent variables are predominantly categorical. The use of the maximum likelihood approach is recommended when sample selection bias is possible [92]. The data analysis was performed using version 25 of IBM SPSS Statistics.

4. Empirical Evidence

Correlations between independent and control variables employed in the study are relatively modest (not exceeding 0.3) (Table 2). Male students in our sample are more likely to have previous professional experience in a technology company. Female students are more likely to be enrolled in an entrepreneurship course and to report that they can count on support from family, partner, friends and acquaintances if they become entrepreneurs after their studies. Students in universities with better research exhibit lower willingness to take risks. Students with lower willingness to take risks tend to perceive lower new technology venture desirability. Students with more positive perceptions of concept and business development support report higher perceived new technology venture feasibility. Students with higher perceived new technology venture desirability also exhibit higher perceived new technology venture feasibility.

A binary logistic regression exploring the effects of university-level factors and individual-level factors related to the university on the likelihood of technopreneurial intentions is presented in Table 3. The model is significant at 99% confidence level according to Chi-square statistics. Therefore, the null hypothesis that all coefficients (except the constant) are zero can be rejected. The variance of inflation factor (VIF) is used as a check on multicollinearity. The VIFs for the variables in the regression indicate that there are no serious multicollinearity problems, as they are all within the acceptable limits (less than 4). The overall predictive ability of the model to correctly classify students by their technopreneurial intentions is 89.1%. The variable RES_EXC significantly and positively influences the odds of technopreneurial intentions. Students in research-oriented universities are more likely to exhibit technopreneurial intentions than other entrepreneurial intentions ($p < 0.01$). Hypothesis 4 cannot be rejected. The variable BUS_DEV positively affects the odds of technopreneurial intentions. Students perceiving greater university support for business development are more likely to exhibit technopreneurial intentions ($p < 0.10$). These results provide weak support for Hypothesis 3. The coefficients of the variables ENTR_EDU and CONCEPT_DEV are not significant. Participation in entrepreneurship education and perceptions of university support with concept development are not related to the likelihood of technopreneurial intentions. Hypotheses 1 and 2 can be rejected.

Table 2. Correlation matrix.

	Variables	1	2	3	4	5	6	7	8	9	10
1	GENDER	1									
2	RISK	0.02	1								
3	TECH_EXP	0.16 **	0.00	1							
4	ROLE_MODELS	0.05	−0.05	−0.06	1						
5	SOC_NET_SUP	−0.14 *	0.02	0.04	−0.07	1					
6	ENTR_EDU	−0.22 **	−0.1	0.05	−0.02	0.06	1				
7	CONCEPT_DEV	−0.12 *	0.01	0.05	−0.04	−0.00	0.15 **	1			
8	BUS_DEV	−0.08	0.1	0.00	−0.05	−0.04	0.09	0.62 **	1		
9	RES_EXC	−0.07	−0.14 **	−0.02	0.13 *	0.03	0.07	−0.05	−0.06	1	
10	DESIRABILITY_TE	−0.01	−0.14 **	−0.05	0.06	0.08	0.02	0.1	0.05	0.01	1
11	FEASIBILITY_TE	−0.06	−0.01	−0.08	−0.007	0.09	0.1	0.22 **	0.20 **	−0.12 *	0.30 **

* Correlation is significant at 0.01. ** Correlation is significant at 0.05.

Table 3. Results from a binary logistic regression (dependent variable = TECH_INT) [a].

Variable	B	S.E.	Wald	Sig.	Exp(B)
RES_EXC	0.134	0.045	8.693	0.003	1.143
ENTR_EDU	0.475	0.435	1.190	0.275	1.607
CONCEPT_DEV	−0.020	0.039	0.267	0.605	0.980
BUS_DEV	0.104	0.058	3.160	0.075	1.109
FEASIBILITY_TE	0.012	0.043	0.081	0.776	1.012
DESIRABILITY_TE	0.206	0.057	13.206	0.000	1.229
GENDER	0.513	0.417	1.514	0.219	1.670
RISK	−0.003	0.038	0.008	0.929	0.997
ROLE_MODELS	0.394	0.410	0.922	0.337	1.483
SOC_NET_SUP	−0.032	0.479	0.004	0.947	0.969
TECH_EXP	−0.369	0.342	1.166	0.280	0.691
−2 Log likelihood	196.848				
Nagelkerke R-Square	0.217				
Model Chi-square	38.628				
Overall correct predictions	89.1%				

[a] A constant has been estimated, but it is not included in the table.

The control variables DESITABILITY_TE exert a significant influence on the likelihood of technopreneurial intentions ($p < 0.001$). Students with greater new technology venture desirability are more likely to have technopreneurial intentions. The variables GENDER, SOC_NET_SUP,

ROLE_MODELS, RISK, FEASIBILITY_TE and EXP_TE have no significant effects on the dependent variable.

5. Discussion and Conclusions

In a knowledge-based economy, entrepreneurship is a driving force for economic development [93]. In particular, technology entrepreneurship, which is at the crossroad of entrepreneurship and technology-based innovation [8], may play a vital role in enhancing sustainable development. Universities need to operate more entrepreneurially and to create favourable conditions for entrepreneurship and innovation at all its levels [94]. To date, a large number of studies have focused on students' entrepreneurial intentions to gain understanding of their future entrepreneurial behaviour. This study examined factors related to the university at both the individual and organizational levels, which are associated with the incidence of technopreneurial intentions using a sample of 337 Bulgarian STEM students with entrepreneurial intentions. The proposed conceptual model posits that participation in entrepreneurship education, university support for entrepreneurship and university research excellence determine the formation of technopreneurial intentions among STEM students. The study controlled for other individual-level differences related to the support from social networks, positive entrepreneurial role models, gender, willingness to take risks, perceived new technology venture feasibility and desirability, previous experience in a technology company.

Our findings reveal that the university environment influences the odds of technopreneurial intentions. University research excellence has a strong positive impact on the odds of technopreneurial intentions, while university support with business development weakly and positively affects the odds of technopreneurial intentions. University support with concept development has no significant effect on the likelihood of technopreneurial intentions. These findings are in line with previous empirical evidence about the role of university support in entrepreneurial intentions among students [56,65–67,95] and reinforce the direct contextual link with entrepreneurial intentions [96]. It seems that students with entrepreneurial intentions in universities providing greater business development support may gain confidence for technopreneurial venturing [56]. The weak statistical significance of the effect of the university support with business development and the lack of statistical significance of the effect of the university support with concept development on the odds of students' techopreneurial intentions may be explained by the fact that these variables do not account particularly for the university support with concept development for technology entrepreneurship and university support for technology business development. University support specifically targeted at encouraging technology entrepreneurship may influence more strongly the formation of technopreneurial intentions among STEM students.

As expected, students in universities with better research in their scientific field of study are more likely to exhibit technopreneurial intention, which is in contradiction with previous findings that university research orientation negatively influences students' self-employment intentions [64] and is in line with previous research demonstrating that technology entrepreneurs often have a "research" background [97]. Although Bulgarian universities encounter significant internal and external barriers to their transformation into entrepreneurial universities [98], our study demonstrates that Bulgarian universities offering educational programs in STEM fields have the capacity to foster technology entrepreneurship among students and thus to contribute to entrepreneurial activity and innovation in the economy. Our empirical results support previous claims that universities may play an important role in the regional entrepreneurship and innovation ecosystem [99].

Surprisingly, participation in entrepreneurship education is not related to a higher likelihood of technopreneurial intentions. The lack of significant effect of entrepreneurship education on the dependent variable may be due to the general nature of entrepreneurship education in the studied universities. Such general entrepreneurship educations may not be suited to the specific needs of STEM students and may not be able to encourage the formation of technopreneurial intentions. These results contradict previous evidence about the role of entrepreneurship education for entrepreneurial intentions [61,99–101] and highlight the need to tailor entrepreneurship education to suit a particular target group [102]. They raise the question about what contents and teaching methods are used in entrepreneurship courses for STEM students at the studied universities and to what extent they are conductive for the formation of technopreneurial intentions. Past research results conceive of entrepreneurial learning as an experiential process [103]. Hence, entrepreneurship education should go beyond promoting awareness and providing knowledge [104] and should focus on real-world experience, action, and reflection to increase technopreneurial intentions and eventually to enhance entrepreneurial performance [105]. Entrepreneurship courses and programs specifically targeted at STEM students should help them to overcome specific constraints and obstacles such as lack of entrepreneurial skills and skills for start-up development, aversion to financial and bureaucracy issues, lack of knowledge in the management field, and lack of market focus [69].

The control variables employed in the study do not have a statistically significant effect on the odds of technopreneurial intentions, except for the perceived desirability for technology entrepreneurship. Previous research reveals the importance of entrepreneurial role models for the formation of entrepreneurial intentions [106]. The key functions of entrepreneurial role models include inspiration and motivation to get started, increased entrepreneurial self-efficacy, learning by example, and learning by support [106]. The present study demonstrates that entrepreneurial role models are not more conductive for the formation of technopreneurial intentions. Students with higher perceived desirability for technology entrepreneurship exhibit higher odds of technopreneurial intentions. Perceived feasibility has no effect on the odds of technopreneurial intentions versus other intentions. Although willingness to take risk is an important trait related to entrepreneurship [107,108], the evidence suggests that willingness to take risk does not affect the odds of technopreneurial intentions in our sample. An important finding is that gender is not associated with the odds of technopreneurial intentions. It seems that factors that constrain women's entrepreneurial intentions compared to men do not affect in the same way the formation of technopreneurial intentions among women. Contrary to prior claims that stronger social network support may provide needed resources for entrepreneurship, and thus is associated with entrepreneurial choice [64], this study did not find a link between social network support and the odds of technopreneurial intentions. Although previous research suggests that specific knowledge acquired through previous work experience facilitates the discovery of new opportunities [58], this study demonstrates that previous experience in a technology company is not related to the likelihood of technopreneurial intentions. These results imply that determinants of entrepreneurial intentions identified in the literature such as entrepreneurial role models, perceived support from social networks, willingness to take risks and gender may not be relevant specifically for technopreneurial intentions.

Our cross-level study on university-level and individual-level factors related to the university associated with the incidence of technopreneurial intentions among Bulgarian STEM students extends the literature in several ways. Our study contributes to the literature on the role of university for student entrepreneurship by providing evidence about the influence of various factors related to the university at two levels (university level and individual level), which helps to understand the context in which some pre-venture processes occur [64]. Given the increasing importance of student entrepreneurship, this research enhances our understanding about how universities can better stimulate and support the entrepreneurial efforts of students. The findings of this research highlight the importance of the university research excellence and university support for entrepreneurship development. The presented empirical evidence reinforces the view of university as a driver of entrepreneurship and innovation [1] and contributes to the debate about the role of institutional underpinnings in national innovation systems [109]. This study extends the literature on technology entrepreneurship, which lacks significant research attention to pre-venture processes, by identifying determinants of technopreneurial intentions. It contributes to the literature on entrepreneurial intentions by underscoring the need not to regard this construct as homogeneous and to direct research attention to different types of entrepreneurial intentions.

The study has several limitations that should be discussed explicitly before outlining recommendations for future research. A major limitation of this research is the cross-sectional study design, which does not allow to make inferences about causal relationships. The use of cross-sectional data does not allow to control for unobservable fixed effects that may affect both the dependent and the independent variables such as student's capabilities [84]. The study does not address possible endogeneity in the proposed regression model. The size of the sample is not large, and the variability of the dependent variable is rather low. The data were collected through a self-reported survey, and thus may be subjected to cognitive biases and errors. The findings may be influenced by specific features of the Bulgarian cultural and institutional environment, and therefore may not be applicable to other countries and contexts.

The reported empirical findings and the outlined limitations of the study open several new directions for future research. First, the presented conceptual model should be modified to include the constructs university support with concept development for technology entrepreneurship and university support with technology business development. Thus, future research should test the modified conceptual model in large, representative samples drawn from Bulgaria and other countries, in order to determine to what extent our findings are applicable to both the Bulgarian context and other contexts. Second, future research should investigate the role of entrepreneurship education for fostering entrepreneurial intentions and behaviour among STEM students and should provide greater understanding of the impact of various educational variables related to entrepreneurship education such as teaching methods, learning outcomes, educator teaching beliefs, etc., on students' technopreneurial intentions and behaviour. Third, future research should devote greater attention to different types of entrepreneurial intentions and should clarify which determinants of entrepreneurial intentions identified in the literature are relevant particularly for technopreneurial intentions. Forth, future research should provide greater understanding about the role of universities for knowledge and technology transfer, focusing, in particular, on student involvement in the early stages of the entrepreneurial process. Future studies need to investigate the impact of the university ecosystem for entrepreneurship and innovation and to identify effective entrepreneurship support services and activities that stimulate students' technopreneurial intentions. The identification of cases of success can help to formulate best practices that can guide universities into developing their internal ecosystems for entrepreneurship and innovation. Future research with longitudinal design is necessary to provide insights about university-related factors at the individual and organizational levels that contribute to technopreneurial intentions.

The findings from this study have important policy implications. Policy makers involved in national and regional development issues should be aware of the key role universities may play in regional entrepreneurship and innovation systems. University managers and policy makers concerned with enhancing the 'third mission' of universities should be aware of the important role of university research excellence and specific university support for stimulating technopreneurial intentions and behaviour of students. They should adopt the view of universities as entrepreneurship and innovation ecosystems and make specific efforts to promote diverse entrepreneurship and innovation support structures and activities within universities for stimulating and supporting technology entrepreneurship, involve a variety of entrepreneurs such as students, faculty members and alumni, and attract internal and external support actors and investors. Policies aimed at increasing and improving research output in STEM fields of Bulgarian universities may contribute to fostering entrepreneurship and innovation at the regional level in the long term [77]. Educators teaching entrepreneurship to STEM students should adopt more tailored approaches to entrepreneurship education that suit the specific characteristics, needs and obstacles experienced by this specific group of students. There is a need for introducing contents and teaching methods which allow for gaining real-world experience and promote experimentation, risk-taking, action and reflection in entrepreneurship courses and programs for STEM students.

Author Contributions: Conceptualization, D.Y., J.A.F. and M.P.C.; Methodology, D.Y., J.A.F. and M.P.C.; Software, D.Y. and J.A.F.; Validation, D.Y., J.A.F. and M.P.C.; Formal analysis, D.Y., J.A.F. and M.P.C.; Investigation, D.Y., J.A.F. and M.P.C.; Resources, D.Y., J.A.F. and M.P.C.; Data curation, D.Y. and J.A.F.; Writing—review and editing, D.Y. and J.A.F.; Visualization, D.Y. and M.P.C.; Supervision, D.Y.; Project administration, D.Y. and M.P.C.; Funding acquisition, D.Y. and M.P.C. All authors have read and agreed to the published version of the manuscript.

Funding: The first author acknowledges financial support from the Scientific Research Fund of Sofia University "St. Kliment Ohridski" (contract N 80-10-232/13.05.2020). This research was also funded by FCT (Fundação para a Ciência e Tecnologia), grant number UIDB/04521/2020.

Acknowledgments: The authors thank the Sofia University "St. Kliment Ohridski", the Iscte-Instituto Universitário de Lisboa, ISTAR-IUL and ISEG/Universidade de Lisboa for their support, which are acknowledged. They finally thank three anonymous reviewers for their valuable comments.

Conflicts of Interest: The authors declare no conflict of interest.

References

1. Guerrero, M.; Urbano, D.; Fayolle, A.; Klofsten, M.; Mian, S. Entrepreneurial universities: Emerging models in the new social and economic landscape. *Small Bus. Econ.* **2016**, *47*, 551–563. [CrossRef]
2. Markkula, M.; Kune, H. Making smart regions smarter: Smart specialization and the role of universities in regional innovation ecosystems. *Technol. Innov. Manag. Rev.* **2015**, *5*, 7–15. [CrossRef]
3. Åstebro, T.; Bazzazian, N.; Braguinsky, S. Startups by recent university graduates and their faculty: Implications for university entrepreneurship policy. *Res. Policy* **2012**, *41*, 663–677. [CrossRef]
4. Martin, B.C.; McNally, J.J.; Kay, M.J. Examining the formation of human capital in entrepreneurship: A meta-analysis of entrepreneurship education outcomes. *J. Bus. Ventur.* **2013**, *28*, 211–224. [CrossRef]
5. Dickson, P.H.; Solomon, G.T.; Weaver, K.M. Entrepreneurial selection and success: Does education matter? *J. Small Bus. Enterp. Dev.* **2008**, *15*, 239–258. [CrossRef]
6. Pittaway, L.; Cope, J. Entrepreneurship Education: A Systematic Review of the Evidence. *Int. Small Bus. J.* **2007**, *25*, 479–510. [CrossRef]
7. Wright, M.; Siegel, D.S.; Mustar, P. An emerging ecosystem for student start-ups. *J. Technol. Transf.* **2017**, *42*, 909–922. [CrossRef]
8. Beckman, C.M.; Eisenhardt, K.; Kotha, S.; Meyer, A.; Rajagopalan, N. The role of the entrepreneur in technology entrepreneurship. *Strateg. Entrep. J.* **2012**, *6*, 203–206. [CrossRef]
9. UN General Assembly. *Entrepreneurship for Sustainable Development*; Resolution Adopted by the General Assembly on 21 December 2016; United Nations: New York, NY, USA, 2016.
10. Hall, J.K.; Daneke, G.A.; Lenox, M.J. Sustainable development and entrepreneurship: Past contributions and future directions. *J. Bus. Ventur.* **2010**, *25*, 439–448. [CrossRef]
11. Filser, M.; Kraus, S.; Roig-Tierno, N.; Kailer, N.; Fischer, U. Entrepreneurship as catalyst for sustainable development: Opening the black box. *Sustainability* **2019**, *11*, 4503. [CrossRef]
12. UN General Assembly. *Transforming our World: The 2030 Agenda for Sustainable Development*; Resolution Adopted by the General Assembly on 25 September 2015; United Nations: New York, NY, USA, 2015.
13. Kardos, M. The Relationship between Entrepreneurship, Innovation and Sustainable Development. Research on European Union Countries. *Procedia Econ. Financ.* **2012**, *3*, 1030–1035. [CrossRef]
14. Ratinho, T.; Harms, R.; Walsh, S. Structuring the Technology Entrepreneurship publication landscape: Making sense out of chaos. *Technol. Forecast. Soc. Chang.* **2015**, *100*, 168–175. [CrossRef]
15. Shane, S.; Venkataraman, S. Guest editors' introduction to the special issue on technology entrepreneurship. *Res. Policy* **2003**, *32*, 181–184. [CrossRef]
16. Mosey, S.; Guerrero, M.; Greenman, A. Technology entrepreneurship research opportunities: Insights from across Europe. *J. Technol. Transf.* **2017**, *42*, 1–9. [CrossRef]
17. Mosey, S. Teaching and research opportunities in technology entrepreneurship. *Technovation* **2016**, *57*, 43–44. [CrossRef]
18. Bailetti, T. Technology entrepreneurship: Overview, definition, and distinctive aspects. *Technol. Innov. Manag. Rev.* **2012**, *2*, 5–12. [CrossRef]
19. Lee, S.H.; Wong, P.K. An exploratory study of technopreneurial intentions: A career anchor perspective. *J. Bus. Ventur.* **2004**, *19*, 7–28. [CrossRef]

20. Carey, T.A.; Flanagan, D.J.; Palmer, T.B. An examination of university student entrepreneurial intentions by type of venture. *J. Dev. Entrep.* **2010**, *15*, 503–517. [CrossRef]

21. Liñán, F.; Fayolle, A. A systematic literature review on entrepreneurial intentions: Citation, thematic analyses, and research agenda. *Int. Entrep. Manag. J.* **2015**, *11*, 907–933. [CrossRef]

22. Fayolle, A.; Liñán, F. The future of research on entrepreneurial intentions. *J. Bus. Res.* **2014**, *67*, 663–666. [CrossRef]

23. Burgelman, R.A.; Christensen, C.M.; Wheelwright, S.C. *Strategic Management of Technology and Innovation*; International Edition; McGraw Hill: New York, NY, USA, 2004.

24. Petti, C.; Zhang, S. Factors influencing technological entrepreneurship capabilities: Towards an integrated research framework for Chinese enterprises. *J. Technol. Manag. China* **2011**, *6*, 7–25. [CrossRef]

25. Petti, C. *Cases in Technological Entrepreneurship: Converting Ideas into Value*; Edward Elgar: Northampton, MA, USA, 2009.

26. Garud, R.; Karnøe, P. Bricolage versus breakthrough: Distributed and embedded agency in technology entrepreneurship. *Res. Policy* **2003**, *32*, 277–300. [CrossRef]

27. Hsu, D.H. *Technology-Based Entrepreneurship. Handbook of Technology and Innovation Management*; Blackwell Publishers, Ltd.: Oxford, UK, 2008; pp. 367–387.

28. Gans, J.S.; Stern, S. The Product Market and the Market for Ideas: Commercialization Strategies for Technology Entrepreneurs. *Res. Policy* **2003**, *32*, 333–350. [CrossRef]

29. Antoncic, B.; Prodan, I. Alliances, corporate technological entrepreneurship and firm performance: Testing a model on manufacturing firms. *Technovation* **2008**, *28*, 257–265. [CrossRef]

30. Colovic, A.; Lamotte, O. Technological Environment and Technology Entrepreneurship: A Cross-Country Analysis. *Creat. Innov. Manag.* **2015**, *24*, 617–628. [CrossRef]

31. Spiegel, M.; Marxt, C. Defining Technology Entrepreneurship. In Proceedings of the 2011 IEEE International Conference on Industrial Engineering and Engineering Management, Singapore, 6–9 December 2011; IEEE: Piscataway, NJ, USA, 2011; pp. 1623–1627.

32. Pathak, S.; Xavier-Oliveira, E.; Laplume, A.O. Influence of intellectual property, foreign investment, and technological adoption on technology entrepreneurship. *J. Bus. Res.* **2013**, *66*, 2090–2101. [CrossRef]

33. Phan, P.H.; Foo, M. Technological entrepreneurship in emerging regions. *J. Bus. Ventur.* **2004**, *19*, 1–5. [CrossRef]

34. McPhee, C.; Bailetti, T. Editorial: Technology Entrepreneurship. *Technol. Innov. Manag. Rev.* **2012**, *2*, 3–4.

35. Bird, B. Implementing entrepreneurial ideas: The case for intention. *Acad. Manag. Rev.* **1988**, *13*, 442–453. [CrossRef]

36. Krueger, N.; Reilly, M.; Carsrud, A. Competing models of entrepreneurial intentions. *J. Bus. Ventur.* **2000**, *15*, 411–432. [CrossRef]

37. Krueger, N.F.; Carsrud, A.L. Entrepreneurial intentions: Applying the theory of planned behaviour. *Entrep. Reg. Dev.* **1993**, *5*, 315–330. [CrossRef]

38. Krueger, N.F. The impact of prior entrepreneurship exposure on perception of new venture feasibility and desirability. *Entrep. Theory Pract.* **1993**, *18*, 5–21. [CrossRef]

39. Kolvereid, L. Organizational employment versus self-employment: Reasons for career choice intentions. *Entrep. Theory Pract.* **1996**, *20*, 23–31. [CrossRef]

40. MacMillan, I.; Katz, J. Idiosyncratic milieus of entrepreneurship research: The need for comprehensive theories. *J. Bus. Ventur.* **1992**, *7*, 1–8. [CrossRef]

41. Shapero, A.; Sokol, L. The social dimensions of entrepreneurship. In *Encylclopedia of Entrepreneurship*; Kent, C.A., Sexton, D.L., Vesper, K.H., Eds.; Prentice-Hall: Englewood Cliffs, NJ, USA, 1982; pp. 72–90.

42. Ajzen, I. Theory of planned behaviour. *Organ. Behav. Hum. Decis. Process.* **1991**, *50*, 179–211. [CrossRef]

43. Veciana, J.M.; Aponte, M.; Urbano, D. University students' attitudes towards entrepreneurship: A two countries comparison. *Int. Entrep. Manag. J.* **2005**, *1*, 165–182. [CrossRef]

44. Lüthje, C.; Franke, N. The 'making' of an entrepreneur: Testing a model of entrepreneurial intent among engineering students at MIT. *R&D Manag.* **2003**, *33*, 135–147.

45. Martínez-González, J.A.; Kobylinska, U.; García-Rodríguez, F.J.; Nazarko, L. Antecedents of Entrepreneurial Intention among Young People: Model and Regional Evidence. *Sustainability* **2019**, *11*, 6993. [CrossRef]

46. Tkachev, A.; Kolvereid, L. Self-employment intentions among Russian students. *Entrep. Reg. Dev.* **1999**, *11*, 269–280. [CrossRef]

47.   Souitaris, V.; Zerbinati, S.; Al-Laham, A. Do entrepreneurship programmes raise entrepreneurial intention of science and engineering students? The effect of learning, inspiration and resources. *J. Bus. Ventur.* **2007**, *22*, 566–591. [CrossRef]

48.   Hmieleski, K.M.; Corbett, C. Proclivity for improvisation as a predictor of entrepreneurial intentions. *J. Small Bus. Manag.* **2006**, *44*, 45–63. [CrossRef]

49.   De Pillis, E.; Reardon, K. The influence of personality traits and persuasive messages on entrepreneurial intention: A cross-cultural comparison. *Career Dev. Int.* **2007**, *12*, 382–396. [CrossRef]

50.   Etzkowitz, H. Research groups as 'quasi firms': The invention of the entrepreneurial university. *Res. Policy* **2003**, *32*, 109–121. [CrossRef]

51.   Graham, R. Creating University-Based Entrepreneurial Ecosystems: Evidence from Emerging World Leaders. MIT Skoltech Initiative. 2014. Available online: http://www.rhgraham.org/RHG/Recent_publications_files/MIT%3ASkoltech%20entrepreneurial%20ecosystems%20report%202014%20_1.pdf (accessed on 15 June 2020).

52.   Etzkowitz, H.; Webster, A.; Gebhardt, C.; Terra, B.R.C. The future of the university and the university of the future: Evolution of ivory tower to entrepreneurial paradigm. *Res. Policy* **2000**, *29*, 313–330. [CrossRef]

53.   Rothaermel, F.T.; Agung, S.D.; Jiang, L. University Entrepreneurship: Taxonomy of the literature. *Ind. Corp. Chang.* **2007**, *16*, 691–791. [CrossRef]

54.   Breznitz, S.M.; Zhang, Q. Fostering the growth of student start-ups from university accelerators: An entrepreneurial ecosystem perspective. *Ind. Corp. Chang.* **2019**, *28*, 855–873. [CrossRef]

55.   Guenther, J.; Wagner, K. Getting out of the ivory tower–new perspectives on the entrepreneurial university. *Eur. J. Int. Manag.* **2008**, *2*, 400–417. [CrossRef]

56.   Kraaijenbrink, J.; Bos, G.; Groen, A. What do students think of the entrepreneurial support given by their universities? *Int. J. Entrep. Small Bus.* **2010**, *9*, 110–125. [CrossRef]

57.   Johannisson, B. University training for entrepreneurship: A Swedish approach. *Entrep. Reg. Dev.* **1991**, *3*, 67–82. [CrossRef]

58.   Falkäng, J.; Alberti, F. The assessment of entrepreneurship education. *Ind. High. Educ.* **2000**, *14*, 101–108. [CrossRef]

59.   Shane, S. Prior knowledge and the discovery of entrepreneurial opportunities. *Organ. Sci.* **2000**, *11*, 448–469. [CrossRef]

60.   Hahn, D.; Minola, T.; Bosio, G.; Cassia, L. The impact of entrepreneurship education on university students' entrepreneurial skills: A family embeddedness perspective. *Small Bus. Econ.* **2020**, *55*, 257–282. [CrossRef]

61.   Rideout, E.C.; Gray, D.O. Does entrepreneurship education really work? A review and methodological critique of the empirical literature on the effects of university-based entrepreneurship education. *J. Small Bus. Manag.* **2013**, *51*, 329–351. [CrossRef]

62.   Bae, T.J.; Qian, S.; Miao, C.; Fiet, J.O. The relationship between entrepreneurship education and entrepreneurial intentions: A meta-analytic review. *Entrep. Theory Pract.* **2014**, *38*, 217–254. [CrossRef]

63.   Hoque, A.S.M.M.; Awang, Z.; Siddiqui, B.A. Technopreneurial intention among university students of business courses in Malaysia: A structural equation modeling. *Int. J. Entrep. Small Medium Enterp.* **2017**, *4*, 1–16.

64.   Walter, S.G.; Parboteeah, K.P.; Walter, A. University departments and self-employment intentions of business students: A cross-level analysis. *Entrep. Theory Pract.* **2013**, *37*, 175–200. [CrossRef]

65.   Trivedi, R. Does university play significant role in shaping entrepreneurial intention? A cross-country comparative analysis. *J. Small Bus. Enterp. Dev.* **2016**, *23*, 790–811. [CrossRef]

66.   Mustafa, M.J.; Hernandez, E.; Mahon, C.; Chee, L.K. Entrepreneurial intentions of university students in an emerging economy: The influence of university support and proactive personality on students' entrepreneurial intention. *J. Entrep. Emerg. Econ.* **2016**, *8*, 162–179. [CrossRef]

67.   Saeed, S.; Yousafzai, S.Y.; Yani De Soriano, M.; Muffatto, M. The role of perceived university support in the formation of students' entrepreneurial intention. *J. Small Bus. Manag.* **2015**, *53*, 1127–1145. [CrossRef]

68.   Minola, T.; Donina, D.; Meoli, M. Students climbing the entrepreneurial ladder: Does university internationalization pay off? *Small Bus. Econ.* **2016**, *47*, 565–587. [CrossRef]

69.   Paço, A.; Ferreira, J.; Raposo, M. How to foster young scientists' entrepreneurial spirit? *Int. J. Entrep.* **2017**, *21*, 47–60.

70. Meoli, M.; Vismara, S. University support and the creation of technology and non-technology academic spin-offs. *Small Bus. Econ.* **2016**, *47*, 345–362. [CrossRef]
71. D'Este, P.; Mahdi, S.; Neely, A.; Rentocchini, F. Inventors and entrepreneurs in academia: What types of skills and experience matter? *Technovation* **2012**, *32*, 293–303. [CrossRef]
72. Di Gregorio, D.; Shane, S. Why do some universities generate more start-ups than others? *Res. Policy* **2003**, *32*, 209–227. [CrossRef]
73. Van Looy, B.; Landoni, P.; Callaert, J.; Van Pottelsberghe, B.; Sapsalis, E.; Debackere, K. Entrepreneurial effectiveness of European universities: An empirical assessment of antecedents and trade-offs. *Res. Policy* **2011**, *40*, 553–564. [CrossRef]
74. Rasmussen, E.; Mosey, S.; Wright, M. The influence of university departments on the evolution of entrepreneurial competencies in spin-off ventures. *Res. Policy* **2014**, *43*, 92–106. [CrossRef]
75. Acs, Z.J.; Audretsch, D.B.; Lehmann, E.E. The knowledge spillover theory of entrepreneurship. *Small Bus. Econ.* **2013**, *41*, 757–774. [CrossRef]
76. Audretsch, D.B.; Lehmann, E.E. Does the knowledge spillover theory of entrepreneurship hold for regions? *Res. Policy* **2005**, *34*, 1191–1202. [CrossRef]
77. Fritsch, M.; Aamoucke, R. Regional public research, higher education, and innovative start-ups: An empirical investigation. *Small Bus. Econ.* **2013**, *41*, 865–885. [CrossRef]
78. Ghio, N.; Guerini, M.; Lehmann, E.E.; Rossi-Lamastra, C. The emergence of the knowledge spillover theory of entrepreneurship. *Small Bus. Econ.* **2015**, *44*, 1–18. [CrossRef]
79. Ghio, N.; Guerini, M.; Rossi-Lamastra, C. The creation of high-tech ventures in entrepreneurial ecosystems: Exploring the interactions among university knowledge, cooperative banks, and individual attitudes. *Small Bus. Econ.* **2019**, *52*, 523–543. [CrossRef]
80. Amoroso, S.; Audretsch, D.B.; Link, A.N. Sources of knowledge used by entrepreneurial firms in the European high-tech sector. *Eurasian Bus. Rev.* **2018**, *8*, 55–70. [CrossRef]
81. Barbosa, N.; Faria, A.P. The effect of entrepreneurial origin on firms' performance: The case of Portuguese academic spinoffs. *Ind. Corp. Chang.* **2020**, *29*, 25–42. [CrossRef]
82. Bonaccorsi, A.; Colombo, M.G.; Guerini, M.; Rossi-Lamastra, C. The impact of local and external university knowledge on the creation of knowledge-intensive firms: Evidence from the Italian case. *Small Bus. Econ.* **2014**, *43*, 261–287. [CrossRef]
83. Beyhan, B.; Findik, D. Student and graduate entrepreneurship: Ambidextrous universities create more nascent entrepreneurs. *J. Technol. Transf.* **2018**, *43*, 1346–1374. [CrossRef]
84. Singhry, H.B. The effect of technology entrepreneurial capabilities on technopreneurial intention of nascent graduates. *Eur. J. Bus. Manag.* **2015**, *7*, 8–20.
85. Allen, J.C. *Starting a Technology Business*; Pitman: London, UK, 1992.
86. Peterman, N.; Kennedy, J. Enterprise education: Influencing students' perceptions of entrepreneurship. *Entrep. Theory Pract.* **2003**, *8628*, 129–144. [CrossRef]
87. Hair, F.J.; Anderson, E.R.; Tathan, L.R.; Black, C. *Multivariate Data Analysis*, 5th ed; Prentice Hall: Upper Saddle River, NJ, USA, 1998.
88. Drennan, J.; Kennedy, J.; Renfrow, P. Impact of childhood experiences on the development of entrepreneurial intentions. *Int. J. Entrep. Innov.* **2005**, *6*, 231–238. [CrossRef]
89. Gomez-Mejia, L.R.; Balkin, D.B. Effectiveness of individual and aggregate compensation strategies. *Ind. Relat.* **1989**, *28*, 431–445. [CrossRef]
90. Greene, W.H. *Econometric Analysis*; Prentice Hall: Upper Saddle River, NJ, USA, 1997.
91. Maddala, G. *Limited Dependent and Qualitative Variables in Econometrics*; Cambridge University Press: New York, NY, USA, 1983.
92. Nawata, K. Estimation of sample selection bias models by the maximum likelihood estimator and Heckman's two-step estimator. *Econ. Lett.* **1994**, *45*, 33–40. [CrossRef]
93. Audretsch, D.; Thurik, R. What's new about the new economy? Sources of growth in the managed and entrepreneurial economies. *Ind. Corp. Chang.* **2001**, *10*, 267–315. [CrossRef]
94. Kirby, D.A. Entrepreneurship education: Can business schools meet the challenge? International Entrepreneurship Education. In *International Entrepreneurship Education: Issues and Newness*; Fayolle, A., Klandt, H., Eds.; Edward Elgar Publishing: Northampton, MA, USA, 2006; pp. 35–54.

95. Arrighetti, A.; Caricati, L.; Landini, F.; Monacelli, N. Entrepreneurial intention in the time of crisis: A field study. *Int. J. Entrep. Behav. Res.* **2016**, *22*, 835–859. [CrossRef]
96. Nabi, G.; Liñán, F. Considering business start-up in recession time. *Int. J. Entrep. Behav. Res.* **2013**, *19*, 633–655. [CrossRef]
97. Jones-Evans, D. A typology of technology-based entrepreneurs: A model based on previous occupational background. *Int. J. Entrep. Behav. Res.* **1995**, *1*, 26–47. [CrossRef]
98. Yordanova, D.; Filipe, J.A. Towards entrepreneurial universities: Barriers, facilitators, and best practices in Bulgarian and Portuguese universities. *Int. J. Econ. Bus. Adm.* **2019**, *7*, 213–227. [CrossRef]
99. Portuguez Castro, M.; Ross Scheede, C.; Gómez Zermeño, M.G. The Impact of Higher Education on Entrepreneurship and the Innovation Ecosystem: A Case Study in Mexico. *Sustainability* **2019**, *11*, 5597. [CrossRef]
100. Asimakopoulos, G.; Hernández, V.; Peña Miguel, J. Entrepreneurial intention of engineering students: The role of social norms and entrepreneurial self-efficacy. *Sustainability* **2019**, *11*, 4314. [CrossRef]
101. Vodă, A.I.; Florea, N. Impact of personality traits and entrepreneurship education on entrepreneurial intentions of business and engineering students. *Sustainability* **2019**, *11*, 1192. [CrossRef]
102. Maresch, D.; Harms, R.; Kailer, N.; Wimmer-Wurm, B. The impact of entrepreneurship education on the entrepreneurial intention of students in science and engineering versus business studies university programs. *Technol. Forecast. Soc. Chang.* **2016**, *104*, 172–179. [CrossRef]
103. Secundo, G.; Del Vecchio, P.; Schiuma, G.; Passiante, G. Activating entrepreneurial learning processes for transforming university students' idea into entrepreneurial practices. *Int. J. Entrep. Behav. Res.* **2017**, *23*, 465–485. [CrossRef]
104. Ahmed, T.; Chandran, V.G.R.; Klobas, J. Specialized entrepreneurship education: Does it really matter? Fresh evidence from Pakistan. *Int. J. Entrep. Behav. Res.* **2017**, *23*, 4–19. [CrossRef]
105. Kassean, H.; Vanevenhoven, J.; Liguori, E.; Winkel, D.E. Entrepreneurship education: A need for reflection, real-world experience and action. *Int. J. Entrep. Behav. Res.* **2015**, *21*, 690–708. [CrossRef]
106. Bosma, N.; Hessels, J.; Schutjens, V.; Van Praag, M.; Verheul, I. Entrepreneurship and role models. *J. Econ. Psychol.* **2012**, *33*, 410–424. [CrossRef]
107. Stewart, W.H., Jr.; Roth, P.L. Risk propensity differences between entrepreneurs and managers: A meta-analytic review. *J. Appl. Psychol.* **2001**, *86*, 145–153. [CrossRef]
108. Zhao, H.; Seibert, S.E.; Lumpkin, G.T. The relationship of personality to entrepreneurial intentions and performance: A meta-analytic review. *J. Manag.* **2010**, *36*, 381–404. [CrossRef]
109. Audretsch, D.B.; Seitz, N.; Rouch, K.M. Tolerance and innovation: The role of institutional and social trust. *Eurasian Bus. Rev.* **2018**, *8*, 71–92. [CrossRef]

Review

Exploring Mission-Oriented Innovation Ecosystems for Sustainability: Towards a Literature-Based Typology

Malte Jütting [1,2]

[1] Fraunhofer IAO, Fraunhofer Institute for Industrial Engineering, Center for Responsible Research and Innovation (CeRRI), 13086 Berlin, Germany; malte.juetting@iao.fraunhofer.de

[2] Faculty of Economics and Management, Technical University Berlin, 10587 Berlin, Germany

Received: 10 July 2020; Accepted: 14 August 2020; Published: 18 August 2020

Abstract: With mounting sustainability challenges, policy makers have embraced the idea of transformative, mission-oriented innovation policies, to direct innovation objectives towards the 'grand challenges' in recent years. Against this backdrop, the discourse on innovation ecosystems, bringing together actors from science, industry, government and civil society for collaborative research and innovation, has increasingly gained traction. Yet, their role and architectural set-up in a sustainability context remains rather poorly understood. Complementing a systematic literature review with methods of bibliometric analysis and typology building, this paper introduces a typology of mission-oriented innovation ecosystems. It finds that, depending on the type of mission they are trying to address, ecosystems differ, with both a view to the actors involved, and the specific role taken on by them throughout the innovation process. In particular, it points to an increasingly important role of the state for realizing system-level transformations, underlines the importance of civil society involvement, and highlights research organizations' need to adapt to new requirements.

Keywords: grand challenges; innovation ecosystems; mission-oriented innovation; SDGs; sustainable innovation; systematic literature review; SLR; transformative innovation; typology

---

1. Introduction

How do we ensure health—both mental and physical—in an aging society? How do we improve quality of life within and beyond urbanizing spaces? How do we proceed towards a less resource-intensive circular economy? How do we arrive at net carbon zero whilst progressing towards gender equality, ending hunger and reducing global disparities? Confronted with these and other grand challenges of our time, research and innovation faces new demands by politics and society. Having been scrutinized with a view to its ability to deliver originality, prove commercially viable and thus contribute to economic growth over the last decades, research and innovation is now expected to not only provide prosperity, but also deliver solutions to society's most pressing questions. As pointed out by Schot and Steinmueller [1], this is reflected in a tremendous change in science, technology and innovation (STI) policy, from World War II to the present day. "Entering a new era of innovation policy" [2] (p. 76), policy makers, such as the European Commission (EC), are embracing the idea of 'transformative' innovation policies that direct innovation objectives at sustainability challenges. Drawing on the ideas of Mazzucato [3–5], the grand challenges, such as the United Nation's Sustainable Development Goals (SDGs), are translated into concrete, achievable steps—so-called missions, to be achieved through research and innovation. The European Union's programs 'Horizon 2020' [6] and 'Horizon Europe' [7] or Germany's 'High-Tech Strategy 2025' [8] are only three among many examples illustrating this mission-oriented turn in innovation policy making over the last years. Given their complex and often wicked nature, these missions cannot be solved by one actor—be it politics, science, industry or civil society—alone. Instead, "they require different sectors to come together in new ways" [4] (p. 3). Against this backdrop, a shift from linear, largely bilateral innovation processes to non-linear, collaborative forms, involving multiple actors can be observed [9] (p. 4). Yet, the question of how these collaborations for mission-oriented, sustainable innovation might work exactly remains underexplored and unanswered so far.

Describing and analyzing processes of joint value creation towards a common innovation objective, the innovation ecosystem literature [10–14] offers a promising frame of reference for scrutinizing these collaborations (see Section 2). Thereby, the particular focus of the innovation ecosystem perspective complements related research on mission-oriented innovation policy [3,5], sustainability transitions [15,16] and broader system perspectives therein [2,17]. However, an initial scoping review reveals that, while the discourse on innovation ecosystems has grown significantly over the last years, their role in light of today's sustainability challenges is yet to be explored [18].

Addressing said research gap, this paper is the first to systematically explore and conceptualize innovation ecosystems from a mission-oriented perspective. As suggested by Liu and Stephens [18], a systematic literature review (SLR) based on Tranfield et al. [19] is conducted for this purpose. To this end, the paper is guided by the following research questions:

1. What characterizes mission-oriented innovation ecosystems as an emerging area of research and how does it relate to similar research strands?
2. How can mission-oriented innovation ecosystems and their sub-types be conceptualized?
3. What should an agenda for future research look like to improve our understanding of mission-oriented innovation ecosystems?

To answer these questions, the remainder of this paper is structured as follows: Section 2 provides the conceptual background by defining innovation ecosystems as key frame of reference. While the Section 3 describes the methodological approach consisting of systematic literature review, bibliometric analysis and typology building, the Section 4 displays the paper's findings. In accordance with the guiding research questions, it explores the research field as such before conceptualizing mission-oriented innovation ecosystems (MOIEs) by developing a typology of sub-types. Before concluding, Section 5 discusses the paper's contribution, reflects upon its limitations, and sketches an agenda for further research.

2. Conceptual Background: Innovation Ecosystems

Aiming to conceptualize mission-oriented innovation ecosystems and analytically distinguish them from similar approaches requires examining the innovation ecosystem concept and related constructs. However, as numerous SLRs on innovation ecosystems have been conducted for precisely this purpose over the last three years [10–14,20], only a brief summary is provided.

With mounting popularity of the innovation ecosystem concept over the last years, countless—and at times contradictory—definitions have been developed. Some of these are vague and terms have been used interchangeably, however, all definitions identify joint innovation activities, collaboration towards a common goal and value co-creation as central to the innovation ecosystem concept [20]. As summarized by Suominen et al. [13] (p. 16)

> The central literature on ecosystems highlights the capability of ecosystems to create value larger than that which any single organization could create. This process of value creation requires co-evolution, where actors enhance each other's capabilities, but also governance of the dynamics of the endeavour

Drawing on Carayannis and Campbell [21], ecosystems bring together actors from different sectors, such as politics, science, industry or civil society, and are best described by the principles of co-creation, co-evolution, co-specialization and co-opetition. It is these key ideas and principles that form this paper's underlying ecosystem definition (for a more comprehensive definition, see [11]).

The innovation ecosystem concept has developed alongside a number of related concepts. A conclusive differentiation between these is difficult and inconsistent across publications, causing fragmentation and ambiguity in the research field [20,22]. Drawing on previous studies [9–14,20,23,24], this paper distinguishes between discourses on the innovation ecosystem, entrepreneurial ecosystem, innovation system, innovation network, open innovation, supply chain collaboration and public-private partnerships (see Table 1). Here, these concepts' applications in the sphere of sustainable innovation are of particular interest (see e.g., 'sustainable entrepreneurship ecosystems' [25]; 'dedicated' [17] resp. 'mission-oriented' [2] innovation systems; 'sustainable innovation networks' [26] or 'Public-Private Partnerships for Sustainable Development' [27]).

Table 1. Ecosystems and related concepts (author's own elaboration based on [9–14,20,23,24]).

Concept	Description	Analytical Focus	Authors
Innovation Ecosystem	Innovation ecosystems can be understood as collaborative multi-sector arrangements, in which organizations co-create value to achieve a shared innovation target.	• joint value-creation • co-creation, co-evolution, co-specialization, co-opetition • ecosystem orchestration and governance	• Adner 2006 [28] • Adner/Kapoor 2010 [29] • Gomes et al. 2018 [10]
Entrepreneurial Ecosystem	Centered on the entrepreneur, the concept shines a light on the environment and institutions indispensable for nurturing new ventures, hereby generating economic wealth and new jobs.	• entrepreneurial activity, start-up success and wealth creation • necessary support structures and intermediaries to nurture new ventures	• Autio et al. 2014 [30] • Isenberg 2010 [31] • Spigel 2017 [32]
Innovation System	The systems view on innovation analyzes the different factors (political, economic, social, institutional, organizational) shaping development, diffusion and use of innovations.	• contextual, institutional and organizational factors influencing creation and use of innovation • national, regional, sectoral or technological reference points	• Edquist 1997 [33] • Lundvall 1992 [34] • Nelson 1993 [35]
Innovation Network	Focusing on the relational ties and networks between different organizations (mostly firms), the network perspective does not look at their co-evolution.	• within- and across-industry collaboration of (mostly) businesses • direct and indirect ties between actors	• DeBresson/Amesse 1991 [36] • Powell et al. 1996 [37] • Powell et al. 2005 [38]
Open Innovation	The open innovation concept describes the use of external knowledge sources from the perspective of a focal firm. Innovation ecosystems could be understood as a specific form of open innovation arrangements.	• firms incorporating external knowledge sources, e.g., users • usually bilateral partnerships between a focal firm and its partners	• Bogers 2018 [39] • Chesbrough 2003 [40] • von Hippel 1986 [41]

Table 1. *Cont.*

Concept	Description	Analytical Focus	Authors
Supply Chain Collaboration	Closely related to the concept of open innovation, supply chain collaboration describes firms' cooperation with supply chain partners to improve products or optimize resource use.	▪ firms collaborating along the supply chain to refine offers or optimize resource use ▪ effectiveness vs. efficiency	▪ Cao/Zhang 2011 [42] ▪ Holweg et al. 2005 [43] ▪ Stank et al. 2001 [44]
Public-Private Partnership (PPP)	Coming from a new public management tradition, the PPP literature analyzes prospects and pitfalls of governmental collaboration with private sector partners for joint solution resp. service provision.	▪ collaboration between governments and businesses and/or NGOs ▪ effectiveness and efficiency of solution/service provision (not necessarily innovation)	▪ Bäckstrand 2006 [45] ▪ Glasbergen et al. 2007 [46] ▪ Pattberg et al. 2013 [27]

Throughout the fast-growing literature on innovation ecosystems, different research themes, pertaining to concept [11–13], life cycle [47], evaluation [48], as well as actors and roles [23,49], have emerged. However, as of now, innovation ecosystems are predominantly understood as a vehicle for technology, product or service development, and only rarely discussed within the context of mission-oriented innovation policy or the sustainability agenda [18].

3. Methodology

In accordance with the research interest, this paper builds upon a systematic literature review (SLR) [19,50,51]. To complement the analysis, it also borrows methodological approaches from bibliometric analysis [52,53] and typology building [54,55]. The next sections outline the methodological approach in detail, while its potential limitations will be resumed and discussed in this paper's discussion section.

3.1. Systematic Literature Review

Systematic reviews can be defined as "a methodology for rigorous and extensive synthesis of research findings, using transparent, explicit and replicable procedures" [56] (p. 237). Providing a detailed analysis about what is or else what is not known so far and how the existing knowledge was empirically generated [57] (p. 3), this approach is particularly appropriate in light of the research aim's exploratory nature. As suggested by Tranfield et al. [19] (pp. 214–215), as well as Arksey and O'Malley [58], an initial scoping review of existing SLRs in the field of innovation ecosystems was conducted, to ensure the planned review's feasibility and relevance. After having identified and screened 35 SLRs, 13 were reviewed in detail. While numerous authors have puzzled their heads over the conceptualization of 'standard'—commercially oriented—innovation ecosystems, particularly as distinct from similar concepts such as business and entrepreneurial ecosystems or innovation systems (see Section 2) [10,11,13,14], little systematic attention has been paid to them against the backdrop of today's mission-oriented innovation policy. Having specified and refined the research aim based on the scoping review, the actual SLR proceeded in four steps (also see Figure 1).

The first step aimed to identify all relevant papers. Following a configurative rather than an aggregative approach, as suggested by the research interest [59] (p. 63), it is not stringently required to identify every single study, but to ensure a sufficient breadth of different concepts [60] (p. 100). Nevertheless, three distinct strategies were implemented to identify relevant literature irrespective of its disciplinary background. Through a keyword-based search, four key bibliographic databases for academic literature—Scopus, Web of Science, EBSCO (among others also including the sub-database EconLit) and ProQuest—were systematically screened. Accounting for potential conceptual overlaps (as illustrated in Section 2), the two keyword blocks—innovation ecosystems and mission-oriented innovation—comprised a broad range of synonyms derived from previous studies in both fields (see Appendix A Table A1 for review protocol) [10,12,61]. The search was limited to results published from 1987 onwards, as "the idea of 'innovating for sustainability' can be traced back" [62] (p. 6) to the publication of the Brundlandt report in the same year. Establishing a base line in terms of scientific quality, only articles published in peer-reviewed scientific journals were included in the search. To complement the keyword-based database search, the ten most relevant journals, as identified by previous systematic reviews on innovation ecosystems and sustainable innovation [10,12–14,61], were screened for additional papers (see Appendix A Table A1 for details). Lastly, reference snowballing was applied throughout the review process, to further identify promising articles. The complete search was conducted in April 2020, with the manual search repeated on 15th of June 2020 to ensure the inclusion of recently published articles. After removing duplicates and cleaning the data (e.g., excluding records with central bibliographic information missing), this tripartite search strategy yielded 1984 results.

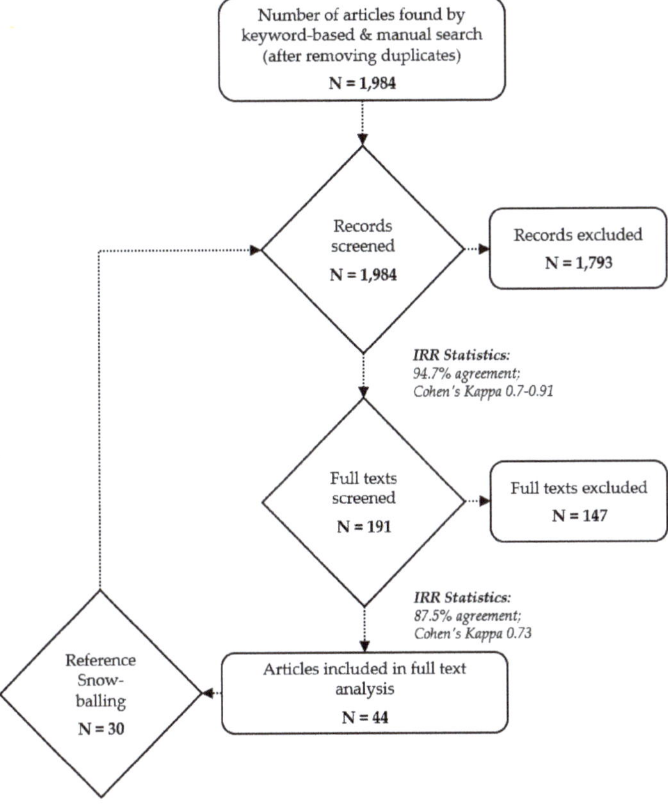

Figure 1. PRISMA flow chart of research process (author's own illustration).

In a second step, the corresponding 1984 abstracts were screened, with view to the following inclusion criteria deducted from the research interest: (a) whether the article deals with a case of mission-oriented innovation, and if yes, (b) whether a specific type of collaboration within this innovation process is scrutinized. For the latter purpose, abstracts were categorized based on their reference to innovation ecosystems or related constructs. Drawing on the conceptual considerations outlined in this paper's second section, articles either explicitly referring to innovation ecosystems or describing similar characteristics (such as a collaboration's multi-sector, co-creative nature) without using the term were included in the subsequent full text screening. The abstract screening procedure was performed by two researchers, who—after completing two training rounds with smaller samples—independently screened all 1984 abstracts. The overall agreement rate of

94.7% in their inclusion-exclusion decisions and the Kappa values of 0.70 (=substantial agreement) for articles describing innovation ecosystem-like characteristics more vaguely respectively 0.91 (=almost perfect agreement), for those explicitly using innovation ecosystem language indicate a high inter-rater reliability of the screening process [63,64]. After resolving remaining disagreements through discussion, 191 records were selected for the subsequent full text screening. While the rate of excluded articles may appear to be high initially, it may be explained by the wide and general nature of the set of keywords employed throughout the search process. Although the majority of articles was ultimately rendered irrelevant to this paper's particular research interest, this approach was necessary due to the ecosystem concept's fuzziness (see Section 2).

The systematic review's third step comprised the full text screening of the 191 remaining records against the backdrop of refined inclusion criteria, as outlined above. Again, two researchers independently conducted the screening process in parallel. After two training rounds with smaller samples, an overall agreement of 87.5% resp., a Cohen's Kappa of 0.73 (=substantial agreement) was achieved. Remaining differences were resolved through discussion. Excluding 26 articles due to technical issues (e.g., because they were not accessible or despite the English search written in another language), and 121 articles on content-related grounds, 44 articles (see Appendix B Table A2 for full list) were selected for subsequent bibliometric (see Section 3.2) and full text analyses (see Section 3.3).

3.2. Bibliometric Analysis

In order to explore and map the research landscape around the emerging field of mission-oriented innovation ecosystems, a bibliometric analysis similar to that performed on innovation ecosystems in general by Gomes et al. [10] was conducted. Coined by Pritchard [53], the term bibliometric analysis refers to the quantitative study of a certain stock of literature and its respective bibliographies. Bibliometric analyses may adopt two different foci: choosing a rather descriptive productivity count, the publication behavior of authors and journals is examined over time. On the other hand, a more evaluative usage count considers the number of citations of a certain author or journal [52].

In accordance with this paper's research aim, different techniques from bibliometric analysis were applied at two stages of the systematic literature review. Firstly, productivity counting was used to explore differences in publication behavior between different research streams, based on the categorization made during the abstract screening ($n = 603$ records, in which some form of collaboration for mission-oriented innovation is mentioned). Secondly, the 44 full texts eligible for content analysis were assessed both from productivity as well as usage points of view. For this purpose, their complete bibliographic datasets were retrieved from Scopus (available for 41 of the 44 articles) and analyzed using the software VOSviewer (Nees Jan van Eck and Ludo Waltman at Leiden University's Centre for Science and Technology Studies (CWTS), version 1.6.15) [65].

3.3. Content Analysis and Typology Building

Concluding the review process, those 44 papers identified as relevant throughout the rigorous procedure outlined above were analyzed in detail, incorporating elements of framework and narrative synthesis [66,67]. For this purpose, each article was read several times, in order to scrutinize its content. Using MaxQDA as a software tool for qualitative data analysis, the papers' key features were coded based on a codebook derived from the underlying research questions [68] (p. 124). Meta-data, such as information on articles' research aims, their methodological approaches or perspective, were gathered. Furthermore, studies' findings were analyzed, paying particular attention to the conceptualization of ecosystems and their characteristics, e.g., target focus, architectural set-up or collaboration formats.

Aiming to explore and ultimately systemize the differences between several forms of mission-oriented innovation ecosystems, typology building was used as methodological approach drawing on Kluge [54,55], Collier et al. [69] and Schütz [49]. In a first step, an ecosystem's sustainability focus (on one of the three dimensions people, prosperity, planet resp. an integrated understanding) and its solution approach (focus on single solutions vs. focus on system level transformation) were set as the relevant dimensions, opening up the property space. Following Kluge's [54,55] four step procedure, cases were then grouped, empirical regularities analyzed, and types constructed. Additional variables which had been coded as part of content analysis but not used for type construction were then utilized to characterize the constructed types in greater detail.

4. Findings

In line with the two-fold research interest, the systematic review's findings are presented in two separate sections. While the first section focuses on mapping the research landscape by highlighting selected descriptive statistics and results from bibliometric analysis, the subsequent section dives deeper into the content-related analysis, ultimately introducing a typology of mission-oriented innovation ecosystems.

4.1. Overview of the Research Field

Drawing on the conceptual differentiation between innovation ecosystems and related concepts outlined above, abstracts were categorized according to their reference to one of these constructs during the screening process. In total, 603 of the 1984 abstracts focus on some form of collaborative, mission-oriented resp. sustainable innovation. Throughout those articles, the leading discourses are open innovation (20.3%), supply chain (13.3%) and entrepreneurial ecosystems (11.1%). The concept of interest—innovation ecosystems—is present in 7.5% of the 603 abstracts, while remaining constructs figure at roughly 5% (see Figure 2).

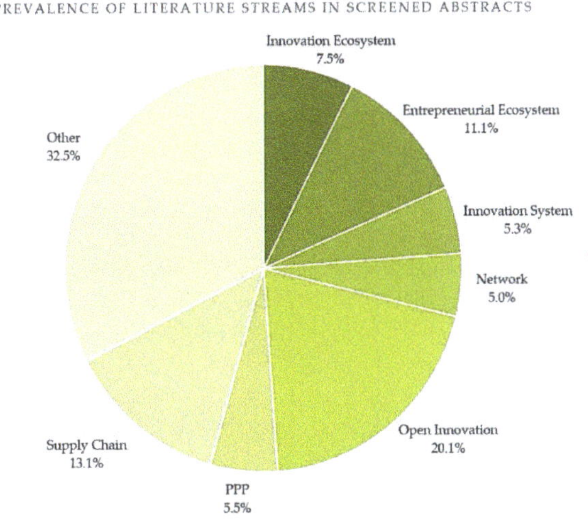

Figure 2. Prevalence of different literature streams in screened abstracts (author's own illustration).

Introducing a time axis to observe publication behavior over time, it becomes apparent that the topic of collaborative innovation for sustainability only gained momentum from 2011 onwards. The research field's growth has significantly accelerated since 2015. As visualized by Figure 3, this general trend can be observed across different streams of research. However, a certain time lag is visible with the open innovation discourse leading the field, while the supply chain and entrepreneurial ecosystem discourses followed a bit later. The prevalence of the innovation ecosystem terminology within the mission-oriented innovation discourse only gained traction from 2017 onwards, jumping up the ladder in 2019, with yearly publications more than doubling in comparison to the previous year. Although not included in Figure 3, a sneak peek into preliminary 2020 data suggests that the innovation ecosystem stream might even outpace other streams. These observations are in line with findings of previous SLRs on innovation ecosystems, pointing to a significant increase of publications in recent years, with the majority stemming from the last three years alone [12,22].

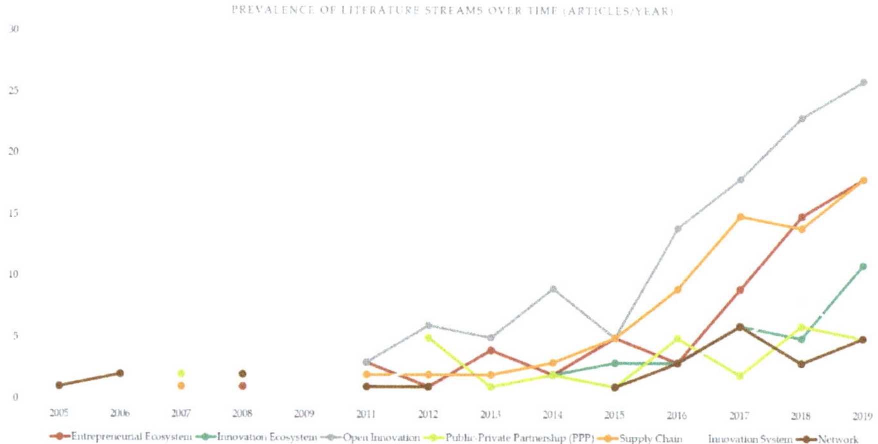

Figure 3. Prevalence of literature streams found in abstracts over time (author's own illustration).

Investigating the papers' source of publication, the impression of a scattered field emerges at first sight, with the 603 articles stemming from 328 different journals. Taking a closer look, however, the analysis reveals an uneven distribution, with two journals—*Journal of Cleaner Production* and *Sustainability*—leading the field by far with 65 resp. 50 records each, compared to just 11 publications in the third-ranked *Business Strategy and the Environment*. Differentiating between the various conceptual streams (see Section 2) shows that the former two are characterized by their broad approach, providing a platform to almost all conceptual traditions. Beyond that, each research stream is dominating one or two important journals, as is the case with *Small Business Economics* for the entrepreneurship-focused literature, the *International Journal of Production Economics* for the supply chain discourse, or *Technological Forecasting & Social Change* and *European Planning Studies* for the proponents of sustainable innovation systems. Looking specifically at the 44 articles selected for full-text analysis, the findings emphasize the importance of the *Journal of Cleaner Production* and *Sustainability* for the discourse on mission-oriented innovation ecosystems. These two journals are the only ones with more than two publications, dominating this niche from 2017 onwards (see Figure 4). Interestingly, however, these are not the journals of high relevance for the general debate on innovation ecosystems (such as *International Journal of Technology Management*, *Research Policy*, *R&D Management*, *Strategic Management Journal*, *Technovation* or *Technological Forecasting & Social Change* [10,12–14,22]), but rooted in the sustainable innovation discourse [61].

Looking beyond the journal covers, Table 2 provides an overview of the eleven most cited articles within the full text sample. Strikingly, there are no papers specifically dedicated to the phenomenon of mission-oriented innovation ecosystems included in this list. Instead, articles focus on certain sub-types, such as Carayannis and Campbell's [70] influential work on quadruple and quintuple helix structures for achieving sustainability, as well as the former author's conceptualization of social innovation ecosystems [71].

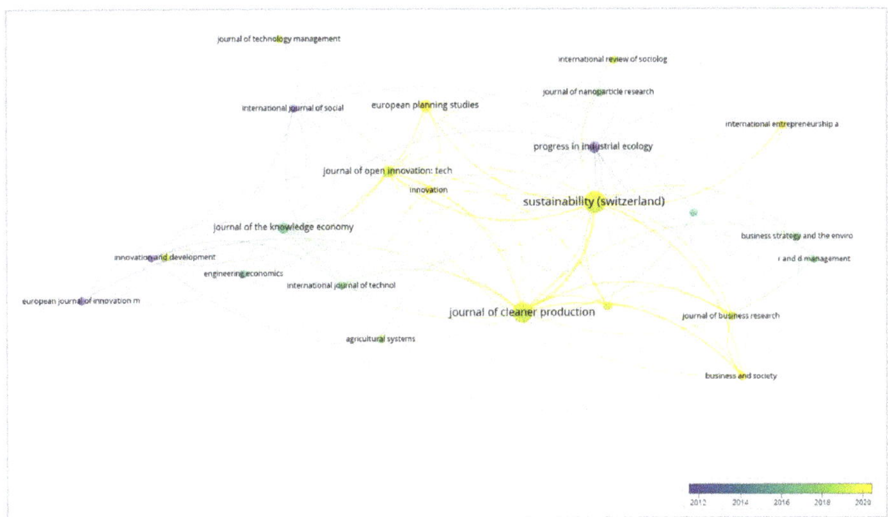

Figure 4. Development of journal importance over time (based on bibliographic coupling using VOSviewer 1.6.15).

Table 2. Overview of most-cited articles within the sample of full texts.

Rank	Authors (Year)	Title	No. of Citations
1	Carayannis and Campbell (2010) [70]	Triple Helix, Quadruple Helix and Quintuple Helix and How Do Knowledge, Innovation and the Environment Relate to Each Other?	196
2	Carayannis et al. (2019) [71]	Social Business Model Innovation: A Quadruple-/Quintuple- Helix-Based Social Innovation Ecosystem.	113
3	Pigford et al. (2018) [72]	Beyond Agricultural Innovation Systems? Exploring an Agricultural Innovation Ecosystems Approach for Niche Design and Development in Sustainability Transitions.	41
4	Goodman et al. (2017) [73]	Our Collaborative Future: Activities and Roles of Stakeholders in Sustainability-Oriented Innovation.	37
5	Yarahmadi and Higgins 2012 [74]	Motivations towards Environmental Innovation: A Conceptual Framework for Multiparty Cooperation.	30
6	Behnam et al. (2018) [75]	How should Firms Reconcile their Open Innovation Capabilities for Incorporating External Actors in Innovations Aimed at Sustainable Development?	26
7	Yun and Liu (2019) [76]	Micro- and Macro-Dynamics of Open Innovation with a QH-Model.	24
8	Brown et al. (2019) [77]	Why Do Companies Pursue Collaborative Circular Oriented Innov?	11
9	Hossain et al. (2019) [78]	A Systematic Review of Living Lab Literature.	10
9	Barrie et al. (2017) [79]	Leveraging Triple Helix and System Intermediaries to Enhance Effectiveness of Protected Spaces and Strategic Niche Management for Transitioning to Circular Economy.	10
9	Yang et al. (2012) [80]	What can Triple Helix Frameworks Offer to the Analysis of Eco-Innovation Dynamics? Theo. and Method. Considerations.	10

Against this backdrop, it is no wonder that E.G. Carayannis is also leading the ranking of the ten most important authors within the sample (see Table 3), co-authoring not only three different papers, but also having the most citations by far. Two more authors with a high publication output, pushing the research frontiers forward, are Zheng Liu and Nancy Bocken. Since their articles have been published more recently (4 in 2019; 2 in 2020), their citation count is inevitably lower—up to this point. This also highlights the more general point that citation statistics should be used with caution, specifically within a research field that is relatively young and still evolving.

Table 3. Overview of most important authors within the sample of full texts.

Rank	Author	No. of Documents within Sample	No. of Citations	Author's h-Index (Source: Google Scholar)
1	E.G. Carayannis	3	310	57
2	Z. Liu	3	31	*n.a.*
3	N. Bocken	3	13	36
4	J. Barrie	2	16	*n.a.*
5	E. João	2	16	17
6	G. Zawdie	2	16	*n.a.*
7	R. Balkenende	2	11	25
8	P. Brown	2	11	3
9	Y. Yang	2	11	*n.a.*
10	M. Grimaldi	2	1	*n.a.*

Although the research field is only just coming of age, its conceptual roots can be traced back to discourses that have been around for much longer. Running a co-citation analysis of our sample to better understand the intellectual structure of the field, six main conceptual traditions or research clusters influencing mission-oriented innovation ecosystem thinking were revealed (see Figure 5). These are: (1) the open innovation discourse around authors such as Chesbrough, Bogers and von Hippel; (2) the innovation systems and helix debates, going back to e.g., Nelson, Carayannis, Campbell or Etzkowitz; (3) the living lab concept put forward by Leminen and Schuurman; (4) the sustainable innovation literature with authors such as Bocken, Evans or Lüdeke-Freund; (5) the social innovation concept—sometimes viewed from a more sociological perspective—established by Howaldt and Terstriep, and finally, (6) the debates around sustainability transitions, transformative change and mission-oriented innovation policies, dominated by authors such as Mazzucato, Schot or Klerkx.

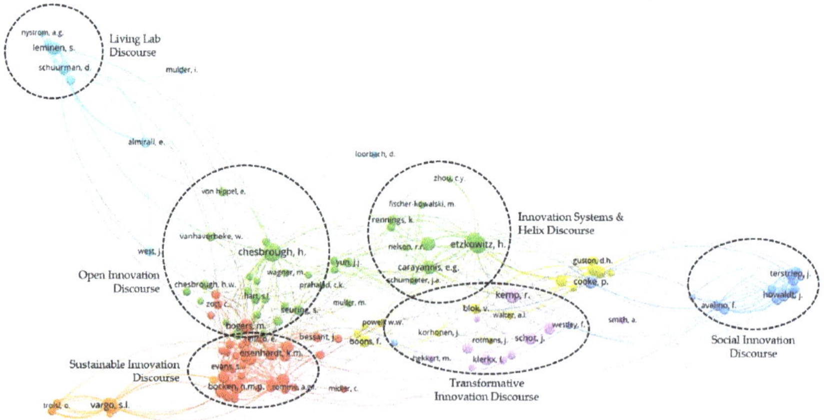

Figure 5. Conceptual roots of MOIE thinking (based on co-citation analysis using VOSviewer 1.6.15).

Building on these conceptual roots, about one third of the 44 articles are of a conceptual nature, whereas about two thirds apply an empirical approach. Zooming in, Table 4 displays that half of the empirical papers build on single case studies, while another 21.4% utilize a small-n comparative approach (2–5 cases). On the contrary, only a quarter of the empirical papers are based on the analysis of more than five cases. This lack of medium- and large-n research reinforces findings from previous SLRs on innovation ecosystems [10]. By far the most common methodological approaches are qualitative interviews (57.1% of empirical papers) and document analyses (also 57.1%). While social network analysis (SNA) is used in at least four papers, other methods are scarce.

Table 4. Overview of methodological approaches used within the sample of full texts.

Nature of Paper	No. of Studies (% Total)	Research Approach	No. of Studies (% Empirical)	Research Method * (* Note: One Study can Use Several Methods)	No. of Studies (% Empirical)
Conceptual	16 (36.4%)	—	—	—	—
Empirical	28 (63.6%)	Single Case Study	14 (50.0%)	Interviews	16 (57.1%)
		Small-N Case Study	6 (21.4%)	Document Analysis	16 (57.1%)
		Medium-N Case Study	7 (25.0%)	Social Network Analysis	4 (14.3%)
		Other	1 (3.6%)	Survey	2 (7.1%)
				Participant Observations	1 (3.6%)
				Action Research	1 (3.6%)

4.2. Content Analysis: Towards a Typology of Mission-Oriented Innovation Ecosystems

Having provided an overview of the research field in response to the first research question, the following section focuses on the second research question by laying out the conceptual foundations of the mission-oriented innovation ecosystem idea. As stated in the introduction and background section, a mission-oriented innovation ecosystem is characterized by its distinct target focus. As opposed to 'standard' innovation ecosystems that have been dealt with in the business strategy and innovation management literature, and predominantly aim to bring innovative technologies, products and services to the market [28,29], mission-oriented innovation ecosystem pit themselves against the grand challenges. Aiming to direct innovation towards sustainability challenges, as for example represented by the United Nation's Sustainable Development Goals (SDGs), a mission-oriented innovation ecosystem brings together all relevant actors from politics, science, industry and civil society for joint value creation. Directionality and value co-creation, as well as the principles of 'co-evolution' and 'co-specialization' [21] mark key characteristics of, and dynamics within, these ecosystems (see Figure 6).

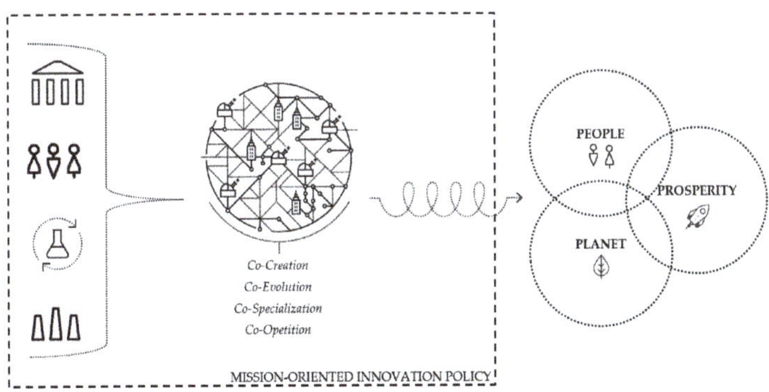

Figure 6. Mission-oriented innovation ecosystems (author's own illustration inspired by Schütz [49]).

Starting from this relatively broad definition of mission-oriented innovation ecosystems, several sub-types could be identified based on the systematic literature review. On the one hand, the 'new' missions, as defined by Mazzucato [5], equally encompass all three dimensions of sustainability—economic, social, environmental. Hence, mission-oriented innovation ecosystems could specifically target one of the three dimensions—people, prosperity, planet—or apply a more holistic understanding that aims to integrate these elements. On the other hand, mission-oriented innovation policy leaves room for two different solution approaches towards the grand challenges: Answers to pressing societal challenges can be sought at the immediate solution level (e.g., electric instead of fuel-based vehicles), or at the system level (e.g., transforming mobility systems as such). As mission-oriented innovation ecosystems could potentially emerge at the intersection of all four sustainability foci and take on either one of the two solution approaches, the typology developed as part of this article is based upon a 4 × 2 property space. Figure 7 visualizes the eight potential sub-types (of which one was not empirically found within the analyzed papers, and one is excluded based on conceptual considerations), which will be outlined in more detail in the following.

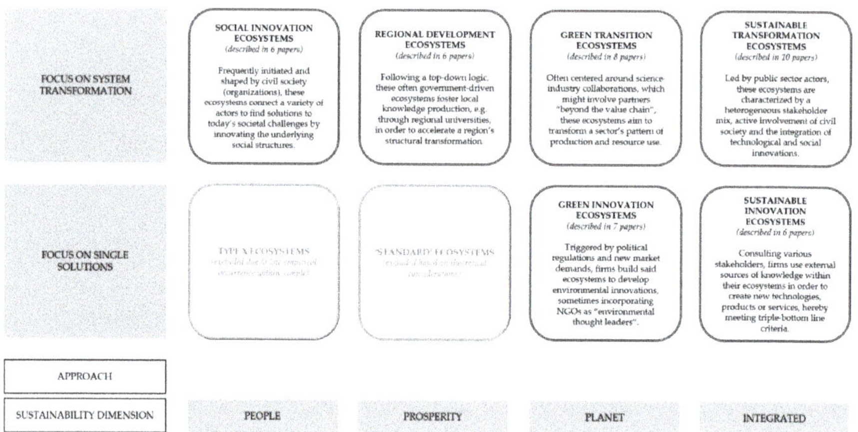

Figure 7. A typology of mission-oriented innovation ecosystems (author's own elaboration).

Social Innovation Ecosystems: The first sub-type of mission-oriented innovation ecosystems describes ecosystems which aim to give "answers […] to social needs that will lead to better results for the entire society" [71] (p. 2). As such, these ecosystems focus on sustainability's 'people dimension', while trying to tackle challenges at a system level. The six papers clustered under this heading all refer to the 'social innovation ecosystem' terminology. Acknowledging the ambiguity of the term 'social innovation' [81,82], this paper uses the terminology explicitly to describe innovation ecosystems leveraging social impact. Social, in the sense of behavioral, non-technological innovation, cuts across all four different target dimensions and may—some might even say must—be part of any innovation ecosystem. With view to their architectural set-up, these innovation ecosystems are characterized by a high variety of actors [71,82,83]. All six papers are based upon a quadruple-helix perspective, and describe ecosystems as involving actors from science, industry, politics and civil society. Alcaide Lozano et al. [83] consider stakeholder variety a key success factor of social innovation ecosystems, stating that "the more heterogeneous an ecosystem is […], the greater are its possibilities of gaining access to new ideas and strategies". Despite this quest for stakeholder diversity, public sector actors and civil society (organizations) in particular take on a central role in social innovation ecosystems [81,83]. In contrast, actors from science seem to play a less prominent role compared to other innovation processes [81,83,84]. As trust and relational ties are perceived to be particularly relevant to social innovation ecosystems, their local embedding is utterly important [81–83]. Still, they "are often supported in their attempts at social change through translocal, international collaborations with like-minded local initiatives" [84] (p. 5).

Regional Development Ecosystems: Within this sub-type of mission-oriented innovation ecosystems, a primary focus on economic sustainability is combined with a system level perspective. An illustrative example that features prominently in most of the six underlying papers are missions aiming at regional development and structural transformation, for instance within the context of smart specialization strategies driven forward in EU innovation policy. According to Lopes and Franco [85] (p. 278), these ecosystems' "final objective is the creation of territorial added value in the form of increased wealth, employment and well-being, leading to regional competitiveness". Five out of the six papers refer to a quadruple-helix set-up; however, the empirical findings allude to a dominance of triple-helix patterns, and a marginalized role of civil society in some cases [86]. This may be due to the often government-driven nature of regional development initiatives [87]. Although hardly an easy endeavor, combining this top-down logic with bottom-up elements and ensuring meaningful citizen participation is critical for success [87,88]. To this end, the 'social foresight lab approach' developed by Schroth et al. [89] depicts a promising methodological approach. Reflected in the particular university perspective taken by two of the six papers, a key role within regional development ecosystems is ascribed to academia, especially regional universities [88,90]. Not only do they provide a region with a highly skilled workforce, but generate new scientific knowledge, which can be transferred and commercialized through science-industry collaboration or academic spin-offs [88]. At this point, a blurring of innovation and regional development policies and a conceptual overlap with regional innovation system (RIS), cluster and entrepreneurial ecosystem discourses can be observed. Inherent to the concept is the ecosystem's highly local nature, with the geographical region as the central point of reference [85–88,91], nevertheless, aiming at global markets [86,88].

Green Transition Ecosystems: Green transition ecosystems—at the intersection of planet focus and system approach—can be defined as innovation ecosystems, with a clear focus on achieving environmental sustainability, through the transformation of the underlying patterns of production and resource use. An excellent example for these ecosystems addressed by several of the eight underlying papers [e.g., 231; 232; 901; 1953] is the circular economy. This is because the

> Transition to a circular economy can be achieved through multiple protected spaces targeted—for example, at key circular economy growth markets such as renewable energy, biorefinery, remanufacturing, sustainable mobility and the sharing economy, to co-evolve, paving the way for smooth transition within a governance framework that is capable of mitigating tensions and conflicts that are likely to arise in the transition process [79] (p. 3)

An alternative example could be the discourse around agricultural innovation ecosystems [72,92]. With view to these ecosystems' architectural set-up, it is notable that, while actors from science and industry are discussed in almost all papers, the role of governments and especially of civil society is less prominent. Throughout the analyzed papers, the leading role of industry and science and the importance of the two sectors' collaboration is emphasized [92–95]. If considered relevant by these key actors, additional partners "beyond traditional value-chains" [93] (p. 11) may be included. Although said ecosystems "may be driven by firms only", it "may often require a push from policy organizations", for example "trough financing mechanisms" [94] (p. 14), to accelerate the ecosystems' initial emergence. In some cases, these early investment costs are also borne by large keystone companies in order to nurture the emerging ecosystem [95]. The strong science-industry collaboration at the heart of green transition ecosystems is also reflected in typical collaboration formats. In contrast to some of the other ecosystem types, they predominantly focus on classical knowledge and technology transfer channels, such as patents and licenses [96], as well as the establishment of common standards [95,97,98]. Contrary to the previous two sub-types, green transition ecosystems can have a "decentralized" [79] (p. 7) nature, and do not necessarily rely on a specific geographical point of reference.

Sustainable Transformation Ecosystems: The fourth sub-type of mission-oriented innovation ecosystems distinguish itself by their holistic perspective on sustainability, seeking to integrate and balance the three dimensions people, planet and prosperity. Such an approach can, for example, be found in smart city projects, where positioning the city competitively in the region and reducing CO_2 footprints and increasing citizens' well-being may be three (among other) aims pursued at the same time. Such an integrated perspective combined with a system level approach is described in ten papers of the full text sample. While all papers refer to the three sectors politics, industry and civil society, science is mentioned once less. A heterogeneous stakeholder set-up is perceived as central to these ecosystems [78], however, taking a closer look reveals a leading role of public sector actors and the importance of active citizen participation [70,76,99]. As ensuring the latter is challenging at times, it is particularly interesting that—despite the steering role of the public sector—"there is usually an intermediary, like a design or architecture firm, taking care of the co-creative process" [99] (p. 12). While design-based methods' ability to translate between different stakeholder groups and make potential future(s) tangible is a promising pathway to ensure wider participation in innovation processes [100], citizens' preferences towards participatory approaches must be considered [101]. Besides various co-creation formats, for example in living lab environments [78,99,102], the literature highlights the important role of capacity building [102,103] and the opportunities of open government platforms for increasing government-citizen interaction [99,102]. Compared to other types of ecosystems, their aspiration to pursue different goals simultaneously, hereby adding complexity, and their heterogeneous stakeholder setting, might make sustainable transformation ecosystems especially prone to tensions, which may arise with regard to competing priorities, value creation vs. value capture mechanisms, or potential hi-jacking through interest groups [99,104].

Type X Ecosystems: Type X ecosystems focus on sustainability's people dimension, and adopt a solution level approach. A recent practice example of such ecosystems is the development of the COVID-19 tracing app in Germany, where actors from government, industry, science and civil society joined forces to make a concrete step towards the containment of the corona pandemic. However, as this sub-type was identified in only one paper, no further characterization is developed here.

Standard Innovation Ecosystems: Given the two dimensions spanning the typology's property space—sustainability focus and solution approach—there could potentially be a type of ecosystem at the intersection of prosperity focus and solution level. Based on the research aim to conceptualize mission-oriented innovation ecosystems and drawing on Mazzucato's [3–5] mission definition, this type is nevertheless excluded due to theoretical considerations. Introducing a new solution—in other words a product or service—whilst aiming at economic sustainability is what all commercial innovation ecosystems strive for. Yet, such focus does not qualify as a dedicated mission from a policy perspective. Instead, this ecosystem type can be defined as a 'standard innovation ecosystem', representing the research focus of the general innovation ecosystem discourse (see Section 2). Most of what has been written on ecosystems so far characterizes this non-mission-oriented ecosystem type.

Green Innovation Ecosystems: Described in seven papers, green innovation ecosystems are defined by their strong focus on environmental sustainability, while applying a solution level approach. For such ecosystems, numerous examples aiming at developing 'green' or 'cleaner' technologies, for instance in the areas of energy production [105,106], construction [107] or mobility [108], could be listed. Within green innovation ecosystems, firms are considered the ecosystem leaders and primary solution drivers, not least reflected by the fact that all seven papers are analyzing ecosystems from a focal firm perspective. Over the course of the innovation process, these ecosystem leaders consult and incorporate various stakeholders, for two reasons [107,108]: according to Yarahmadi and Higgins [74], they aim at leveraging external competences on the one hand, whilst trying to comply with (environmental) regulations and obtain legitimacy on the other. With regard to the latter cooperation rationale, the literature particularly highlights the importance of governments, as regulators, and in some cases of civil society organizations and NGOs, as environmental thought leaders [80,105]. On the contrary, science seems to play a less prominent role, and is only considered as an important stakeholder by three of the seven papers. More often, competency is sought within the firm's own supply chain. The open innovation and supply chain literature, as neighboring strands of research, which have already inquired this topic in detail, serve as important points of reference here. The two main motivations for collaboration within this ecosystem sub-type may also help to explain the collaboration formats, which are mentioned by the texts: While knowledge exchange among firms and the involvement of users are meant to build competencies [106,108], political lobbying and awareness campaigns are used for reasons of compliance [80].

Sustainable Innovation Ecosystems: Sustainable innovation ecosystems combine a holistic view on sustainability with a focus on the immediate solution level. Six papers apply this understanding. However, a closer look at their methodology and sampling in particular shows that some of them define sustainable innovation as "a new or significantly improved product or service whose implementation in the market solves or alleviates an environmental OR a social problem" [73] (p. 732; own accentuation). Thus, sustainable innovation ecosystems encompass both cases with a more singular people or planet focus, as well as with a more integrated approach. Therefore, the following characterization based upon the underlying texts must be viewed with caution. Generally, said ecosystems show significant similarity with green innovation ecosystems, as the former appear to be equally firm-driven and are once again analyzed from the focal firm perspective in all cases. Perceiving the grand challenges "as a potential future market" [109] (p. 5), in which a first mover advantage may still be achieved, firms try to incorporate them into their business models. Applying a 'boundary-spanning business model' [110] perspective, different stakeholders are consulted and incorporated at certain stages or for certain tasks during the innovation process [18,73,109,111].

5. Discussion and Agenda for Further Research

Having presented the SLR's findings above, the subsequent sections aim to embed them into the broader discourse to derive their theoretical and practical contribution (see Section 5.1), outline avenues for further research (see Section 5.2) and reflect upon this study's limitations (see Section 5.3).

5.1. Ecosystems in Times of Mission-Oriented Innovation Policy

The grand challenges of our time cannot be solved by one actor—be it politics, science or industry—alone. Yet, an appropriate analytical framework for conceptualizing and ultimately analyzing these processes of collaborative innovation is lacking. As argued within this paper, the innovation ecosystem construct depicts a particular promising frame of reference in this regard. For this purpose, the paper integrates two different strands of research—the discourse on mission-oriented or transformative innovation policy on the one, as well as the innovation ecosystem literature on the other hand. Following a rigorous approach of systematic literature review, this study is the first to map adjacent fields of research and conceptualize mission-oriented innovation ecosystems with view to both the scope and target of their proposed solution and the underlying structure of multi-sector, collaborative innovation.

While the question of how such mission-oriented innovation ecosystems might work remains underexplored, the SLR's content-related findings presented in Section 4.2 indicate that the overarching mission or innovation target typically implies the prioritization of one sustainability dimension over others, influencing ecosystems' strategy and architectural set-up in turn. Compared to the broader innovation ecosystem literature, where firms usually constitute the keystone players [10], the analysis of mission-oriented innovation ecosystems reveals a more central role of the public sector. Within many sub-types of mission-oriented innovation ecosystems, governments move beyond their purely regulatory role, in order to actively create and orchestrate ecosystems. This finding is in line with the general shift in the role of governments within mission-oriented innovation policy [5]. However, the sub-types social innovation ecosystem, regional development ecosystem and sustainable transformation ecosystem show that—in order to be successful, this top-down approach is in need of societal participation and bottom-up experimentation, as argued by Schot and Steinmueller [1].

Whereas single solutions are driven by firms within ecosystem set-ups, which are partly similar to 'standard' innovation ecosystems, real, system-level transformation does not only require fresh ideas and innovative solutions, but political commitment and societal acceptance. With view to the academic sector, the analysis reveals a less central role within several ecosystem types, despite mission-oriented policy's initial claim 'to solve the grand challenges with the means of research and innovation'. This observation does certainly not imply a general loss of importance, but rather points to the challenge of overcoming linear innovation processes and strategically positioning research organizations within these new forms of collaboration.

This bears important implications for mission-oriented policy practices: on the one hand, politics should realize its formative power, especially with view to system-level transformations. For this purpose, new collaboration formats, enabling the active involvement of policy makers in innovation processes beyond merely providing financial resources and setting the regulatory framework, should be tested. On the other hand, both public and private actors need to recognize that civil society involvement is indispensable with regard to system level transformations. Moving beyond the narrow 'user-centered' focus often found at the solution level, broader societal involvement is necessary. This is especially true given that for technological solutions to leverage significant effects, they must go hand in hand with social, behavioral innovations. The academic sector should accelerate these transformational processes, by not only providing the necessary contextual and specialist knowledge, but also by taking on a mediating role, enabling the co-production of knowledge and innovation as such. In order to adapt to these new roles and requirements, research organizations are advised to constantly rethink and reinvent their organizational practices, reputational mechanisms and business models [49].

5.2. Agenda for Further Research

Having conceptualized mission-oriented innovation ecosystems and explored the surrounding field(s) of research, a number of potential avenues for further inquiry emerge. Those were identified based on (a) the transfer of the general innovation ecosystem discourse based on the scoping review and (b) more than 60 open research questions raised within the analyzed full texts. While this paper does not aim to provide an exhaustive list, Table 5 groups and aggregates these questions along six broader themes: (1) concept; (2) architectural set-up, actors and roles; (3) emergence and life cycle; (4) collaboration formats; (5) success factors; (6) impact.

Table 5. Opportunities for further research (author's own elaboration).

Theme	Description	Potential Research Questions
Concept	Starting from this paper, providing a first literature-based conceptualization of MOIEs, the concept should be empirically tested and refined.	• How can today's mission-orientated innovation policy and sustainability issues be integrated in our understanding of innovation ecosystems? • Which of the sub-types of MOIEs can be found empirically? How can they be characterized?
Architectural Set-Up, Actors and Roles	Knowing that different sub-types of MOIEs are also characterized by differences within their architectural set-up, different configurations of actors and roles should be scrutinized. The works of Dedehayir et al. [23], Jacobides et al. [112] and Schütz [49] depict particularly promising starting points.	• What are architectural characteristics of MOIEs? How do they differ from 'standard' innovation ecosystems? • (How) do targeted mission and the ecosystem's architectural set-up interrelate? • Which actors resp. which functional roles are needed to achieve certain goals? • How does the role of governments change from 'standard' to mission-oriented innovation ecosystems? What are the prospects and pitfalls in this regard? • How can the architectural set-up account for the transnational nature of today's sustainability challenges?

Table 5. *Cont.*

Theme	Description	Potential Research Questions
Emergence and Life Cycle	Building upon Dedehayir et al. [23], understanding emergence and evolution of MOIEs is not only an urgent question among scholars, but equally relevant for policy makers and practitioners.	▪ What are necessary conditions, resources and circumstances for MOIEs to emerge? What inspires their creation? ▪ What is the role of innovation policy and politics in the emergence of MOIEs? ▪ How do MOIEs evolve over time? Which life cycle phases do exist?
Collaboration Formats	As collaboration faces a wide range of barriers, suitable formats for overcoming these should be explored.	▪ Which collaboration formats are used by (different sub-types of) MOIEs? ▪ Which methods can facilitate the integration of sustainability goals within collaborative innovation processes?
Success Factors	Failure is inherent not only to innovation but also to innovation ecosystems. Identifying the adjusting screws is of high strategic priority for innovation managers.	▪ What are necessary and sufficient conditions for successful MOIEs? How do these factors interact? ▪ Which intra-organizational prerequisites need to be fulfilled in order to engage in MOIEs? ▪ How can mechanisms of value creation (also considering external effects) and value capture be balanced?
Impact	As measuring the performance of ecosystems has always been far from trivial [48], assessing the impact of MOIEs depicts a particular challenge.	▪ How should the impact of MOIEs be conceptualized and measured? ▪ Which KPIs could be used to evaluate MOIE's impact? ▪ How can MOIEs balance different (sometimes conflicting) targets, for example the three dimensions of sustainability?

However, the analysis also reveals that the largest, most commonly mentioned research gap is a methodological one. As shown in Section 4.1, new methods beyond single and small-n case studies and a diversified sampling are urgently needed to proceed over the course of the coming years. This is in line with the current state of general innovation ecosystem research [10].

5.3. Limitations

In order to appraise this paper's contribution to the evolving area of research around mission-oriented innovation ecosystems, its limitations must be considered as well. For this purpose, two methodological and one conceptual challenge potentially influencing results' validity shall be discussed. Firstly, although systematic reviews depict the most rigorous approach towards existing literature, not all of the previous research is necessarily covered, due to flaws within the search strategy, technically unavailable sources or language barriers [60] (p. 97). A second challenge inherent to the chosen methodological approach arises from publication bias. As not all forms of research are equally submitted to, and published in, academic journals, studies addressing the failure of innovation ecosystems might be less prevalent within the sample [60] (p. 101). In addition, one conceptual limitation should be pointed out with regard to the development of the typology in Section 4.2 of this paper. As it is the aim of typologies to reduce complexity by providing a relatively simple model of real-life phenomena, the potential loss of case-specific granularity must be acknowledged.

6. Conclusions

Collaborative innovation within innovation ecosystems is critical for proceeding towards the SDGs. Following a rigorous review approach, and supplementing it with elements from bibliometric analysis and typology building, this paper explores and conceptualizes the idea of mission-oriented innovation ecosystems. In doing so, it closes the research gap detected through the scoping analysis of previous SLRs on innovation ecosystems and similarly identified by Liu and Stephens [18]. By differentiating mission-oriented innovation ecosystems from related streams of research, such as innovation systems, entrepreneurial ecosystems or supply chain collaboration, conceptual boundaries towards similar constructs are drawn. Analyzing 44 articles in depth, and developing a typology of mission-oriented innovation ecosystem sub-types, allows for an internal sharpening of the concept. Ultimately, the study has shown that the actual mission and respectively prioritization of sustainability dimensions influence ecosystems' strategy and architectural set-up. In a next step, the concept of mission-oriented innovation ecosystems, as well as the literature-based typology, should be tested and refined through empirical studies.

Funding: This research was funded by the German Federal Ministry for Education and Research (Bundesministerium für Bildung und Forschung (BMBF)) under grant number 01IO1915. The APC was funded by Fraunhofer-Gesellschaft zur Förderung der angewandten Forschung e.V.

Acknowledgments: The author would like to express his deep gratitude to Svenja Lemke for supporting the systematic literature review, as well as to her, Florian Schütz and Anja Müller for their valuable feedback on earlier versions of this paper.

Conflicts of Interest: The author declares no conflict of interest.

Appendix A.

Table A1. Review protocol (author's own elaboration).

AIM	Objective	Exploring and conceptualizing mission-oriented innovation ecosystems by aggregating and systemizing previous research (based on scoping review and [18]). Research questions: • What characterizes mission-oriented innovation ecosystems as an emerging area of research and how does it relate to similar research strands? • How can mission-oriented innovation ecosystems and their sub-types be conceptualized? • What should an agenda for future research look like to improve our understanding of mission-oriented innovation ecosystems?
IDENTIFICATION	Information Sources	Bibliographic databases: • EBSCO (including among others EconLit) • ProQuest • Scopus • Web of Science Manual search within 10 most important journals (based on previous SLRs [10,12–14,61]): • International Journal of Sustainability • International Journal of Technology Management • Journal of Cleaner Production • Journal of Technology Transfer • R&D Management • Research Policy • Strategic Management Journal • Sustainability • Technological Forecasting and Social Change • Technovation Reference checking and literature snowballing

Table A1. *Cont.*

	Key Words	Search of all possible combinations of key words in block A ("ecosystem") and key words in block B ("mission-oriented innovation"). Key words were developed based on previous SLRs in both fields [10,12,61]. Block A ("ecosystem"): • alliance • collaboration • ecosystem • helix • network • open innovation • partnership • platform Block B ("mission-oriented innovation"): • circular innovation • eco innovation • green innovation • inclusive innovation • innovation AND bottom of the pyramid • innovation AND triple-bottom line • mission-oriented innovation • responsible innovation • social innovation • societal innovation • sustainable innovation • value-based innovation
	Filter	Publication period: 1987–2020 (based on [62]) Publication type: peer-reviewed journal articles
SCREENING and ELIGIBILITY	Inclusion vs. Exclusion Criteria	Technical exclusion, if: • not peer-reviewed journal article • central bibliographic information, such as author, title or abstract missing • article is not written in English Content-based inclusion, if: • article deals with case of mission-oriented innovation • article scrutinizes the underlying innovation process, hereby directly referring to ecosystem terminology or describing similar characteristics (such as a collaboration's multi-sector, co-creative nature)
PROCESSING	Data Management	The data were stored and processed using Excel (general data management and descriptive statistics), Citavi (bibliographic records), MaxQDA (coding and content analysis) and VOSViewer (bibliometric analysis).
	Data Collection Process	Abstract and full text screening were independently conducted by two researchers after two training rounds, inter-rater reliability checked and remaining differences solved through discussion. Abstract screening: • overall agreement: 94.7% • Cohen's Kappa: 0.70 (=substantial agreement) for articles describing innovation ecosystem-like characteristics more vaguely resp. 0.91 (=almost perfect agreement) for those explicitly using innovation ecosystem language. Full text screening: • overall agreement: 87.5% • Cohen's Kappa: 0.73 (=substantial agreement)

Table A1. *Cont.*

	Data Items	The following data items were coded as part of the full text analysis: Meta data: • conceptual vs. empirical paper • method • sample • perspective • context • further research • literature snowballs Content: • name • definition • goal • regional scope • actors • roles • motivation (for ecosystem participation) • collaboration formats • emergence and life cycle • success factors • barriers and risks
	Synthesis	Integration of elements of framework and narrative synthesis [66,67]; subsequent typology building based on Kluge [54,55], Collier et al. [69] and Schütz [49].
CHALLENGES and LIMITATIONS	Technical Issues	26 out of the 191 articles identified for full text screening were not accessible through the channels available to the researchers (portals of Fraunhofer IAO, FU Berlin, HU Berlin, TU Berlin).
	Potential Bias	Coverage: despite the broad search strategy, not all research is necessarily covered due to potential flaws in the search strategy, technically unavailable sources (see technical issues) or language barriers [60]. Publication bias: as not all forms of research are equally submitted to and published in academic journals, studies addressing the failure of innovation ecosystems might be less prevalent within the sample [60].

Appendix B.

Table A2. Overview of the 44 articles included in the full text analysis.

Author (Year)	Title	Journal
Alcaide et al. (2019) [83]	Understanding the Effects of Social Capital on Social Innovation Ecosystems in Latin America through the Lens of Social Network Approach.	International Review of Sociology
Barile et al. (2020) [111]	Technology, Value Co-Creation and Innovation in Service Ecosystems: Toward Sustainable Co-Innovation.	Sustainability
Barrie et al. (2017) [79]	Leveraging Triple Helix and System Intermediaries to Enhance Effectiveness of Protected Spaces and Strategic Niche Management for Transitioning to Circular Economy.	Int. J. of Techn. Mgm. and Sust. Development
Barrie et al. (2019) [96]	Assessing the Role of Triple Helix System Intermediaries in Nurturing an Industrial Biotechnology Innovation Network.	Journal of Cleaner Production
Behnam et al. (2018) [75]	How Should Firms Reconcile their Open Innovation Capabilities for Incorporating External Actors in Innovations Aimed at Sustainable Development?	Journal of Cleaner Production

Table A2. *Cont.*

Author (Year)	Title	Journal
Borowska/Osborne (2018) [103]	Locating the Fourth Helix: Rethinking the Role of Civil Society in Developing Smart Learning Cities.	International Review of Education
Brown et al. (2019) [77]	Why Do Companies Pursue Collaborative Circular Oriented Innovation?	Sustainability
Brown et al. (2020) [93]	How Do Companies Collaborate for Circular Oriented Innovation?	Sustainability
Callaghan/Herselman (2015) [113]	Applying a Living Lab Methodology to Support Innovation in Education at a University in South Africa.	TD—The Journal for Transdisciplinary Research in S.A.
Carayannis/Campbell (2010) [70]	Triple Helix, Quadruple Helix and Quintuple Helix and How Do Knowledge, Innovation and the Environment Relate to Each Other? A Proposed Framework for a Transdisciplinary Analysis of Sustainable Development and Social Ecology.	Int. J. of Social Ecology and Sust. Development
Carayannis/Rakhmatullin (2014) [87]	The Quadruple/Quintuple Innovation Helixes and Smart Specialisation Strategies for Sustainable and Inclusive Growth in Europe and Beyond.	Journal of the Knowledge Economy
Carayannis et al. (2019) [71]	Social Business Model Innovation: A Quadruple and Quintuple Helix-Based Social Innovation Ecosystem.	IEEE Transactions on Engineering Mgm.
Ceicyte/Petraite (2018) [109]	Networked Responsibility Approach for Responsible Innovation: Perspective of the Firm.	Sustainability
Chaminade/Randelli (2020) [92]	The Role of Territorially Embedded Innovation Ecosystems Accelerating Sustainability Transformations: A Case Study of the Transformation to Organic Wine Production in Tuscany (Italy).	Sustainability
Ciasullo et al. (2020) [102]	Multi-Level Governance for Sustainable Innovation in Smart Communities: An Ecosystems Approach.	Int. Entrepreneurship and Mgm. Journal
Domanski et al. (2020) [81]	A Comprehensive Concept of Social Innovation and its Implications for the Local Context: On the Growing Importance of Social Innovation Ecosystems and Infrastructures.	European Planning Studies
Fliaster/Kolloch (2017) [105]	Implementation of Green Innovations: The impact of Stakeholders and their Network Relations.	R&D Management
Foley/Wiek (2017) [86]	Bridgework Ahead! Innovation Ecosystems vis-à-vis Responsible Innovation.	J. of Nanoparticle Research
Goodman et al. (2017) [73]	Our Collaborative Future: Activities and Roles of Stakeholders in Sustainability-Oriented Innovation.	Business Strategy and the Environment
Hossain et al. (2019) [78]	A Systematic Review of Living Lab Literature.	Journal of Cleaner Production
Jucevicius et al. (2016) [91]	The Emerging Innovation Ecosystems and "Valley of Death": Towards the Combination of Entrepreneurial and Institutional Approaches.	Engineering Economics
Kirschten (2005) [26]	Sustainable Innovation Networks: Conceptual Framework and Institutionalisation.	Progress in Industrial Ecology
Koch-Ørvad et al. (2019) [107]	Transforming Ecosystems: Facilitating Sustainable Innovations Through the Lineage of Exploratory Projects.	Project Management Journal

Table A2. *Cont.*

Author (Year)	Title	Journal
Konietzko et al. (2020) [94]	Circular Ecosystem Innovation: An Initial Set of Principles.	Journal of Cleaner Production
Liu et al. (2019) [108]	An Investigation on Responsible Innovation in the Emerging Shared Bicycle Industry: Case Study of a Chinese Firm.	J. of Open Innovation: Technology, Market, and Complexity
Liu/Stephens (2019) [18]	Exploring Innovation Ecosystem from the Perspective of Sustainability: Towards a Conceptual Framework.	J. of Open Innovation: Technology, Market, and Complexity
Lopes/Franco (2019) [85]	Review About Regional Development Networks: An Ecosystem Model Proposal.	Journal of the Knowledge Economy
Madsen (2020) [114]	Business Model Innovation and the Global Ecosystem for Sustainable Development.	Journal of Cleaner Production
Markkula/Kune (2015) [88]	Making Smart Regions Smarter: Smart Specialization and the Role of Universities in Regional Innovation Ecosystems.	Technology Innovation Management Review
Mejia et al. (2019) [90]	A Hub-Based University Innovation Model.	Journal of Technology Mgm. and Innovation
Oskanen/Hautamäki (2015) [115]	Sustainable Innovation: A Competitive Advantage for Innovation Ecosystems.	Technology Innovation Management Review
Oskam et al. (2020) [104]	Valuing Value in Innovation Ecosystems: How Cross-Sector Actors Overcome Tensions in Collaborative Sustainable Business Model Development.	Business and Society
Parida et al. (2019) [95]	Orchestrating Industrial Ecosystem in Circular Economy: A Two-Stage Transformation Model for Large Manufacturing Companies.	Journal of Business Research
Pel et al. (2019) [84]	Unpacking the Social Innovation Ecosystem: An Empirically Grounded Typology of Empowering Network Constellations.	Innovation: The European J. of Social ScienceResearch
Pigford et al. (2018) [72]	Beyond Agricultural Innovation Systems? Exploring an Agricultural Innovation Ecosystems Approach for Niche Design and Development in Sustainability Transitions.	Agricultural Systems
Reficco et al. (2018) [110]	Collaboration Mechanisms for Sustainable Innovation.	Journal of Cleaner Production
Schuurman et al. (2016) [116]	Living Labs as Open Innovation Systems for Knowledge Exchange: Solutions for Sustainable Innovation Development.	Int. J. of Business Innovation and Research
Terstriep et al. (2020) [82]	Favourable Social Innovation Ecosystem(s)? An Explorative Approach.	European Planning Studies
Van Genuchten et al. (2019) [99]	Open Innovation Strategies for Sustainable Urban Living.	Sustainability
Walter/Scholz (2006) [117]	Sustainable Innovation Networks: An Empirical Study on Interorganisational Networks in Industrial Ecology.	Progress in Industrial Ecology
Yang et al. (2012) [80]	What Can Triple Helix Frameworks Offer to the Analysis of Eco-Innovation Dynamics? Theoretical and Methodological Considerations.	Science and Public Policy
Yang et al. (2019) [106]	Dynamics of Triple Helix Relations in the Development of Cleaner Technologies: Case of a Chinese Power Equipment Manufacturer.	Innovation and Development

Table A2. *Cont.*

Author (Year)	Title	Journal
Yarahmadi/Higgins (2012) [74]	Motivations towards Environmental Innovation: A Conceptual Framework for Multiparty Cooperation.	European Journal of Innovation Management
Yun/Liu (2019) [76]	Micro- and Macro-Dynamics of Open Innovation with a Quadruple-Helix Model.	Sustainability

References

1. Schot, J.; Steinmueller, W.E. Three frames for innovation policy: R&D, systems of innovation and transformative change. *Res. Policy* **2018**, *47*, 1554–1567. [CrossRef]
2. Hekkert, M.P.; Janssen, M.J.; Wesseling, J.H.; Negro, S.O. Mission-oriented innovation systems. *Environ. Innov. Soc. Transit.* **2020**, *34*, 76–79. [CrossRef]
3. Mazzucato, M. From market fixing to market-creating: A new framework for innovation policy. *Ind. Innov.* **2016**, *23*, 140–156. [CrossRef]
4. Mazzucato, M. Mission-Oriented Innovation Policy: Challenges and Opportunities. Available online: https://www.ucl.ac.uk/bartlett/public-purpose/publications/2017/sep/mission-oriented-innovation-policy-challenges-and-opportunities (accessed on 10 July 2020).
5. Mazzucato, M. Mission-oriented innovation policies: Challenges and opportunities. *Ind. Corp. Chang.* **2018**, *27*, 803–815. [CrossRef]
6. European Commission. Horizon. 2020. Available online: https://www.eur-lex.europa.eu/legal-content/EN/ALL/?uri=CELEX:52011DC0808 (accessed on 3 July 2020).
7. European Commission. Implementation Strategy for Horizon Europe: Version 1.0. Available online: https://ec.europa.eu/info/sites/info/files/research_and_innovation/strategy_on_research_and_innovation/documents/ec_rtd_implementation-strategy_he.pdf (accessed on 3 July 2020).
8. Bundesministerium für Bildung und Forschung. Forschung und Innovation für die Menschen. Die Hightech-Strategie 2025. 2018. Available online: https://www.bmbf.de/upload_filestore/pub/Forschung_und_Innovation_fuer_die_Menschen.pdf (accessed on 3 July 2020).
9. Yaghmaie, P.; Vanhaverbeke, W. Identifying and describing constituents of innovation ecosystems: A systematic review of the literature. *EuroMed J. Bus.* **2019**. [CrossRef]
10. De Vasconcelos Gomes, L.A.; Facin, A.L.F.; Salerno, M.S.; Ikenami, R.K. Unpacking the innovation ecosystem construct: Evolution, gaps and trends. *Technol. Forecast. Soc. Chang.* **2018**, *136*, 30–48. [CrossRef]
11. Granstrand, O.; Holgersson, M. Innovation ecosystems: A conceptual review and a new definition. *Technovation* **2020**, *90*, 102098. [CrossRef]
12. Scaringella, L.; Radziwon, A. Innovation, entrepreneurial, knowledge, and business ecosystems: Old wine in new bottles? *Technol. Forecast. Soc. Chang.* **2018**, *136*, 59–87. [CrossRef]
13. Suominen, A.; Seppänen, M.; Dedehayir, O. A bibliometric review on innovation systems and ecosystems: A research agenda. *Eur. J. Innov. Manag.* **2019**, *22*, 335–360. [CrossRef]
14. Tsujimoto, M.; Kajikawa, Y.; Tomita, J.; Matsumoto, Y. A review of the ecosystem concept: Towards coherent ecosystem design. *Technol. Forecast. Soc. Chang.* **2018**, *136*, 49–58. [CrossRef]
15. Schot, J.; Geels, F.W. Strategic niche management and sustainable innovation journeys: Theory, findings, research agenda, and policy. *Technol. Anal. Strateg. Manag.* **2008**, *20*, 537–554. [CrossRef]
16. Schot, J.; Kanger, L. Deep transitions: Emergence, acceleration, stabilization and directionality. *Res. Policy* **2018**, *47*, 1045–1059. [CrossRef]
17. Pyka, A. Dedicated innovation systems to support the transformation towards sustainability: Creating income opportunities and employment in the knowledge-based digital bioeconomy. *J. Open Innov.* **2017**, *3*, 385. [CrossRef]
18. Liu, Z.; Stephens, V. Exploring innovation ecosystem from the perspective of sustainability: Towards a conceptual framework. *J. Open Innov. Technol. Mark. Complex.* **2019**, *5*, 48. [CrossRef]
19. Tranfield, D.; Denyer, D.; Smart, P. Towards a methodology for developing evidence-informed management knowledge by means of systematic review. *Br. J. Manag.* **2003**, *14*, 207–222. [CrossRef]

20. Hakala, H.; O'Shea, G.; Farny, S.; Luoto, S. Re-Storying the business, innovation and entrepreneurial ecosystem concepts: The model-narrative review method. *Int. J. Manag. Rev.* **2020**, *22*, 10–32. [CrossRef]
21. Carayannis, E.G.; Campbell, D.F.J. 'Mode 3' and 'Quadruple Helix': Toward a 21st century fractal innovation ecosystem. *Int. J. Technol. Manag.* **2009**, *46*, 201–234. [CrossRef]
22. Khademi, B. Ecosystem value creation and capture: A systematic review of literature and potential research opportunities. *Technol. Innov. Manag. Rev.* **2020**, *10*, 16–34. [CrossRef]
23. Dedehayir, O.; Makinen, S.J.; Ortt, J.R. Roles during innovation ecosystem genesis: A literature review. *Technol. Forecast. Soc. Chang.* **2018**, *136*, 18–29. [CrossRef]
24. Russo-Spena, T.; Tregua, M.; Bifulco, F. Searching through the jungle of innovation conceptualisations: System, network and ecosystem perspectives. *J. Serv. Theory Pract.* **2017**, *27*, 977–1005. [CrossRef]
25. Simatupang, T.M.; Schwab, A.; Lantu, D. Introduction: Building sustainable entrepreneurship ecosystems. *Int. J. Entrep. Small Bus.* **2015**, *26*, 389–398. [CrossRef]
26. Kirschten, U. Sustainable innovation networks: Conceptual framework and institutionalisation. *Prog. Ind. Ecol.* **2005**, *2*, 132–147. [CrossRef]
27. Pattberg, P.H.; Biermann, F.; Chan, S.; Mert, A. *Public-Private Partnerships for Sustainable Development: Emergence, Influence and Legitimacy*; Pattberg, P.H., Biermann, F., Chan, S., Mert, A., Eds.; Edward Elgar: Cheltenham, UK, 2013; ISBN 1781006520.
28. Adner, R. Match your innovation strategy to your innovation ecosystem. *Harv. Bus. Rev.* **2006**, *84*, 98. [PubMed]
29. Adner, R.; Kapoor, R. Value creation in innovation ecosystems: How the structure of technological interdependence affects firm performance in new technology generations. *Strateg. Manag. J.* **2010**, *31*, 306–333. [CrossRef]
30. Autio, E.; Kenney, M.; Mustar, P.; Siegel, D.; Wright, M. Entrepreneurial innovation: The importance of context. *Res. Policy* **2014**, *43*, 1097–1108. [CrossRef]
31. Isenberg, D. The big idea: How to start an entrepreneurial revolution. *Harv. Bus. Rev.* **2010**, *88*, 40–50.
32. Spigel, B. The relational organization of entrepreneurial ecosystems. *Entrep. Theory Pract.* **2017**, *41*, 49–72. [CrossRef]
33. Edquist, C. *Systems of Innovation. Technologies, Institutions and Organizations*; Pinter Publishers/Cassell Academic: London, UK, 1997; ISBN 9781855674523.
34. Lundvall, B.A. *National Systems of Innovation: Towards a Theory of Innovation and Interactive Learning*; Pinter Publishers: London, UK, 1992.
35. Nelson, R.R. *National Innovation Systems: A Comparative Analysis*; Oxford University Press: New York, NY, USA, 1993; ISBN 9781601298904.
36. De Bresson, C.; Amesse, F. Networks of innovators: A review and introduction to the issue. *Res. Policy* **1991**, *20*, 363–379. [CrossRef]
37. Powell, W.W.; Koput, K.W.; Smith-Doerr, L. Interorganizational collaboration and the locus of innovation: Networks of learning in biotechnology. *Adm. Sci. Q.* **1996**, *41*, 116. [CrossRef]
38. Powell, W.W.; White, D.R.; Koput, K.W.; Owen-Smith, J. Network dynamics and field evolution: The growth of interorganizational collaboration in the life sciences. *Am. J. Sociol.* **2005**, *110*, 1132–1205. [CrossRef]
39. Bogers, M.; Chesbrough, H.; Moedas, C. Open innovation: Research, practices, and policies. *Calif. Manag. Rev.* **2018**, *60*, 5–16. [CrossRef]
40. Chesbrough, H. *Open Innovation: The New Imperative for Creating and Profiting from Technology*; Harvard Business School Press: Boston, MA, USA, 2003.
41. Von Hippel, E. Lead users: A source of novel product concepts. *Manag. Sci.* **1986**, *32*, 791–805. [CrossRef]
42. Cao, M.; Zhang, Q. Supply chain collaboration: Impact on collaborative advantage and firm performance. *J. Oper. Manag.* **2011**, *29*, 163–180. [CrossRef]
43. Holweg, M.; Disney, S.; Holmström, J.; Småros, J. Supply chain collaboration: Making sense of the strategy Continuum. *Eur. Manag. J.* **2005**, *23*, 170–181. [CrossRef]
44. Stank, T.P.; Keller, S.B.; Daugherty, P.J. Supply chain collaboration and logistical service performance. *J. Bus. Logist.* **2001**, *22*, 29–48. [CrossRef]
45. Bäckstrand, K. Multi-stakeholder partnerships for sustainable development: Rethinking legitimacy, accountability and effectiveness. *Eur. Environ.* **2006**, *16*, 290–306. [CrossRef]

46. Glasbergen, P.; Biermann, F.; Mol, A.P.J. *Partnerships, Governance and Sustainable Development. Reflections on Theory and Practice*; Edward Elgar: Cheltenham, UK, 2007; ISBN 9781847204059.

47. Rabelo, R.J.; Bernus, P. A holistic model of building innovation ecosystems. *IFAC Pap. Online* **2015**, *48*, 2250–2257. [CrossRef]

48. Galvão, A.; Mascarenhas, C.; Gouveia Rodrigues, R.; Marques, C.S.; Leal, C.T. A quadruple helix model of entrepreneurship, innovation and stages of economic development. *Rev. Int. Bus. Strategy* **2017**, *27*, 261–282. [CrossRef]

49. Schütz, F. Das Geschäftsmodell Kollaborativer Innovation: Eine Empirische Analyse zu Funktionalen Rollen in Quadruple-Helix-Innovationsprozessen. Ph.D. Thesis, Technische Universität Berlin, Berlin, Germany, 2020.

50. Gough, D.; Oliver, S.; Thomas, J. (Eds.) *An Introduction to Systematic Reviews*, 2nd ed.; SAGE: Los Angeles, CA, USA; London, UK; New Delhi, India; Singapore; Washington, DC, USA; Melbourne, Australia, 2017; ISBN 9781473929432.

51. Petticrew, M.; Roberts, H. *Systematic Reviews in the Social Sciences. A Practical Guide*; Blackwell: Oxford, UK, 2006.

52. Osareh, F. Bibliometrics, citation analysis and co-citation analysis: A review of Literature I. *Libri* **1996**, *46*, 149–158. [CrossRef]

53. Pritchard, A. Statistical bibliography or bibliometrics? *J. Doc.* **1969**, *25*, 348–349.

54. Kluge, S. *Empirisch Begründete Typenbildung. Zur Konstruktion von Typen und Typologien in der Qualitativen Sozialforschung*; Leske+Budrich: Opladen, Germany, 1999.

55. Kluge, S. Empirically grounded construction of types and typologies in qualitative social research. *Forum Qual. Soc. Res.* **2000**, *1*. [CrossRef]

56. Langer, L.; Stewart, R. What have we learned from the application of systematic review methodology in international development: A thematic overview. *J. Dev. Eff.* **2014**, *6*, 236–248. [CrossRef]

57. Gough, D.; Oliver, S.; Thomas, J. Introducing systematic reviews. In *An Introduction to Systematic Reviews*, 2nd ed.; Gough, D., Oliver, S., Thomas, J., Eds.; SAGE: Los Angeles, CA, USA; London, UK; New Delhi, India; Singapore; Washington, DC, USA; Melbourne, Australia, 2017; pp. 1–18. ISBN 9781473929432.

58. Arksey, H.; O'Malley, L. Scoping studies: Towards a methodological framework. *Int. J. Soc. Res. Methodol.* **2005**, *8*, 19–32. [CrossRef]

59. Gough, D.; Thomas, J. Commonality and diversity in reviews. In *An Introduction to Systematic Reviews*, 2nd ed.; Gough, D., Oliver, S., Thomas, J., Eds.; SAGE: Los Angeles, CA, USA; London, UK; New Delhi, India; Singapore; Washington, DC, USA; Melbourne, Australia, 2017; pp. 43–70. ISBN 9781473929432.

60. Brunton, G.; Stansfield, C.; Caird, J.; Thomas, J. Finding relevant studies. In *An Introduction to Systematic Reviews*, 2nd ed.; Gough, D., Oliver, S., Thomas, J., Eds.; SAGE: Los Angeles, CA, USA; London, UK; New Delhi, India; Singapore; Washington, DC, USA; Melbourne, Australia, 2017; pp. 93–122. ISBN 9781473929432.

61. Cillo, V.; Petruzzelli, A.M.; Ardito, L.; Del Giudice, M. Understanding sustainable innovation: A systematic literature review. *Corp. Soc. Resp. Environ. Manag.* **2019**, *26*, 1012–1025. [CrossRef]

62. Lupova-Henry, E.; Dotti, N.F. Governance of sustainable innovation: Moving beyond the hierarchy-market-network trichotomy? A systematic literature review using the 'who-how-what' framework. *J. Clean. Prod.* **2019**, *210*, 738–748. [CrossRef]

63. Landis, J.R.; Koch, G.G. The measurement of observer agreement for categorical data. *Biometrics* **1977**, *33*, 159. [CrossRef]

64. Belur, J.; Tompson, L.; Thornton, A.; Simon, M. Interrater reliability in systematic review methodology. *Sociol. Methods Res.* **2018**. [CrossRef]

65. Van Eck, N.J.; Waltman, L. Software survey: VOSviewer, a computer program for bibliometric mapping. *Scientometrics* **2010**, *84*, 523–538. [CrossRef]

66. Thomas, J.; O'Mara-Eves, A.; Harden, A.; Newman, M. Synthesis methods for combining and configuring textual or mixed methods data. In *An Introduction to Systematic Reviews*, 2nd ed.; Gough, D., Oliver, S., Thomas, J., Eds.; SAGE: Los Angeles, CA, USA; London, UK; New Delhi, India; Singapore; Washington, DC, USA; Melbourne, Australia, 2017; pp. 181–210. ISBN 9781473929432.

67. Snilstveit, B.; Oliver, S.; Vojtkova, M. Narrative approaches to systematic review and synthesis of evidence for international development policy and practice. *J. Dev. Eff.* **2012**, *4*, 409–429. [CrossRef]

68. Sutcliff, K.; Oliver, S.; Richardson, M. Describing and analysing studies. In *An Introduction to Systematic Reviews*, 2nd ed.; Gough, D., Oliver, S., Thomas, J., Eds.; SAGE: Los Angeles, CA, USA; London, UK; New Delhi, India; Singapore; Washington, DC, USA; Melbourne, Australia, 2017; pp. 123–144. ISBN 9781473929432.

69. Collier, D.; LaPorte, J.; Seawright, J. Putting typologies to work: Concept formation, measurement, and analytic rigor. *Political Res. Q.* **2012**, *65*, 217–232. [CrossRef]

70. Carayannis, E.G.; Campbell, D.F.J. Triple helix, quadruple helix and quintuple helix and how do knowledge, innovation and the environment relate to each other? *Int. J. Soc. Ecol. Sustain. Dev.* **2010**, *1*, 41–69. [CrossRef]

71. Carayannis, E.G.; Grigoroudis, E.; Stamati, D.; Valvi, T. Social business model innovation: A quadruple/quintuple Helix-Based Social innovation ecosystem. *IEEE Trans. Eng. Manag.* **2019**, 1–14. [CrossRef]

72. Pigford, A.-A.E.; Hickey, G.M.; Klerkx, L. Beyond agricultural innovation systems? Exploring an agricultural innovation ecosystems approach for niche design and development in sustainability transitions. *Agric. Syst.* **2018**, *164*, 116–121. [CrossRef]

73. Goodman, J.; Korsunova, A.; Halme, M. Our collaborative future: Activities and roles of stakeholders in sustainability-oriented innovation. *Bus. Strat. Environ.* **2017**, *26*, 731–753. [CrossRef]

74. Yarahmadi, M.; Higgins, P.G. Motivations towards environmental innovation. *Eur. J. Innov. Manag.* **2012**, *15*, 400–420. [CrossRef]

75. Behnam, S.; Cagliano, R.; Grijalvo, M. How should firms reconcile their open innovation capabilities for incorporating external actors in innovations aimed at sustainable development? *J. Clean. Prod.* **2018**, *170*, 950–965. [CrossRef]

76. Yun, J.J.; Liu, Z. Micro- and Macro-Dynamics of open innovation with a Quadruple-Helix Model. *Sustainability* **2019**, *11*, 3301. [CrossRef]

77. Brown, P.; Bocken, N.; Balkenende, R. Why do companies pursue collaborative circular oriented innovation? *Sustainability* **2019**, *11*, 635. [CrossRef]

78. Hossain, M.; Leminen, S.; Westerlund, M. A systematic review of living lab literature. *J. Clean. Prod.* **2019**, *213*, 976–988. [CrossRef]

79. Barrie, J.; Zawdie, G.; João, E. Leveraging triple helix and system intermediaries to enhance effectiveness of protected spaces and strategic niche management for transitioning to circular economy. *Int. J. Technol. Manag. Sustain. Dev.* **2017**, *16*, 25–47. [CrossRef]

80. Yang, Y.; Holgaard, J.E.; Remmen, A. What can triple helix frameworks offer to the analysis of eco-innovation dynamics? Theoretical and methodological considerations. *Sci. Public Policy* **2012**, *39*, 373–385. [CrossRef]

81. Domanski, D.; Howaldt, J.; Kaletka, C. A comprehensive concept of social innovation and its implications for the local context–On the growing importance of social innovation ecosystems and infrastructures. *Eur. Plan. Stud.* **2020**, *28*, 454–474. [CrossRef]

82. Terstriep, J.; Rehfeld, D.; Kleverbeck, M. Favourable social innovation ecosystem(s)?–An explorative approach. *Eur. Plan. Stud.* **2020**, *28*, 881–905. [CrossRef]

83. Alcaide Lozano, V.; Moliner, L.A.; Murillo, D.; Buckland, H. Understanding the effects of social capital on social innovation ecosystems in Latin America through the lens of Social Network Approach. *Int. Rev. Sociol.* **2019**, *29*, 1–35. [CrossRef]

84. Pel, B.; Wittmayer, J.; Dorland, J.; Søgaard Jørgensen, M. Unpacking the social innovation ecosystem: An empirically grounded typology of empowering network constellations. *Innov. Eur. J. Soc. Sci. Res.* **2019**, *37*, 1–26. [CrossRef]

85. Lopes, J.; Franco, M. Review about regional development networks: An ecosystem model proposal. *J. Knowl. Econ.* **2019**, *10*, 275–297. [CrossRef]

86. Foley, R.; Wiek, A. Bridgework ahead! Innovation ecosystems vis-à-vis responsible innovation. *J. Nanopart. Res.* **2017**, *19*, 216. [CrossRef]

87. Carayannis, E.G.; Rakhmatullin, R. The quadruple/quintuple innovation Helixes and smart specialisation strategies for sustainable and inclusive growth in Europe and beyond. *J. Knowl. Econ.* **2014**, *5*, 212–239. [CrossRef]

88. Markkula, M.; Kune, H. Making smart regions smarter: Smart specialization and the role of universities in regional innovation ecosystems. *Technol. Innov. Manag. Rev.* **2015**, *5*, 7–15. [CrossRef]

89. Schroth, F.; Glatte, H.; Kaiser, S. Integrating civil society into regional innovation systems: A social foresight lab approach. *Int. J. Innov. Reg. Dev.* **2020**, *8*, 6.

90. Mejia, S.P.; Hincapie, J.M.M.; Giraldo, J.A.T. A Hub-based university innovation model. *J. Technol. Manag. Innov.* **2019**, *14*, 11–17. [CrossRef]
91. Jucevicius, G.; Juceviciene, R.; Gaidelys, V.; Kalman, A. The emerging innovation ecosystems and "Valley of Death": Towards the combination of entrepreneurial and institutional approaches. *Eng. Econ.* **2016**, *27*. [CrossRef]
92. Chaminade, C.; Randelli, F. The role of territorially embedded innovation ecosystems accelerating sustainability transformations: A Case study of the transformation to organic wine production in tuscany (In Italy). *Sustainability* **2020**, *12*, 4621. [CrossRef]
93. Brown, P.; Bocken, N.; Balkenende, R. How do companies collaborate for circular oriented innovation? *Sustainability* **2020**, *12*, 1648. [CrossRef]
94. Konietzko, J.; Bocken, N.; Hultink, E.J. Circular ecosystem innovation: An initial set of principles. *J. Clean. Prod.* **2020**, *253*, 119942. [CrossRef]
95. Parida, V.; Burström, T.; Visnjic, I.; Wincent, J. Orchestrating industrial ecosystem in circular economy: A two-stage transformation model for large manufacturing companies. *J. Bus. Res.* **2019**, *101*, 715–725. [CrossRef]
96. Barrie, J.; Zawdie, G.; João, E. Assessing the role of triple helix system intermediaries in nurturing an industrial biotechnology innovation network. *J. Clean. Prod.* **2019**, *214*, 209–223. [CrossRef]
97. Wurster, S.; Ladu, L. Bio-Based products in the automotive industry: The need for ecolabels, standards, and regulations. *Sustainability* **2020**, *12*, 1623. [CrossRef]
98. Majer, S.; Wurster, S.; Moosmann, D.; Ladu, L.; Sumfleth, B.; Thrän, D. Gaps and research demand for sustainability certification and standardisation in a sustainable bio-based economy in the EU. *Sustainability* **2018**, *10*, 2455. [CrossRef]
99. Van Genuchten, E.; Calderón González, A.; Mulder, I. Open innovation strategies for sustainable urban living. *Sustainability* **2019**, *11*, 3310. [CrossRef]
100. Heidingsfelder, M.; Kimpel, K.; Best, K.; Schraudner, M. Shaping future: Adapting design know-how to reorient innovation towards public preferences. *Technol. Forecast. Soc. Chang.* **2015**, *101*, 291–298. [CrossRef]
101. Schütz, F.; Heidingsfelder, M.L.; Schraudner, M. Co-shaping the future in quadruple Helix innovation systems: Uncovering public preferences toward participatory research and innovation. *J. Des. Econ. Innov.* **2019**, *5*, 128–146. [CrossRef]
102. Ciasullo, M.V.; Troisi, O.; Grimaldi, M.; Leone, D. Multi-level governance for sustainable innovation in smart communities: An ecosystems approach. *Int. Entrep. Manag. J.* **2020**, *21*, 1. [CrossRef]
103. Borkowska, K.; Osborne, M. Locating the fourth helix: Rethinking the role of civil society in developing smart learning cities. *Int. Rev. Educ.* **2018**, *64*, 355–372. [CrossRef]
104. Oskam, I.; Bossink, B.; de Man, A.-P. Valuing value in innovation ecosystems: How cross-sector actors overcome tensions in collaborative sustainable business model development. *Bus. Soc.* **2020**. [CrossRef]
105. Fliaster, A.; Kolloch, M. Implementation of green innovations: The impact of stakeholders and their network relations. *RD Manag.* **2017**, *47*, 689–700. [CrossRef]
106. Yang, Y.; Etzkowitz, H.; Yin, S.; Luo, Y. Dynamics of triple helix relations in the development of cleaner technologies: Case of a Chinese power equipment manufacturer. *Innov. Dev.* **2019**, *9*, 65–84. [CrossRef]
107. Koch-Ørvad, N.; Thuesen, C.; Koch, C.; Berker, T. Transforming ecosystems: Facilitating sustainable innovations through the lineage of exploratory projects. *Proj. Manag. J.* **2019**, *50*, 602–616. [CrossRef]
108. Liu, M.; Zhu, J. An investigation on responsible innovation in the emerging shared bicycle industry: Case study of a chinese firm. *J. Open Innov. Technol. Mark. Complex.* **2019**, *5*, 42. [CrossRef]
109. Ceicyte, J.; Petraite, M. Networked responsibility approach for responsible innovation: Perspective of the firm. *Sustainability* **2018**, *10*, 1720. [CrossRef]
110. Reficco, E.; Gutiérrez, R.; Jaén, M.H.; Auletta, N. Collaboration mechanisms for sustainable innovation. *J. Clean. Prod.* **2018**, *203*, 1170–1186. [CrossRef]
111. Barile, S.; Grimaldi, M.; Loia, F.; Sirianni, C.A. Technology, value Co-Creation and innovation in service ecosystems: Toward sustainable Co-Innovation. *Sustainability* **2020**, *12*, 2759. [CrossRef]
112. Jacobides, M.G.; Cennamo, C.; Gawer, A. Towards a theory of ecosystems. *Strateg. Manag. J.* **2018**, *39*, 2255–2276. [CrossRef]
113. Callaghan, R.; Herselman, M. Applying a living lab methodology to support innovation in education at a university in South Africa. *J. Transdiscipl. Res. S. Afr.* **2015**, *11*, 21–38. [CrossRef]

114. Madsen, H.L. Business model innovation and the global ecosystem for sustainable development. *J. Clean. Prod.* **2020**, *247*, 119102. [CrossRef]
115. Oksanen, K.; Hautamäki, A. Sustainable innovation: A competitive advantage for innovation ecosystems. *Technol. Innov. Manag. Rev.* **2015**, *5*, 24–30. [CrossRef]
116. Schuurman, D.; Baccarne, B.; Marez, L.D.; Veeckman, C.; Ballon, P. Living Labs as open innovation systems for knowledge exchange: Solutions for sustainable innovation development. *Int. J. Bus. Innov. Res.* **2016**, *10*, 322. [CrossRef]
117. Walter, A.I.; Scholz, R.W. Sustainable innovation networks: An empirical study on interorganisational networks in industrial ecology. *Prog. Ind. Ecol.* **2006**, *3*, 431. [CrossRef]

Article

Exploring Cultivation Path of Building Information Modelling in China: An Analysis from the Perspective of an Innovation Ecosystem

Yingnan Yang [1], Yidan Zhang [1,2,*] and Hongming Xie [3]

[1] College of Civil Engineering and Architecture, Zhejiang University, Hangzhou 310058, China; yyn@zju.edu.cn
[2] Department of Civil Engineering, The University of Hong Kong, Hong Kong 999077, China
[3] School of Management, Guangzhou University, Guangzhou 510006, China; hmxie@gzhu.edu.cn
* Correspondence: 3130102512@zju.edu.cn or ydzhang@connect.hku.hk; Tel.: +86-157-0007-7750

Received: 15 July 2020; Accepted: 21 August 2020; Published: 25 August 2020

Abstract: Ecosystem theory provides a new perspective for studying the development of the architecture engineering and construction (AEC) industry in the age of information and communication technology (ICT). As an extremely ICT innovation, building information modelling (BIM) not only brings technical benefits to the AEC industry, but changes the innovation paradigm of the AEC industry towards an innovation ecosystem, which improve productivity and sustainability throughout the project life cycle. This article contributes to innovation ecosystem theory by exploring the structure of the BIM ecosystem and deriving its cultivation path. Then, as the leading city in China for developing BIM technologies, Shanghai was selected as the case study to elaborate on the cultivation path of the BIM ecosystem. The results indicate that three layers identified in the structure contribute to the understanding of the boundaries, units, and analytical focus of the BIM ecosystem, with the BIM platform being the core layer. This topology structure, with the BIM platform as the hub, promotes interdependency and symbiosis among participants in the cultivation of the BIM ecosystem, supporting the birth, expansion, maturity, re-innovation (or extinction), and sustainable development of the BIM ecosystem. This research complements and extends literature on the BIM ecosystem, and provides implications as to the construction, cultivation, and sustainability of BIM ecosystems for emerging economy firms.

Keywords: architecture engineering and construction (AEC) industry; building information modelling (BIM); ecosystem; cultivation; sustainability

---

1. Introduction

The architecture, engineering, and construction (AEC) industry, as one of the major contributors to both environmental and socioeconomic issues, is vital in achieving sustainable development [1]. There is an enormous demand for infrastructure construction in developing countries in order to sustain an accelerated economic growth, which comes at a very high cost to the environment [2]. To alleviate the environmental burden and low productivity associated with conventional construction, the recent global trend is to promote modular construction [3–6]. As mentioned in the SmartMarket report of 2011 [7], the reemergence of modular constructions as a "new" trend can be tied to the adoption of a new technology-building information modelling (BIM). The main reason is that BIM has allowed implementing manufacturing concepts such as lean design and modularization into the AEC industry [8]. Also, BIM is a technological innovation that can drive the transition to sustainability in the construction sector [9,10]. This means that BIM needs to be integrated into the whole lifecycle of construction projects. Consequently, BIM contributes to the sustainable development of the AEC industry from the economic, social, and environmental perspectives, but also the organizational one [11].

With the introduction of BIM, the innovation paradigm of the construction industry has been changing and upgrading towards an ecological and organic innovation ecosystem [12]. The sustainable development of the AEC industry is not only derived from technological change or policy support, but also depends on the imbedded innovation ecosystem. As an innovation throughout the life cycle of a project, BIM requires technical, capital, process, organizational, and cultural support. The interaction of these different aspects builds a BIM ecosystem. The ecosystem concept has been adopted as a unit of analysis to capture the structural and functional interrelationships between various actors of the AEC industry [12]. However, the theorizing of a BIM-based innovation ecosystem is still in its infancy and in-depth studies from emerging economy are insufficient.

As the largest AEC market in the world, China is struggling to drive an industrial transformation from a traditional extensive pattern to sustainable development [13]. BIM has become a key solution to various challenges of implementing sustainable construction phases [14]. In this context, the objective of this article was to analyze the formation and evolution of BIM in China from the perspective of an innovation ecosystem. More specifically, this paper sought to examine the basic structure and the cultivation of China's BIM ecosystem. A good understanding of the basic structure is the basis for further BIM ecosystem research. Furthermore, the research on the BIM ecosystem seeks to discover how the communities of government, enterprises, and so on practice symbiotic innovation to nurture the BIM ecosystem to achieve sustainable development of the AEC industry. This study took the BIM ecosystem of Shanghai as the object of case study and tried to discuss how the BIM ecosystem is cultivated in China. The cultivation path of the BIM ecosystem in China will provide valuable references for other countries that are applying or preparing to adopt BIM technology.

After this introduction, the rest of the paper is structured as follows: Section 2 provides the literature review of the BIM ecosystem, followed by the research design and methods in Section 3. Section 4 discusses the structure of the BIM ecosystem, and Section 5 discusses cultivation of BIM ecosystems. Section 6 presents the theoretical and practical implications of this paper, and Section 7 concludes the paper by discussing the limitations and suggestions for future research.

2. The Literature Review of the BIM Ecosystem

Prior to this paper, a search study of BIM literature using knowledge-mapping techniques was conducted, and the results are presented in Yang (2017) [15]. In order to ensure that the selected papers are of reference value, we set the inclusion criteria for the search. For example, only journal papers were selected, and they were required to be in English. The search rule for BIM-related literature was ("BIM" OR "building information modeling" OR "building information modelling" OR "building information model"), the search rule for innovation ecosystem literature was ("innovation" OR "construction innovation" OR "innovation ecosystem"), and the search rule for BIM ecosystem literature was ("ecosystem") AND ("BIM" OR "building information modeling" OR "building information modelling" OR "building information model"). As shown in Table 1, 20 journals with great influence in the field of civil engineering and innovation ecosystem were selected, which were highly ranked by construction management and strategic management researchers (e.g., [16,17]). More than 1000 papers on BIM from 2000 to 2019 were selected and analyzed to explore the current research focus and future research directions. It is noteworthy that past and present research has focused too much on BIM technology itself, neglecting its socioeconomic and organizational issues. However, the interdisciplinary nature of BIM, with its technical and nontechnical potentials and challenges, requires a systematic analysis to understand this paradigm shift in the AEC industry [15].

Table 1. Information of selected journals.

Journal Name
Automation in Construction
International Journal of Project Management
Journal of Construction Engineering and Management
Construction Management and Economics
Journal of Management in Engineering
Construction and Architectural Management
Journal of Cleaner Production
Journal of Computing in Civil Engineering
International Journal of Construction Management
Advanced Engineering Informatics
Harvard Business Review
Strategic Management Journal
Management Decision
Journal of Management
Academy of Management Perspectives
Construction Innovation
Architectural Engineering and Design Management
Journal of Professional Issues in Engineering Education and Practice
International Journal of Technology Management
Journal of Product Innovation Management

BIM has been increasingly adopted in the AEC industry since its inception in the 1970s, particularly in the last few years [18,19]. As an important technology to promote the sustainable development of the AEC industry, the BIM implementation has become a focus of study by scholars from different countries in recent years, particularly with regard to barriers to the implementation of BIM (e.g., Doumbouya et al. [20]; Walasek and Barszcz [21]), potential problems (e.g., Becerik-Gerber and Kensek [2]), critical success factors (e.g., Tsai et al. [22]) and case studies intending to draw lessons and experience from the implementation of BIM (e.g., Eadie et al. [23]; Luth et al. [24]; Cao et al. [25]).

By considering the possible gap among technical feasibility, potential value, and practical adoption, increasing research interests and efforts were presented to examine the degree by which BIM is currently adopted through the life cycle of construction projects in different countries or regions [18–20]. These studies offered a better understanding of the current status, problems, and constraints encountered in BIM implementation. With the implementation of BIM, some research studies (e.g., Aksenova et al. [12]; Singh et al. [26]) noted that the innovation paradigm of the AEC industry has been changing and upgrading towards an ecological and organic innovation ecosystem.

In order to study BIM implementation from the perspective of an innovation ecosystem, the research papers related to innovation and innovation ecosystems were further collected and reviewed. The concept of ecosystem was first introduced by Moore [27] to social science, as an approach to viewing firms, not as a part of an industry, but as an ecosystem where interdependent complementary actors cooperate, compete, and co-evolve around a new innovation to achieve competitive advantages [28,29]. Pulkka et al. [30] introduced the concept of the ecosystem in the context of the construction industry and suggested that the ecosystem concept is applicable and offers a useful analytical lens for understanding value creation in the construction industry. With the advancement of industrialization and informatization, BIM is no longer simply understood as a software or technology, but rather the innovation of the AEC industry, affecting all aspects of the industry [25]. Therefore, following Moore [27,31], Singh [32] defined the BIM ecosystem as the network of interacting technologies, processes, policies, and organizations that collectively determine the development and evolution of BIM-related products and services. Some studies (e.g., Pulkka et al. [24]; Jiang et al. [26]) have adopted the ecosystem concept to understand the imbedded system formed by BIM-related products, processes, and stakeholders. Based on the theory of ecology, Liu, Zeng, and Xu [27] proposed a theoretical framework for the BIM ecosystem, making a preliminary understanding of the BIM ecosystem possible.

The term "BIM ecosystem" is used to identify the economic, political, technical, and organizational systems involved in BIM, through which they can cultivate, maintain, and support an environment conducive to the operation of BIM. Through long-term observation of the development and application of BIM in Finland, Aksenova et al. [12] explored the Finnish innovation journey by capturing and recording historical sequence of key events and actors in the process and elaborated on the reasons why the BIM ecosystem in Finland did not evolve as expected. Although the BIM ecosystem has received much attention from academics in recent years, the theoretical background is still insufficient in the basic structure and cultivation of the BIM ecosystem from an evolutionary perspective.

Existing theoretical research of innovation ecosystem provided significant inspiration and foundation for this study. Tomás et al. [33] identified that not all innovation ecosystems have the same architectural models or internal collaboration, and existing research rarely deconstructs an ecosystem of innovation and examines its structure. This article contributes to innovation ecosystem theory by exploring the structure of the BIM ecosystem. Compared with developed countries such as the USA, the application of BIM in China is still in its infancy [34], and the AEC industry has not yet formed a self-organizing and self-updating ecosystem. However, the ecological phenomenon appearing in this process has attracted the attention of scholars [12,26,32,35]. Therefore, studying the structure and cultivation path of China's BIM ecosystem is more valuable for promoting informatization of the AEC industry in an emerging economy.

3. Research Design and Methods

In terms of the research design, this study took the following research process: literature review, draft design and data collection, data analysis and case study (see Figure 1). First, the study started with literature search which was targeted to confirm the research problems of this paper. Secondly, to address the research objectives, a comprehensive data collection and processing was used. The research data consist of literature review, interviews, marketing data, and policy documents. Thirdly, based on the theory of innovation ecosystem, data were analyzed to explore the basic structure and cultivation path of the BIM ecosystem in China. Finally, as the leading city in China for implementing BIM technologies, Shanghai was selected as the case study to elaborate on the cultivation path of the BIM ecosystem.

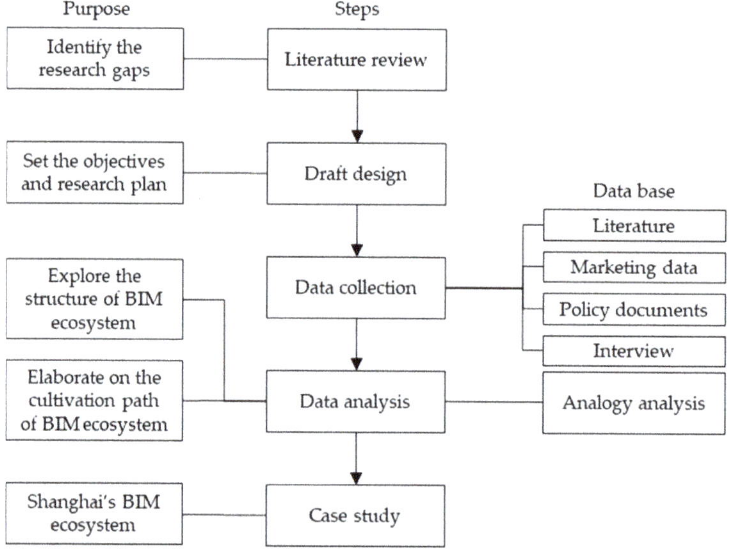

Figure 1. Research flow.

3.1. Interview

In order to get the whole picture of the BIM practices, the interview outlines were designed. The study team conducted formal and informal in-depth interviews with government actors, public agencies, industry, and academics. A semi-structured interview schedule was adopted to collect information by personal interview about the application of BIM, especially in Shanghai. Questions were asked relating to describing the key events, the technological changes, the marketing status, the effects of policies which were likely to affect the BIM practices. McCracken [36] long interview techniques were used to guide the interview and to provide focus using a series of open-ended questions pertaining to the BIM practices in construction projects. The average duration of each interview was one hour. Interviews were recorded and transcribed accordingly. Finally, based on the interview records and multiple data sources collected through the government units, Internet, published papers, and BIM projects reports, the reliable data were used to identify the main players, symbiotic relationships, the socioeconomic environment, and the evolution process of the BIM ecosystem.

Twenty interviews (over 30 hours) were conducted in 2019 and 2020, with representatives distributed among seven key stakeholder and end-user groups: (1) government agency; (2) research institution; (3) property owner; (4) design company; (5) construction company; (6) consulting company; and (7) software company. Table 2 lists the profiles of the interviewees.

Table 2. The profiles of the interviewees.

Interviewee	Average Years of Using BIM	Number of Interviews
government agency	about 5 years	4
research institution	>10 years	4
owner	about 5 years	3
design company	about 8 years	3
construction company	about 8 years	4
consulting company	>10 years	2
software company	>10 years	4

3.2. Marketing Data

In order to understand the current status of BIM applications in Shanghai, market reports provided a great amount of detailed information about BIM technology, policy, enterprises, projects, and so on, especially "Shanghai BIM technology Application & Development Reports" between 2016 and 2019 [37–40] issued by the Shanghai Municipal Commission of Housing Urban-Rural Development and Management and the Shanghai BIM Technology Application Joint Conference Office. In addition, relevant data on the AEC industry were collected from professional networks. Finally, these data on BIM trends and market share were combined in time series.

3.3. Policy Document

In order to understand the evolution process of the BIM ecosystem, this article collected the BIM-related documents issued since 2011. The starting point for China to promote the application of BIM from the policy perspective is 2011, when the 2011–2015 Outline for the Information Development of the Construction industry [41] was issued by the Ministry of Housing and Urban-Rural Development (MOHURD). Subsequently, the MOHURD successively issued a number of important documents to promote the application and development of BIM technology. Around 2014, the number of policies related to building information technology surged. Provincial and municipal entities have issued BIM promotion and application documents in response to the requirements of national and industrial building informatization in accordance with their actual conditions. As for the establishment of the BIM standard system, the plan of developing and revising standards for engineering and construction issued by the MOHURD in 2012 announced that 5 national BIM-related standards would be formulated: Unified standard for building information model application [42], Standard for classification and coding of building information model [43], Standard for storage of building information model [44], Application standard for manufacturing industry design information model [45], Standard for design delivery of building information modelling [46]. Subsequently, the Standard for building information modelling in construction [47] and the Presentation standard for building information modelling [48] were also included in the standard development plan in 2013 and 2014, respectively. Since the first national standard Unified standard for building information model application came out in 2016 [42], another four standards have been put into use successively, and the standard named Standard for storage of building information model [44] is under approval. Due to the slow setting progress of BIM-related standards at the national level, some provinces and cities, especially these areas with developed economies and highly developed construction industries, have issued regional BIM standards before the national ones in order to meet their own development needs and to guide the implementation of BIM technology in projects. In addition, the China BIM Union has also carried out the preparation of 21 P-BIM standards, which are based on the specialization in projects and play an important role as a complement to the BIM standard system.

3.4. Case Study

The potential advantage of a case study analysis is to gain a better understanding of "how" and "why" things happen. In single case study research, the opportunity to open a black box arises by looking at deeper causes of the phenomenon [49]. One key point of this paper is to discuss how Shanghai cultivates its BIM ecosystem. Also, the analytic generalization principle from cases to theory is suitable for exploring and explaining the research phenomena above [50].

This research selected Shanghai's BIM ecosystem as the case for the reason that it reflects the following three principles of case selection. First of all, the case selection takes into account both importance and representativeness. Shanghai is the leading region in China for developing BIM technologies and one of the first provincial governments to issue BIM policy in China. Secondly, the selection of research sample follows the principle of theoretical sampling. The choice of case is based on the need to fill existing theoretical gaps or develop new theories rather than statistical sampling [51]. The basic structure and the cultivation of the BIM ecosystem are important topics of academic concern. Existing theoretical research provides a foundation for the basic concepts of the BIM ecosystem. The research on Shanghai's BIM ecosystem tries to discover how the communities of the governments, enterprises, and other participants practice the symbiosis innovation to cultivate an BIM ecosystem for a resource-rich area. It helps to improve BIM ecosystems theory. Thirdly, the case selection should reflect the consistency principle of theory and research object [50]. The author team who attended the field surveys is deeply impressed by promotion from the Shanghai government and rapid growth of BIM projects. The data are relatively sufficient and integrated. Therefore, carrying out case studies grounded in the Shanghai context is urgently needed to provide both theoretical and practical insights for emerging markets.

4. Structure of the BIM Ecosystem

According to Taylor and Bernstein [52], BIM practice paradigms will evolve along a trajectory, from visualization, to coordination, to analysis, and finally to supply chain integration. This process also incorporates changes of the complex socioeconomic relationships among the technology, process, organization, built environment, and others [26,53]. It is necessary to analyze the structure of the BIM ecosystem as it is the basis for further research. Also, previous research can provide us with many insights for the construction of BIM structures. For example, Tansley [54] described natural ecosystem as a community of living organisms in conjunction with the nonliving components of their environment. Ron Adner [55] proposed a clear definition of the ecosystem construct through a structuralist approach and suggested the basic elements that describe the structure of ecosystems, namely actors, activities, locations, and the links between them.

Based on the literature review, this article firstly conducted an analogy analysis between a BIM ecosystem and a natural ecosystem to identify the involved actors and activities (see Figure 2). In an information ecosystem, information producers, information consumers, and information decomposers are people or organizations that generate, utilize, and ultimately process information. Due to the unique nature of the AEC industry and the level of regional development, the current BIM ecosystem in China is project-based, involving a large number of participants. It includes not only owners, designers, contractors, material suppliers, and government officials, which come from the traditional AEC industry, but also the enterprises from the IT industry. Therefore, project stakeholders are identified as the key players in the BIM ecosystem. Based on their information activities, they can be divided into three categories: information producers, information consumers, and information decomposers (see Figure 2). The information producer (e.g., the design firm) creates the initial information and transmits it to the primary consumer (BIM platform) for storage and initial processing. When other consumers (e.g., other participants in the project) request information, the information is read from the BIM platform as needed. The updated information is then transmitted to the BIM platform for data storage and integration. Unlike a natural ecosystem, the roles of the players in a BIM ecosystem are not always fixed. The roles as producers, consumers, or decomposers depend on their functions in different information activities.

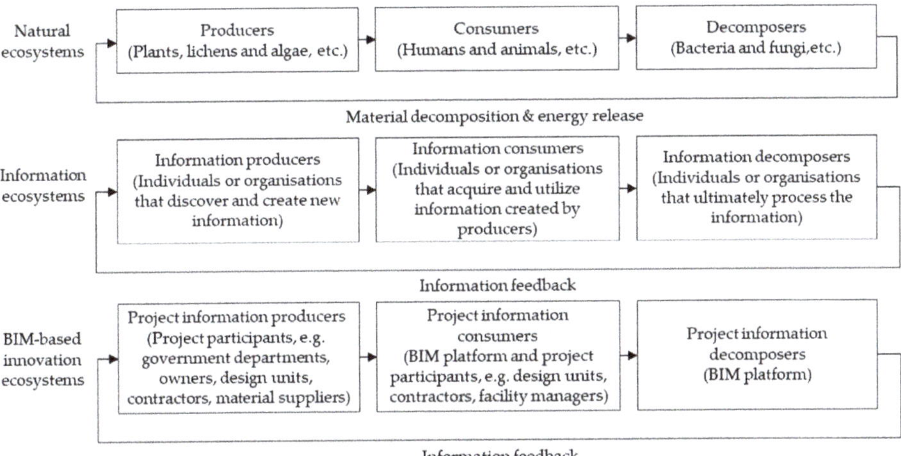

Figure 2. Comparison of natural ecosystems and a BIM ecosystem (biological components).

Singh [32] defined the BIM ecosystem as a network of interacting technologies, processes, policies, and organizations that collectively determine the development and evolution of BIM-related products and services. Based on this definition, Succar [56] and Succar and Kassem [57] identified the sub-domains and constituents that are specific to BIM ecosystems. Of these, technology, process, and policy are three interlocking fields of BIM activity, each of which is followed by two sub-domains (players and deliverables). Drawing on the literatures and interviews, we deconstructed the BIM ecosystem and examined its structure (see Figure 3). The BIM ecosystem can be described as three layers: the core layer, the middle layer, and the outermost layer. The core layer of the BIM ecosystem is the BIM platform, which consists of four main components, namely, databases, the IFC (Industry Foundation Class) engine, the Internet of Things, and big data analytics, to facilitate the information insertion, sharing, processing, and integration [58]. The scope of BIM is expanding from the intra-disciplinary collaboration to multi-disciplinary collaboration through a BIM-server that provides a platform for direct integration, storage, and exchange of data from multiple disciplines [57]. The middle layer represents the project participants in the supply chain, such as owners, consultants, designers, contractors, etc., interacting with each other through the BIM platform. They are the direct beneficiaries of the innovation ecosystem, forming a topology structure with the BIM platform as the hub. The outermost layer is the environment, which supports the collaboration of project participants in terms of technology, hardware device, policies, etc. [56]. As suggested by the interviewees, economic issues and standardization of BIM can be supplemented as supporting environments of the construction. The cost of BIM application hinders the intention of BIM adoption, so it is necessary to include the cost of BIM application in the budget and ensure it is paid to the general contractor at the beginning of the project. Also, new industry standards for BIM have been released in China in 2019. In addition, the biggest obstacle to BIM R&D is the lack of funding. The economic benefits of R&D will be reflected in engineering practices, but few companies are willing to pay for R&D, especially private companies. The main reason is a large initial investment and a long payback period.

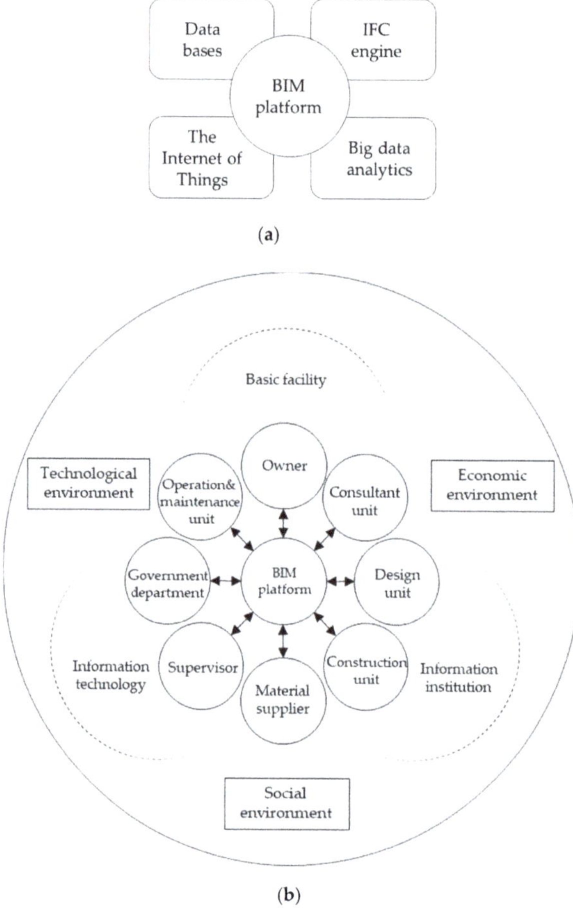

Figure 3. The basic structure of a BIM ecosystem: (**a**) core layer; (**b**) basic structure.

5. Cultivation of BIM Ecosystems

5.1. Cultivation Path of BIM Ecosystems

The AEC industry has a complex sociotechnical environment, with entangled power relationships between different stakeholders [59]. This leads to a complex organization, a fragmented structure of the construction sector, and project-based collaborations, posing critical challenges to change management and diffusion of construction innovation [60–63]. The BIM research team of Tsinghua University identified the tasks in the informatization process from the perspectives of the nation, industry, and enterprises [64]. Based on the ecosystem theory proposed by [31] and taking into account the unique nature of the AEC industry, the cultivation path of BIM ecosystems was derived, as shown in Figure 4. The cultivation path of BIM ecosystems consists of four stages: birth, expansion, maturity, re-innovation or extinction.

In the birth stage, the desire for product and service innovation inspired some pioneer enterprises to invest in R&D of BIM technology. However, the number of such pioneer enterprises is small and is growing slowly. There are several reasons for this phenomenon: As a new innovation, the application of BIM requires enterprises to invest a lot of human, material, and financial resources to support R&D and personnel training [65]. As indicated in the interviews, this requires the enterprises to have strategic vision and courage, because such a huge investment will not bring benefits to the enterprises in the short term; at the same time, lack of support from the market has also prevented some companies from participating. In the expansion stage, enterprises have begun to actively and passively participate in the development and application of BIM technology, gradually forming standardized market-oriented business. One of the salient features of this stage is the policy support from local governments, which greatly stimulated the enthusiasm of related enterprises, thereby fulfilling the potentials of BIM technology and regulating the operating process and market environment. Some new project delivery system, such as the integrated project delivery (IPD) as a natural companion to BIM, have also effectively eased the status of fragmentation and promoted the application of BIM technology [8]. In the maturity stage, the market is becoming stable and mature. Based on the ecosystem theory [31], as the BIM platform has been widely recognized, the status and roles of enterprises in the ecosystem begin to change due to their own innovation capabilities. For those pioneer enterprises that initiated R&D of BIM technology at the early stage, the investment is starting to pay off. These enterprises are gradually becoming leaders in the competition and occupy the core position in the ecosystem. The pioneer enterprises dominate the market, and those that cannot keep up with the trend of building informatization will be merged or eliminated. Although the BIM ecosystem in China is far from being formed, the profits of large design units and construction enterprises are more than those of small enterprises, and in tandem with the increase of application rate [66]. When information and communication technology (ICT) is fully applied throughout the project life cycle, the sustainable development of the AEC industry can be achieved. The fourth stage of the ecosystem is re-innovation or extinction. If it is re-innovation, new innovation points will emerge, and the ecosystem will evolve around this new innovation point to form a new ecosystem [31]. In the final stage, the loose policy environment and the accelerated spread of new technologies will lead to further improvement of the ecosystem [31]. Based on the BIM platform, the pioneer enterprises will gain more market competitive advantages.

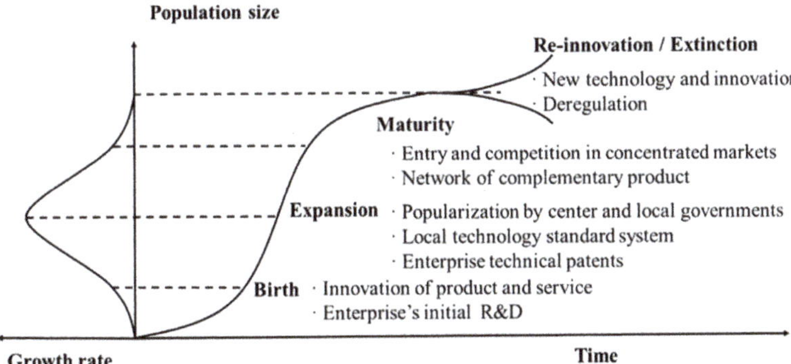

Figure 4. Cultivation path of a BIM ecosystem.

5.2. Case Description: Cultivation of a BIM Ecosystem in Shanghai

China's construction industrialization and informatization is still in its infancy and has not yet formed a self-organizing ecosystem and efficient operation system. As a leading city in China to promote BIM technology, Shanghai was taken as a case study to illustrate the cultivation process of a BIM ecosystem.

Phase 1: The birth of a BIM ecosystem

BIM research in China first appeared in 1995. Initially, it kept a slow pace and mainly focused on the technological issues. In 2011, the MOHURD issued the first document to promote the building informatization reform from the policy perspective. In spite of the potential advantages of BIM, its implementation frequently involves a variety of technical and organizational barriers which may significantly influence the intentions of enterprises to use BIM. According to the report on the application of BIM in China's construction industry [67], 87% of the respondents said they had heard of BIM, but 61% of them had never used it. Due to the lack of BIM talents and enterprises' motivation for implementing BIM, BIM had not been widely used. One of the key benefits that are highly recognized by users is clash detection, but only 6% of respondents have experience in this area. Although Shanghai takes the lead in the promotion of the BIM concept, it is still in the cognitive stage.

Phase 2: The expansion of a BIM ecosystem

One of the developing tasks at the second stage is the policy formulation. In June 2015, the Shanghai housing and urban-rural construction commission issued the Shanghai BIM guide (2015), which is the first regional standard that describes in detail the operational procedures, requirements, and expected results of 23 BIM application items in life-cycle management. In order to meet the actual needs of the BIM implementation and to make the guide more operational, the second version of the guide was revised in 2017. Overall, by the end of 2016, the BIM policy and standard system in Shanghai has basically formed. In 2017, the establishment of the multilevel education and training system, the policy and standard system, and the supporting environment for BIM application was forming a more mature market. In May 2018, the Commission also promulgated the "Evaluation criteria for BIM application in affordable housing projects in Shanghai" to change the mode of government supervision.

The application of BIM technology is gradually being promoted in Shanghai. From September 2015 to June 2016, more than 60 BIM-based projects were selected as pilot projects in 5 batches, including hospitals, schools, affordable housing, rail transit, bridges (tunnels), and other aspects. The Shanghai World Expo Museum project, as the first pilot project, started in December 2013 and ended in 2016, providing valuable practical experience on the construction of the 3D collaboration platform and life-cycle management. Since 2017, the government projects that have invested more than $100 million or have a single building area of more than 20,000 square meters (hereinafter referred to as "above the scale") must use BIM technology, and the society-invested ones are encouraged to use BIM. According to the statistics of BIM projects in Shanghai (see Figure 5), the number of BIM projects grows slowly first and then a bit quicker. Besides, the application rate of BIM projects that are "above the scale" increased from 29% to 88%, but the overall application rate only reached 12% in 2017, far below the goal of the first phase. In addition, it is designed that by 2020 BIM will be fully applied in the planning, design, and construction phases to achieve a goal of reducing project costs by 10% and construction periods by 5%. Overall, Shanghai's BIM ecosystem is still in the immature phase, but gradually stepping to the rapidly developing stage, where Shanghai has made great efforts in terms of the formulation of standards, the implementation of pilot projects and financial support, and learned from big government-invested projects.

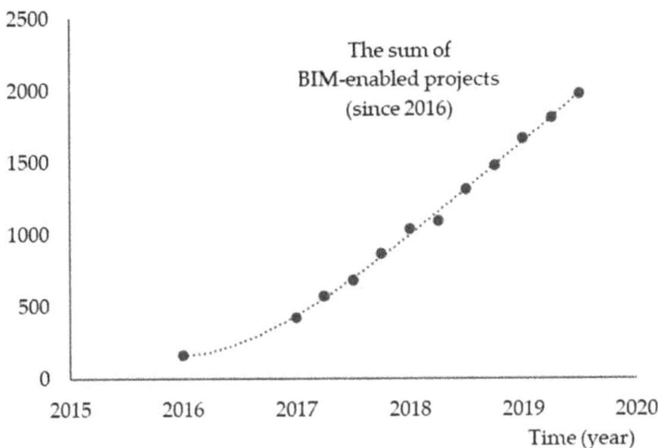

Figure 5. The number of BIM-based projects in Shanghai (since 2016).

Phase 3: The maturity of a BIM ecosystem

The periodic change of a BIM ecosystem is the result of the interaction of enterprises and environment. As the supporting environment in Shanghai matures, the government's role in promoting BIM would become less important, but the profit-seeking nature of enterprises would make them carry out information reform spontaneously to enhance the core competitiveness. The structure of a BIM ecosystem and its communities would become more fixed: large enterprises that take the lead in completing the reform of building informatization may form a monopoly position, while those lagging ones may face a survival crisis. Under the influence of technological, economic, and social environments, the enterprises' transformation to building informatization would be completed in this stage, partly reflected in the stable number of enterprises with information technology.

Phase 4: The re-innovation/extinction of a BIM-based innovation ecosystem

Enterprises also need to constantly discover the value behind BIM and niche markets. Trends like the emergence of new technology and diffusion of policy are predicted to increase the rate of enterprises' transformation and would give them a competitive advantage over the former winner who may fail to keep up with technological progress. As stage three emerges into the next stage, these interdependent but complementary actors would cooperate, compete, and co-evolve around the new innovation, starting a new round of evolution of the established ecosystem or creating an alternative ecosystem.

6. Implications

According to the cultivation path of the BIM ecosystem shown in Figure 4, the current development status of the BIM ecosystem in Shanghai is stepping into the early stage of the second phase, but the application rate of BIM in Shanghai is still at an inferior level. There is a long way to go before the BIM-based project network in China transforms to be a self-updating, self-organizing, and highly efficient ecosystem. Measures should be taken step by step from the following aspects.

Firstly, from the analysis of the structure of the BIM ecosystem, it can be seen that the most important factor influencing the formation of the BIM ecosystem comes from the market. Therefore, the government should implement flexible and diverse policies at an early stage, including pilot projects, financial support, and other preferential policies, promote innovation to benefit enterprises, and encourage enterprises to apply BIM to gain more practical experience. At the same time, in order to promote the construction of a BIM ecosystem, the government and enterprises should increase their investment in R&D at the initial stage, including the construction of hardware facilities and the training of technical personnel. In particular, it is necessary to enhance the core capability of the BIM platform from a technical level to improve the quality of BIM-related products and services.

Secondly, according to the interviewees, many project participants build their own 3D models and do not communicate with each other, which causes a lot of resource waste. The government should help establish a support system to attract more outstanding enterprises to share information through the BIM platform. The support system includes technical standards, application specifications, and user guides. Its application scope covers the entire industry chain and the entire project life cycle.

Thirdly, all the interviewees agreed that the application of BIM to the whole life cycle of a project will maximize the effectiveness of BIM, which will contribute to the formation of the BIM ecosystem and promote the sustainability of the AEC industry. This requires a collaborative effort from all participants in the basic structure of the BIM ecosystem to sustain and cultivate a healthy and sustainable ecosystem, and the government should pay attention to the "networking" practice in the field of engineering projects and take measures to encourage the participants to collaborate. Participants in the networked practices should depend on each other for mutual benefit and co-create value that cannot be achieved by individual firms.

Fourthly, the core layer of the BIM ecosystem is the BIM platform, which can add the analysis module for sustainability assessment. As recommended by Chong, Lee, and Wang [8], the new or revised BIM standards and guidelines should include a set of requirements on the BIM tools to comply with a standard sustainability assessment. Relevant techniques, energy simulation software, and life-cycle assessment (LCA) tools should be integrated into the BIM platform to access the sustainability of building materials and energy consumption in the projects.

7. Conclusions

As a catalyst for sustainable development in the AEC industry, BIM not only brings technical benefits to the AEC industry, but changes the innovation paradigm of the AEC industry towards an innovation ecosystem. However, the theorizing of a BIM-based innovation ecosystem is still in its infancy, and in-depth studies from emerging economy are insufficient. This article contributes to innovation ecosystem theory by exploring the structure of the BIM ecosystem and deriving its cultivation path. The BIM ecosystem is composed of internal systems (BIM platforms and project stakeholders) and external supporting environments (technology, economy and society). The platform is the core of the BIM ecosystem, and each component of the system plays the roles of information producer, information consumer, and information decomposer, respectively. Based on the BIM platform, participants interact in a flexible, networked, and loosely coupled manner, which enables the flow of information and complementary resources across organizational boundaries, and hence ecosystem actors are co-evolving and continuously adapting to the external environment to survive and gain sustainable development. Factors such as policy, technology, and market demand affect the external environment and ultimately affect the evolution of the ecosystem to the next stage.

Mapping the cultivation path is the best way to determine whether the realistic performance expectations have been set for the innovation ecosystem [65]. This study derives the cultivation path of the BIM ecosystem for the long-term sustainability development goals. These goals include resource and energy conservation, whole life cycle, increased productivity, and others. Each stage of the BIM ecosystem has different characteristics and development tasks in terms of technology development and organizational management. With the ups and downs of the cultivation path, the structure of the BIM ecosystem will be adjusted accordingly. In the first stage, as the core layer of the ecosystem, BIM technology is continuously developed and improved, and its application scope is gradually expanded to the whole life cycle of construction projects. Then, with support from the government, the development of technical standards, application specifications, and user guides push the evolution of the BIM ecosystem to the second stage. This stage is mainly the integration of project participants and BIM technology. The transformation of BIM from technology to collaborative platforms is conducive to improving the application environment of the ecosystem. In the third stage, the spontaneous market adjustment and the formation of complementary networks between enterprises promoted the formation of the ecosystem structure. The fourth stage is the re-innovation or decay of the ecosystem, which is characterized by the emergence of new technologies, new innovations, or the deregulation of government regulation, thereby overturning the original structure. The challenge at this stage is to constantly update products and upgrade services, otherwise the ecosystem will face decline. The results show that the interdependent and symbiotic structural relationships among BIM ecosystem participants are formed under the influence of internal and external environments. This relationship structure supports the cultivation path of the birth, expansion, maturation, and reinvention of a BIM ecosystem.

One limitation of our study was that this study is a single exploratory case study from China's experiences. Although China's experiences can be extended to other emerging economies, the building and cultivation of the BIM ecosystem in other countries may show different patterns due to different national conditions. From this background, future research can focus on discussion of BIM ecosystems in different countries and regions, comparing and analyzing the impacts of different national and cultural backgrounds, the current state of the AEC industry, and the level of technological development on the construction and cultivation of BIM ecosystems. Another limitation is that the interviewees come from several different fields, such as civil aviation, transportation, housing construction, etc., which are currently developing standards and specifications for BIM applications appropriate to their fields, but our study does not differentiate between them. Future research could consider the distinction between these industry sectors.

Author Contributions: Conceptualization, Y.Y.; methodology, Y.Y.; data curation, Y.Y.; formal analysis, Y.Y. and Y.Z.; investigation, Y.Y. and Y.Z.; project administration, Y.Y.; supervision, H.X.; writing—original draft, Y.Y. and Y.Z.; writing—review and editing, Y.Y. and Y.Z.; funding acquisition, Y.Y. and H.X. All authors have read and agreed to the published version of the manuscript.

Funding: This research was supported by the National Natural Science Foundation of China (Grant No. 71673240 and 71772163), Zhejiang Provincial Natural Science Foundation of China (Grant No. LY16G020009 and LY17G020024), Guangzhou Philosophy and Social Science Foundation (Grant No.2020GZYB94).

Conflicts of Interest: The authors declare no conflict of interest.

References

1. Roufechaei, K.M.; Bakar, A.H.A.; Tabassi, A.A. Energy-efficient design for sustainable housing development. *J. Clean. Prod.* **2014**, *65*, 380–388. [CrossRef]

2. Becerik-Gerber, B.; Kensek, K. Building information modeling in architecture, engineering, and construction: Emerging research directions and trends. *J. Prof. Issues Eng. Ed. Pract.* **2010**, *136*, 139–147. [CrossRef]

3. Teng, Y.; Mao, C.; Liu, G.; Wang, X. Analysis of stakeholder relationships in the industry chain of industrialized building in China. *J. Clean. Prod.* **2017**, *152*, 387–398. [CrossRef]

4. Aziz, N.D.; Nawawi, A.H.; Ariff, N.R.M. ICT evolution in facilities management (FM): Building information modelling (BIM) as the latest technology. *Procedia Soc. Behav. Sci.* **2016**, *234*, 363–371. [CrossRef]

5. Chong, H.Y.; Lee, C.Y.; Wang, X. A mixed review of the adoption of building information modelling (BIM) for sustainability. *J. Clean. Prod.* **2017**, *142*, 4114–4126. [CrossRef]

6. Elmualim, A.; Gilder, J. BIM: Innovation in design management, influence and challenges of implementation. *Architect. Eng. Des. Manag.* **2014**, *10*, 183–199. [CrossRef]

7. McGraw Hill Construction. *Prefabrication and Modularization: Increasing Productivity in the Construction Industry;* Smart Market Report: Bedford, TX, USA, 2011.

8. Azhar, S.; Khalfan, M.; Maqsood, T. Building information modelling (BIM): Now and beyond. *Aust. J. Constr. Econ. Build.* **2012**, *12*, 15–28. [CrossRef]

9. Zhang, L.; Chu, Z.; Song, H. Understanding the relation between BIM application behavior and sustainable construction: A case study in China. *Sustainability* **2020**, *12*, 306. [CrossRef]

10. Yuan, H.; Yang, Y.; Xue, X. Promoting owners' BIM adoption behaviors to achieve sustainable project management. *Sustainability* **2019**, *11*, 3905. [CrossRef]

11. Hetemi, E.; Ordieres-Meré, J.; Nuur, C. An institutional approach to digitalization in sustainability-oriented infrastructure projects: The limits of the building information model. *Sustainability* **2020**, *12*, 3893. [CrossRef]

12. Aksenova, G.; Kiviniemi, A.; Kocaturk, T.; Lejeune, A. From Finnish AEC knowledge ecosystem to business ecosystem: Lessons learned from the national deployment of BIM. *Constr. Manag. Econ.* **2019**, *37*, 317–335. [CrossRef]

13. Chen, C.; Yang, L.; Tang, L.; Jiang, H. BIM-based design coordination for China's architecture, engineering and construction industry. *WIT Trans. Built Environ.* **2017**, *169*, 211–219.

14. Chen, C.; Tang, L.C.M.; Jin, Y. Development of 5D BIM-based management system for pre-fabricated construction in China. In *Proceedings of the International Conference on Smart Infrastructure and Construction 2019 (ICSIC): Driving Data-Informed Decision-Making, Cambridge, UK, 8–10 July 2019*; De Jong, M., Schooling, J., Viggiani, G., Eds.; ICE Publishing: London, UK, 2019; pp. 215–224.

15. Yang, Y.; Zhang, Y.; Huang, W.T.; Xie, H.; Chen, L.J.; Cai, C.Y. Knowledge mapping of building information modelling research—A visual analysis using CiteSpace. In *Proceedings of the 22nd International Conference on Advancement of Construction Management and Real Estate, Melbourne, Victoria, Australia, 20–23 November 2017*; Zou, P.X.W., Ed.; CRIOCM 2017 Organising Committee: Swinburne University of Technology: Melbourne, Australia, 2017; pp. 608–615.

16. Jin, R.; Zou, P.; Piroozfar, P.; Wood, H.; Yang, Y.; Yan, L.; Han, Y. A science mapping approach based review of construction safety research. *Saf. Sci.* **2019**, *113*, 285–297. [CrossRef]

17. He, Q.; Wang, G.; Luo, L.; Shi, Q.; Xie, J.; Meng, X. Mapping the managerial areas of building information modeling (BIM) using scientometric analysis. *Int. J. Proj. Manag.* **2017**, *35*, 670–685. [CrossRef]

18. Ghaffarianhoseini, A.; Tookey, J.; Ghaffarianhoseini, A.; Naismith, N.; Azhar, S.; Efimova, O.; Raahemifar, K. Building Information Modelling (BIM) uptake: Clear benefits, understanding its implementation, risks and challenges. *Renew. Sustain. Energy Rev.* **2017**, *75*, 1046–1053. [CrossRef]

19. Antwi-Afari, M.F.; Li, H.; Pärn, E.A.; Edwards, D.J. Critical success factors for implementing building information modelling (BIM): A longitudinal review. *Autom. Constr.* **2018**, *91*, 100–110. [CrossRef]

20. Doumbouya, L.; Gao, G.; Guan, C. Adoption of the building information modeling (BIM) for construction project effectiveness: The review of bim benefits. *Neth. Int. Law Rev.* **2016**, *55*, 3–32.

21. Walasek, D.; Barszcz, A. Analysis of the adoption rate of building information modeling [BIM] and its return on investment. *Procedia Eng.* **2017**, *172*, 1227–1234. [CrossRef]

22. Tsai, M.H.; Mom, M.; Hsieh, S.H. Developing critical success factors for the assessment of BIM technology adoption: Part I. Methodology and survey. *J. Chin. Inst. Eng.* **2014**, *37*, 845–858. [CrossRef]

23. Eadie, R.; Browne, M.; Odeyinka, H.; McKeown, C.; McNiff, S. BIM implementation throughout the UK construction project lifecycle: An analysis. *Autom. Constr.* **2013**, *36*, 145–151. [CrossRef]

24. Luth, G.P.; Schorer, A.; Turkan, Y. Lessons from Using BIM to Increase Design-Construction Integration. *Pract. Period. Struct. Des. Constr.* **2014**, *19*, 103–110. [CrossRef]

25. Cao, D.; Li, H.; Wang, G.; Huang, T. Identifying and contextualising the motivations for BIM implementation in construction projects: An empirical study in China. *Int. J. Proj. Manag.* **2017**, *35*, 658–669. [CrossRef]

26. Gu, N.; Singh, V.; London, K. BIM ecosystem: The coevolution of products, processes, and people. In *Building Information Modeling: BIM in Current and Future Practice*; John Wiley & Sons: Hoboken, NJ, USA, 2015; pp. 197–210.

27. Moore, J.F. Predators and prey: A new ecology of competition. *Harv. Bus. Rev.* **1993**, *71*, 75–86. [PubMed]

28. Adner, R.; Kapoor, R. Innovation ecosystems and the pace of substitution: Re-examining technology S-curves. *Strateg. Manag. J.* **2016**, *37*, 625–648. [CrossRef]

29. Clarysse, B.; Wright, M.; Bruneel, J.; Mahajan, A. Creating value in ecosystems: Crossing the chasm between knowledge and business ecosystems. *Res. Policy* **2014**, *43*, 1164–1176. [CrossRef]

30. Pulkka, L.; Ristimäki, M.; Rajakallio, K.; Junnila, S. Applicability and benefits of the ecosystem concept in the construction industry. *Constr. Manag. Econ.* **2016**, *34*, 129–144. [CrossRef]

31. Moore, J.F. *The Death of Competition: Leadership and Strategy in the Age of Business Ecosystems*; HarperBusiness: New York, NY, USA, 1996.

32. Singh, V. BIM ecosystem research: What, why and how? framing the directions for a holistic view of BIM. In *Product Lifecycle Management for Digital Transformation of Industries*; Harik, R., Rivest, L., Bernard, A., Eynard, B., Bouras, A., Eds.; Springer: Cham, Switzerland, 2016; Volume 492.

33. Sant'Ana, T.D.; Bermejo, P.H.; Moreira, M.F.; Souza, W.V. The structure of an innovation ecosystem: Foundations for future research. *Manag. Decis.* **2020**. [CrossRef]

34. McGraw Hill Construction. *The Business Value of BIM for Construction in Major Global Markets: How Contractors around the World are Driving Innovation with Building Information Modeling*; McGraw Hill Construction: Manhattan, NY, USA, 2014.

35. Liu, Y.; Zeng, N.S.; Xu, B. Research on the concept and framework of building information modeling ecosystem. In Proceedings of the 2014 International Conference on Management Science & Engineering 21th Annual Conference Proceedings, Helsinki, Finland, 17–19 August 2014; pp. 1799–1805.

36. McCracken, G. *The Long Interview*; Sage Publications: Newbury Park, CA, USA, 1998.

37. SMCHURDM; SBTAJCO. *Shanghai BIM Technology Application & Development Reports (2016)*; Shanghai Housing and Urban-Rural Construction Administration Commission: Shanghai, China, 2016.

38. SMCHURDM; SBTAJCO. *Shanghai BIM Technology Application & Development Reports (2017)*. Available online: http://www.shbimcenter.org/documents/20552/0/2017%E4%B8%8A%E6%B5%B7%E5%B8%82%E5%BB%BA%E7%AD%91%E4%BF%A1%E6%81%AF%E6%A8%A1%E5%9E%8B%E6%8A%80%E6%9C%AF%E5%BA%94%E7%94%A8%E4%B8%8E%E5%8F%91%E5%B1%95%E6%8A%A5%E5%91%8A.pdf/83792ba0-d329-4661-b31e-7b4bced2ad6c (accessed on 21 March 2019).

39. SMCHURDM; SBTAJCO. *Shanghai BIM Technology Application & Development Reports (2018)*. Available online: http://www.shbimcenter.org/documents/20552/212684/2018%E4%B8%8A%E6%B5%B7%E5%BB%BA%E7%AD%91%E4%BF%A1%E6%81%AF%E6%A8%A1%E5%9E%8B%E6%8A%80%E6%9C%AF%E5%BA%94%E7%94%A8%E4%B8%8E%E5%8F%91%E5%B1%95%E6%8A%A5%E5%91%8Anew.pdf/77f2298c-1ddd-4876-8b75-986e06057a5d (accessed on 21 March 2019).

40. SMCHURDM; SBTAJCO. *Shanghai BIM Technology Application & Development Reports (2019)*. Available online: https://www.shgbc.org/Attach/Attaches/201905/201905131038228857.pdf (accessed on 26 May 2019).

41. MOHURD. The 2011–2015 Outline for the Information Development of the Construction Industry. In *Jianzhi [2011] 67*; 2011. Available online: http://www.mohurd.gov.cn/wjfb/201105/t20110517_203420.html (accessed on 12 September 2018).

42. MOHURD; AQSIQ. *Unified Standard for Building Information Model Application (GB/T51212-2016)*; China Architecture & Building Press: Beijing, China, 2016.

43. MOHURD. *Standard for Classification and Coding of Building Information Model (GB/T51269-2017)*; China Architecture & Building Press: Beijing, China, 2017.

44. MOHURD. *Standard for Storage of Building Information Model*; Under Approval.

45. MOHURD. *Application Standard for Manufacturing Industry Design Information Model (GB/T51362-2019)*; China Planning Press: Beijing, China, 2019.

46. MOHURD. *Standard for Design Delivery of Building Information Modeling (GB/T 51301-2018)*; China Architecture & Building Press: Beijing, China, 2018.

47. MOHURD. *Standard for Building Information Modelling in Construction (GB/T 51235-2017)*; China Architecture & Building Press: Beijing, China, 2017.

48. MOHURD. *Presentation Standard for Building Information Modeling (JGJ/T448-2018)*; China Architecture & Building Press: Beijing, China, 2018.

49. Ridder, H.G. The theory contribution of case study research designs. *Bus. Res.* **2017**, *10*, 281–305. [CrossRef]

50. Ding, L.; Wu, J. Innovation ecosystem of CNG vehicles: A case study of its cultivation and characteristics in Sichuan, China. *Sustainability* **2018**, *10*, 39. [CrossRef]

51. Glaser, B.G.; Strauss, A.L. *The Discovery of Grounded Theory: Strategies for Qualitative Research*; Routledge: Abingdon, UK, 2017.

52. Taylor, J.E.; Bernstein, P. Paradigm trajectories of building information modeling practice in project networks. *J. Manag. Eng.* **2009**, *25*, 69–76. [CrossRef]

53. Holmström, J.; Singh, V.; Främling, K. BIM as infrastructure in a Finnish HVAC actor network: Enabling adoption, reuse, and recombination over a building life cycle and between projects. *J. Manag. Eng.* **2015**, *31*, A4014006. [CrossRef]

54. Tansley, A.G. The use and abuse of vegetational concepts and terms. *Ecol. Environ. Sci.* **1935**, *16*, 284–307. [CrossRef]

55. Adner, R. Ecosystem as structure: An actionable construct for strategy. *J. Manag.* **2017**, *43*, 39–58. [CrossRef]

56. Succar, B. Building information modelling framework: A research and delivery foundation for industry stakeholders. *Autom. Constr.* **2009**, *18*, 357–375. [CrossRef]

57. Succar, B.; Kassem, M. Macro-BIM adoption: Conceptual structures. *Autom. Constr.* **2015**, *57*, 64–79. [CrossRef]

58. Arthur, S.; Li, H.; Lark, R. A Collaborative Unified Computing Platform for Building Information Modelling (BIM). In *Proceedings of the Working Conference on Virtual Enterprises, Cham, Switzerland, 18–20 September 2017*; Springer International Publishing: Cham, Switzerland, 2017; pp. 63–73.

59. Moore, D.R.; Dainty, A.R.J. Integrated project teams' performance in managing unexpected change events. *Team Perform. Manag.* **1999**, *5*, 212–222. [CrossRef]

60. Gu, N.; London, K. Understanding and facilitating BIM adoption in the AEC industry. *Autom. Constr.* **2010**, *19*, 988–999. [CrossRef]

61. Peansupap, V.; Walker, D.H.T. Factors enabling information and communication technology diffusion and actual implementation in construction organisations. *ITcon* **2005**, *10*, 193–218.

62. Taylor, J.E.; Levitt, R.E. Understanding and managing systemic innovation in project-based industries. In *Innovations: Project Management Research*; Project Management Institute: Newtown Square, PA, USA, 2004; pp. 83–99.

63. Dubois, A.; Gadde, L.E. The construction industry as a loosely coupled system: Implications for productivity and innovation. *Constr. Manag. Econ.* **2002**, *20*, 621–631. [CrossRef]

64. Tsinghua University BIM Task Group. *Research on Chinese Building Information Modeling Standard Framework*; China Architecture & Building Press: Beijing, China, 2011.

65. Adner, R. Match your innovation strategy to your innovation ecosystem. *Harv. Bus. Rev.* **2006**, *84*, 98–107.

66. Dodg Data & Analytics. *The Business Value of BIM in China*; Dodge Data and Analytics: Bedford, MA, USA, 2015.

67. He, G. *The Report on the Application of BIM in China's Construction Industry (2011)*; China Real Estate Association-Commercial Real Estate Professional Committee: Beijing, China, 2011.

 *sustainability*

Review

Smart Farming Introduction in Wine Farms: A Systematic Review and a New Proposal

Daniele Sarri *, Stefania Lombardo, Andrea Pagliai, Carolina Perna, Riccardo Lisci, Valentina De Pascale, Marco Rimediotti, Guido Cencini and Marco Vieri

Department of Agricultural, Food, Environment and Forestry, University of Florence, Piazzale Delle Cascine 15, 50144 Florence, Italy; stefania.lombardo@unifi.it (S.L.); andrea.pagliai@unifi.it (A.P.); carolina.perna@unifi.it (C.P.); riccardo.lisci@unifi.it (R.L.); valentina.depascale@unifi.it (V.D.P.); marco.rimediotti@unifi.it (M.R.); guido.cencini@unifi.it (G.C.); marco.vieri@unifi.it (M.V.)
* Correspondence: daniele.sarri@unifi.it

Received: 8 July 2020; Accepted: 28 August 2020; Published: 3 September 2020

Abstract: This study shows a new methodological proposal for wine farm management, as a result of the progressive development of the technological innovations and their adoption. The study was carried out in Italy involving farmers, workers, or owners of wine farms who are progressively introducing or using precision agriculture technologies on their farm. The methodology proposed was divided in four stages (1. understanding the changes in action; 2. identifying the added value of Smart Farming processes; 3. verifying the reliability of new technologies; 4. adjusting production processes) that can be applied at different levels in vine farms to make the adoption of precision agriculture techniques and technologies harmonious and profitable. Data collection was carried out using a participant-observer method in brainstorming sessions, where the authors reflected on the significance of technology adoption means and how to put them in practice, and interviews, questionnaire surveys, diaries, and observations. Moreover, project activities and reports provided auxiliary data. The findings highlighted the issues of a sector which, although with broad investment and finance options, lacks a structure of human, territorial, and organizational resources for the successful adoption of technological innovations. The work represents a basis for the future development of models for strategic scenario planning and risk assessments for farmers, policymakers, and scientists.

Keywords: Technological Readiness Level; smart farming; viticulture; lean; business model canvas

1. Introduction

Agriculture, following the development and modernization paradigm begun with the green revolution (1930–1960) [1], has evolved from extensive production, family, and share farming to intensive production, reliant on chemical products. In this transition, agriculture assumed the characteristics of a productivist and linear system, which aims mainly to increase productivity. This approach leads to industrialized and efficient production whose purpose is profit maximization, usually by increasing farm unit size, specialization, and the reliance on external industrial input [2]. Nevertheless, with the current social, economic, and environmental requirements, this production process is no longer viable, in terms of negative externalities, and the necessity of a more sustainable and circular farming method is emerging. The application of precision agriculture and smart farming approach can be considered a step towards the solution needed to continue feeding the world, sustainably.

The concept of precision agriculture, intended as site-specific crop management, dates back to 1980. Still, this term was used for the first time in 1990, during a workshop in Montana [3], and lately has been associated with the concept of smart farming. The term "smart" came from the definition of Smart Communities: a social unit with commons values and ideals that have made a mindful effort to use information technology to transform life and work in the region where they are located, in meaningful and fundamental, rather than incremental, ways [4]. A "smart" system permits an open, inclusive, systematic, inter and transdisciplinary system vision.

Thus, the concept of "smart" applied in farming includes the employment of new digital and high-tech technology and the creation of a localized community in which these technologies have a meaning. Innovation, intended as a new idea or method, is the cornerstone of smart farming.

New forms of innovation cover all dimensions of the agrarian production cycle, along the entire value chain. Those innovations range from crop, input, and resource management, to organization, marketing, and distribution. New technologies such as sensors, Decision Support Systems (DSS), automation and robotics, collected data, traceability, and blockchain [5] are available to farmers for supporting and enhancing productivity.

Achieving innovation in agriculture is not easy, as it requires diffusion in the early stage of its introduction in a sector. As explained by Rogers [6], diffusion is the process by which an innovation is communicated through various channels over time among the members of a production ecosystem. It is a communication, in that the messages are concerned with new ideas. So, there are four main elements in diffusion: the innovation itself, the communication channel, time, and the social system in which introduce the innovation. As is clear from the above, the efficient application of innovation in a production system needs:

- Those who will adopt and use the innovation (farmers);
- Those who know how to use it (producers, retailers, research institutes);
- Systems and channels of communication between the subject above (instruction, universities);
- Time to achieve communication objective;
- A social system intended as a set of interrelated units that are engaged in joint problem-solving to accomplish a common goal. Following the paradigm of Rogers, the diffusion of innovation appears to be challenging due to the intrinsic characteristic of the agricultural system (distance from urban areas, isolation, difficulties in disseminating information, and farmers' mentality).

Two additional elements should be considered. Firstly, the number of innovations available in precision agriculture and smart farming is large, and the types of new techniques and products are incredibly varied. Secondly, introducing an innovation may lead to an upheaval of a prior structural organization in farms [7], requiring a completely new organization. Those two points may cause confusion and difficulty when farmers have to choose the solution that best suits their needs. Such problems can lead to the wrong choice and to purchasing an underestimated or overestimated set of solutions.

From another perspective, to implement innovation adoption the main economic framework given by common agriculture policy (CAP), funding should be taken into account. It is well known that CAP strategic objectives for the period 2014–2020 mainly aim to invest in rural jobs and growth, improve the sustainable management of natural resources, and roll-out fast and ultra-fast broadband in rural areas. Internet access is closely related to the successful adoption of innovations and technologies. The EU has several tools to finance projects on Information and Communication Technologies (ICT) and broadband in rural areas by 2020. As an example, €21.4 billion for ICT coming from EU structural and Investment funds (ESIF), €1.6–2.0 billion from the CAP's Pillar II Rural development (EAFRD), and €6.4 billion to finance high-speed broadband roll-out. The 2014–2020 timeframe EAFRD budget amounts to approximately €100 billion. Throughout this time, the budget will be invested in implementing rural development programs that run until the end of 2023. The European Agricultural Fund for Rural Development ICT Broadband is in continuous implementation from year to year. Data from ICT Broadband of the rural population potentially benefiting from new or improved ICT services or infrastructures show that if in 2016 the implemented part was only 2% of the total (100%), in 2018 became the 17% of the total. Hopefully, the trend is going to grow [8].

This article aims to create a reference framework to suggest a methodology for farmers, experts, and other actors of the agricultural sector. This methodology allows the orientation towards the multiple-choice offered by precision agriculture and smart farming to find the best entrepreneurial and technological choice and solution. It is possible to summarize this methodology in four steps:

1. Understanding the changes in action;
2. Identifying the added value of smart farming processes;
3. Verifying the reliability of new technologies;
4. Adjusting production processes.

In order to identify drivers and elements of the listed four steps, in the present study both a systematic review of the literature and empirical evidence from on-field wine farm projects are undertaken. The wine sector is chosen as an example for the application of the proposed methodology because it is one of the most important agricultural production in EU. In the period between 2014 and 2018, it accounted for 65% of global production, 60% of consumption and 70% of exports, with 45% of the world's wine-growing regions with 3,215,549 ha in 2018, with an average price of €4 Vol^{-1} Hl^{-1} (red wine) [8–10], making the wine sector one of the most profitable agrarian productions.

The second motivation concerns the background of the research activities carried out, since most of the projects considered focus on wine production and vine growth operations.

This article could be considered as a first guideline in developing a new approach to smart farming, that can lead to further research to gather data necessary to support and increase this theory. Especially regarding lean approach and Technology Readiness Level (TRL) analysis, a minimal number of studies are found.

2. Materials and Methods

2.1. Systematic Literature Review

The method chosen was a systematic literature review which came from a clearly defined objective of the research, following a structured protocol, that minimized subjectivity and allowed the critical evaluation of relevant studies. Furthermore, it permitted the removal of subjectivities derived from traditional literature reviews based on the author's knowledge perspective [11]. The aim was, first, to review and analyze papers that discussed smart farming, focusing on the aforementioned four steps (1,2,3,4). This provided a depth knowledge of such elements that strengthened our postulations derived from empirical findings.

The research database and keywords definitions were built on Scopus, Science Direct (Elsevier), and Google Scholar, the main scientific multi-disciplinary abstract and citation databases of peer-reviewed literature (major journal publishers and conferences in the science and technology fields). Selected keywords were prepared for the databases, chosen coherently with the four steps of the smart farming approach. In the first step we included "Sustainable" and "Multidimensional Analysis". The second step included "Business Model Canvas", the third set involved "Technological Readiness Level", "Market Readiness Levels", and "Local Ecosystem Readiness Level". Finally, the fourth step included "lean management", "manufacturing", and "farm management". When more than one keyword was included in the search, they were linked with the Boolean operator OR, whereas the sets were connected to each other through the AND operator. Specifically, each keyword was followed by the Boolean operator AND with the terms "viticulture" and "agriculture". This phase allowed the authors to achieve the final list of keywords used for the search. The literature search covered the period until May 2020 and all the papers published in Scopus, Science Direct (Elsevier), and Google Scholar in the areas of interest were considered to be screened. The year range limit set was 2000.

Articles which included the selected keywords in their title abstract or keywords were analyzed. In order to provide a significant review, only papers published in peer-reviewed journals, conference proceedings, or books and providing an English version were considered.

In order to select the papers relevant for the scope three screening steps were conducted. The first was the reading of the title and the abstract. In this screening phase articles were classified as included or excluded based on two exclusion criteria (EC). The 1st EC was entire conference proceedings, the 2nd EC out of topic. The latter refers to the papers that clearly showed no relationship with the aim of this paper and its research questions and, then, they have not dealt with smart farming, viticulture, and innovation technology adoption issue. Then, the second step consisted of the reading of the full text of the papers selected and, hence, was a definitive assessment to finally include only relevant papers. During this phase, papers that did not precisely respond to the research topics were excluded, while papers discussing the four steps were considered as included. All the information was extracted and reported in the paper as a specific element of each step, allowing the authors to have a general view and do a thorough evaluation of the present state of the art in the field of research.

The review results derived from the total number of papers searched in the databases with the previously reported keywords. Those papers were 1583. Once the duplicates were deleted, 1112 were used for review. After the screening, 104 papers were included for this study, while the others were omitted according to the two EC previously stated during the first screening phase. Then, 38 articles were chosen after a thorough reading of the 104 documents, while 8 additional articles were explored by scanning of sources, with a total number of 46 documents classified as important, as they specifically suit the intent of the study. The low number of articles selected was a consequence of the highly specific research issues, indicative of a systematic analysis of literature.

2.2. Nonlinear Process Analysis

In parallel with the systematic literature review, a method for empirical findings emerged from a nonlinear process for the introduction of smart farming innovation in wine farms, which was defined as follows. The nonlinear process was developed in periodic brainstorming sessions and interviews in which the authors reflected on the significance of technology adoption means and how it plays out in practice. Our framework was iteratively developed over time, using our individual experiences with bottom-up initiatives, and engaging with the theoretical and practical literature. Our beliefs were presented at various symposiums, and in response to the comments received and our reflections the framework was developed adaptively. The process here described takes its cue from the research and prototyping process. This approach, proven to be very important, provided the comparison between authors and actors in the supply chain, interested in finding a way to orient themselves in the vastness of technological innovation and smart farming. The merit lies in the open opportunities thanks to projects on technological innovations, used not only for the contribution in terms of technology but also in social advancement. The work carried out over years of research-action in the agricultural field, during European, national, or regional funded projects focused on precision agriculture and technology adoption, has made urgent the need to find a way to summarize and spread out the work done. Indeed, we were able to empathise with farmers and the other actors (as consultancy) of the wine sector, and this was very useful to focus a new path, which was needed. For example, it happened with several projects as the INTERREG MARTE+ which had the aim to develop and design technological solutions for "heroic viticulture" with small machinery in the vineyards or with regional funded projects VELTHA, TINIA, OENOSMART, SEMIA, KATTIVO, SMASH, SUSTAIN-BIO, CAMPI CONNESSI, INTRACERT, (Table 1) all focused mainly on viticulture and technology adoption. Moreover, with SPARKLE, we started to test and check the "setting in order" of all the material collected on smart farming with an online e-Learning course for *agripreneurs* (agricultural entrepreneurs) [12]. This permitted to collect information through the practice of focus groups, useful for developing ideas and then defining one or more solutions. The reiterative process adopted helped in receiving feedback and improving solutions for technologies or services provided. Therefore, a set of tools is needed for future agripreneurs and future consultants and technicians.

Table 1. List of research projects from 2012 to 2020.

PROJECTS	YEAR	FOCUS GROUP (EVENTS)	INTERVIEWEES (FARMERS)	MAIN FINDINGS	WEB	FUNDING
TINIA 1	2016–2021	5	40	Technological and Business aspects	[13]	€320,522
INTRACERT [1]	2019–2021	2	20	Agronomical and adoption procedure	[14]	€150,120
KATTIVO [1]	2019–2021	3	34	VRA technologies	[15]	€299,708
CAMPI CONNESSI [1]	2019–2021	3	37	Farming management and technologies	[14]	€322,005
SUSTAIN BIO [1]	2019–2021	1	7	Farming management and technologies	[16]	€153,000
SMASH [2]	2018–2020	2	13	Technologies and Agronomical aspects	[17]	€1,905,000
OENOSMART [1]	2016–2018	2	30	Farm management and business	[18]	€500,000
SPARKLE [4]	2018–2020	3	28 Farmers 19 Researchers 536 Students	Agronomical aspects, technologies, business and farm management	[19]	€775,566
MARTE+ [3]	2012–2013	20	459 Farmers 13 Field day	Technologies and adoption	[20]	€140,000
VELTHA [1]	2016–2018	3	23	Technologies and farm management	[21]	€352,341
SEMIA [1]	2016–2018	2	17	Technologies and farm management	[22]	€499,915

[1] Tuscany Region innovative actions (TR 16.2); [2] Regional Operational Programme (ROP); [3] Interreg Europe (INTERREG); [4] ERASMUS + Key Action 2 (KA2).

Table 1 shows in detail the international and national research projects in which focus groups, field days, and brainstorming sessions were held in order to postulate and build our new proposal through personal experiences and information coming from the actors of the wine farms or farmers. The listed projects provided empirical findings, which represent part of the principles used for the development of the methodological proposal. Information about the name of the research projects involved our research team, the starting year, the number of focus group conducted and interviewees, the main findings regarded by every single project, the webpage, and the total amount funding about every project were listed. The funding resources were of the following types: European Agricultural Fund for Rural Development (EAFRD) through Tuscany Region innovative actions (TR 16.2), European Regional Development Fund (ERDF) through Interreg Europe (INTERREG) and Regional Operational Programme (ROP), ERASMUS + Key Action 2 (KA2). Specifically, 11 running projects from 2012 to 2021 (some are still on-going projects) made possible the organization of 46 sessions, the interviewing of 708 farmers, the conduction of surveys with 536 students (Portugal, Spain, Greece, Italy). Moreover, 19 interviews were carried out with European academic experts on precision agriculture adoption and 19 demo events were conducted on several issues related to innovation and technology adoption. Detailed information is analytically reported in Table 1. Usually, the period for each activity was a working day, but it depended on the kind of project and the target involved.

Each project had questions regarding several issues related mainly to technological, agronomical, economic, environmental, and social issues. The non-linear process adopted was used to collect data and information in the informal part of the event and it is important to emphasize that it is not possible to set standard processes and systems in participation contexts, since each mechanism should be adapted to the group that expresses it. Usually, the results of responses of our projects were more farmers' awareness of the adoption of technologies, and reports tailored for each project.

Questionnaires were usually designed on the basis of a need analysis made before surveying farmers that took into account their needs in terms of economic and social aspects and project objectives, as listed before. Projects and surveys carried out in all partner countries were planned (France, Portugal, Spain, Greece, Italy) and at national level the diversity of the target groups investigated in terms of farm dimension and turnover, etc., was taken into account.

In the following paper, we are going to explain, based on exiting literature and practices, which approach is needed to introduce smart farming in wine farms mainly through four points of view. This method helps farmer (or the technicians, consultancy, etc.) to focus on introducing not only one technology but a set of innovation in the working system as the Multidimensional Method (agronomic), the Business model canvas (economic), the checking of the reliability of new technologies (technological), and then adjusting the production processes (management).

The four points try to simplify and categorize the complexity of the current farm activities. Indeed, these are no longer limited to applying recipes for a single purpose but must necessarily diversify. If they cannot diversify the products, they must diversify the activities, and the limits given by biotic and abiotic factors, as well as human and technological, can be summarized and described through the tools that we are going to describe in detail.

Once this method is defined and assessed, it is important to stress that in this way, more conscious decisions can be taken.

3. Understanding the Changes in Action: The Smart Agricultures Multidimensionality Method

Currently, precision agriculture and smart farming are not very diffuse, as shown by the EU Parliament report of 2016, "Precision Agriculture and the future of farming in Europe". In this article, it was estimated that only 25% of EU farms use technologies applicable to smart farming and precision agriculture. This limited adoption can be caused by encountering some difficulties when proposing or adopting smart farming and precision agriculture. Those obstacles can be summarized in two focal points. Firstly, there is a lack of information on the advantages of applying smart farming instead of a traditional way of production [23], as those advantages are not perceived [24]. Secondly, the significant amount of technology and data collection necessary in smart farming can be challenging to manage.

The conventional way of thinking in agriculture, developed during the 20th century, transformed the agriculture system into an industrialized process that relies on external industrial input for fertilizer, plant protection product, and seeds [2]. However, intensive agricultural practices and agricultural mechanization can have many environmental implications, such as soil erosion and loss of organic matter [25–28], excessive nitrogen application [29], reduction of water reserve [30], and excessive use of pesticides. In particular, the use of pesticides causes many environmental problems (water eutrophication, ecotoxicity, soil degradation, and acidification) [31] and can negatively affect human health [32]. Moreover, the agricultural sector affects climate change, producing approximately 13.5% of global greenhouse gas (GHG) [33]. In particular, methane (CH_4, derived from anaerobic decomposition of organic matter or manure), nitrous oxide (N_2O, mainly due to synthetic fertilizer application), and carbon dioxide (CO_2, resulting from energy use in the farm and the carbon loss due to conventional or excessive tillage) [34]. Specifically, in viticulture, GHG emission is caused by the production and distribution of fertilizers and pesticides, irrigation, pruning, tillage, and pesticide application energy usage, soil emissions, and crop residue management [35], [36]. Some studies [37] show that machinery usage in viticulture accounted for more than 60% of the total warming potential of the wine production process.

In this context, precision viticulture and smart farming could improve the environmental behavior of the viticulture system. By adopting those two forms of management system, it is possible to implement economic, environmental, and social sustainability. Precision agriculture is a circular process which entails data collection, data analysis, decision-making in management, and evaluation of these decisions [38]. In this way, it allows a reduction of agricultural inputs, obtaining the maximum yield and quality of produced grapes [39]. By precisely measuring variations within a field and adapting the strategy accordingly, winegrowers can significantly increase the effectiveness of pesticides and fertilizers, and use them more selectively [40]. Smart farming is a management concept using modern technology (Global Navigation Satellite Systems, soil scanning, data management, or Internet of Things technologies) to increase the quantity and quality of agricultural products.

The application of precision viticulture practices in vineyard field operations could contribute to the reduction of GHG emission thanks to:

- the enhancement of the soil's ability to operate as carbon stock reserve; by less tillage [41] and reduced nitrogen fertilization [42,43];
- the reduction of fuel consumption through fewer in-field operations with tractors (direct GHG decrease);
- the reduction of inputs for the agricultural field operations (indirect GHG decrease) [44].

On the other hand, these practices affect farm productivity by optimizing agricultural inputs producing higher or equal yields with a lower cost than conventional methods and reducing, potentially, the carbon foot print (CFP) of the process by one-quarter [45]. Therefore, the application of smart farming permits to enhance the environmental, economic, and social sustainability of the farming production process. Moreover, precision agriculture and smart farming can give farmers the added value provided by practices that protect and maintain the natural and social environment [46], that is perceived positively by customers and society [47].

One of the critical elements in smart farming is data collection. The current information collection system in agriculture is based almost entirely on using the crop calendar (Figure 1). The crop calendar helps to provide timely crop information. This tool gives details of planting cycles, sowing, and harvesting of locally adapted crops in different agro-ecological areas. It also provides information on the main agricultural practices [48].

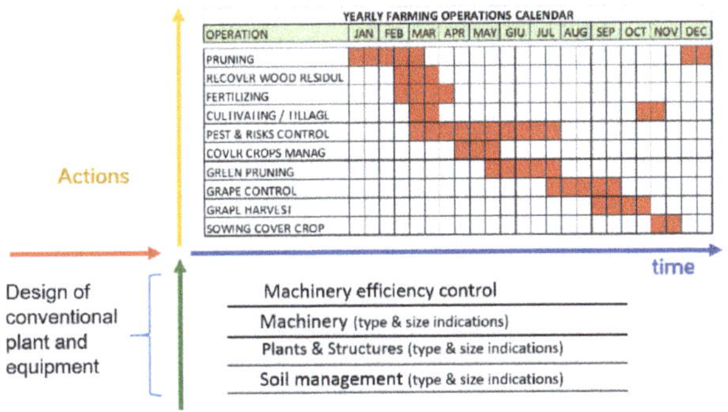

Figure 1. Outdated crop viticulture calendar as a decision-making system. On the x-axis there is time, namely months in a year, and on the y-axis all the action and the cultural operations to carry out are displayed. Those two elements were the only basis for (fourth quarter) designing and choosing the farm equipment and for organizing farm work.

Nowadays, this system is no longer able to support the decision-making process of the farm efficiently. This lack of support is due mainly to the stillness and the lack of comprehensive information collected by this tool.

Technologies, such as sensors, drones, satellites, or intelligent software algorithms, are the primary sources of data in a farm. All information provided by these technologies can generate data, which can be combined and interpreted to give farmers a more comprehensive knowledge, necessary for the optimization of cultivation choices, work, time, and inputs [24]. Therefore, in the smart farming approach, data collection is a more complex and inclusive process that leads to a multiple parametric-specific knowledge, enhanced by contest awareness and enabled by real-time events [49], and a multidimensional approach to data and analysis.

The multidimensional analysis allows us to analyze and categorize several dimensions. Multidimensional data are data that record information related to several different units, called dimensions, for instance, soil, plants, weather, etc. Such a process can help decision-making and planning activities in farms [50].

Therefore, in the methodology proposed here, agronomical choices and objectives (economic, environmental, and social sustainability), all cultivation drivers, and the knowledge needed to accomplish the objectives and use the drivers can be classified in "layers". Risks linked to climate change are additional layers. Risks from climate change in viticulture (late frosts, drought, new pathogens) are to be considered as a key element in the decision-making process in farms. This is because the impact of climate change on crops and the unpredictability of those phenomenon limit farmers, work, and organization in a farm as well as the quality or the quantity of the final product [51]. In creating those layers, there is a transition of knowledge from academic and experimental sources that integrate the previous deploy knowledge.

Referring to Figure 2, those layers are distributed in hierarchical levels, with constraints and available resources at the bottom, and informatics/computing technologies at the top. Each layer and information contained in it must be site-specific and georeferenced. Site-specific information allows optimization of the decision-making process and makes it easy to apply precision agriculture and smart farming protocols. Table 2 shows a reference sequence of layers.

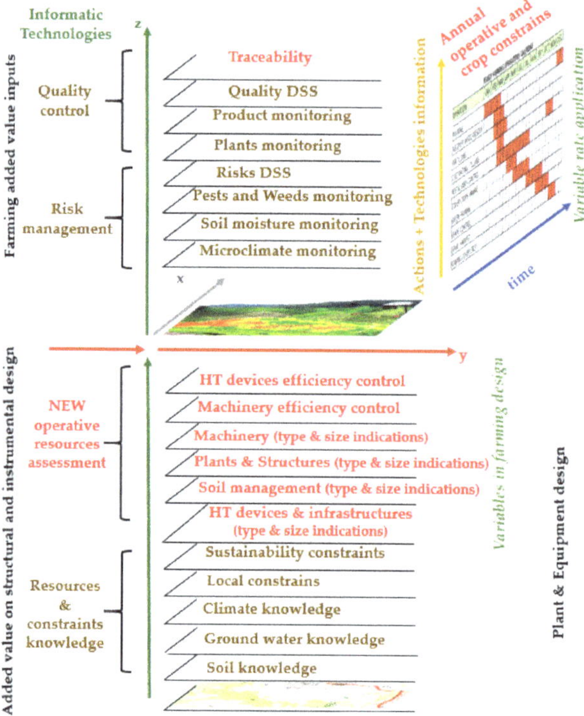

Figure 2. Representation of the future viticulture multidimensional approach for the profitable allocation of new smart technologies in the specific field crop operations. The z-axis specifies the preparatory actions in adopting a new technology. Those actions in the z-axis are summarized in four clusters relative to (1) available resources knowledge, (2) available resources optimization audit, (3) new action in climate change mitigation, (4) new action for quality objective achievement. The x-axis and y-axis show each information and each layer are site-specific georeferenced. Value and costs of each layer of new technology must be allocated on the precise profitable action, for each site-specific area, in the specific time and related farming operations.

As stated earlier, the multidimensional system developed here is based on thorough data collection through various technologies. Some of those technologies are Information and Communication Technologies (ICT), i.e., all techniques that allow storage and exchange of information [52], such as peer-to-peer technologies, broadband. Such technologies are essential in creating some of the previously mentioned layers. The choice between different kinds and levels of technologies depends on various factors that are going to be analyzed in the next points of the presented methodology.

Table 2. Hierarchical levels of the principal wine farm layers, including adopted or necessary knowledge, techniques, and technologies.

Site-Specific Knowledge of Soil and Climate Structural Characteristic	Position, orography, geology, pedology, hydraulic-agrarian managements, ancillary structure Additional resources: biodiversity sites, history, farm culture, and heritage Constraints: risk control, chemical substances, soil conservation, and pollution (Directive 2009/128/EC and 91/67/EEC) Geo-pedologic characteristics obtained through geomatics analysis Groundwater characteristics Historical microclimate site-specific data		
Resources Measurement for Real-Time Variability Management	Agronomic efficiency		Vegetation growth monitoring during season
			Biotic and abiotic damage risk control
		Real-time monitoring	Microclimate, water availability Single plant parameter monitoring (stomatal conductance, canopy growth, hydric stress, plant vigor) IoT, fog and cloud network, field sensor Decision Support System (DSS)
	Technological efficiency		Single device monitoring Registration and monitoring with telemetry of work safety thresholds Automatic setting devices
	Operative efficiency		Automatic or assisted navigation systems in cultural operation Automatic control of machineries and devices VRA variable-rate application Telemetry Operative control and work capability optimization Digital communication between physical connected devices (ISOBUS protocols)
Quality Traceability and Constraints and Confidence Indicators	Policy Farm Management—Digital Integrated System Operation remote control Traceability: QRCode, TAG, telemetry, blockchain		

4. Identifying the Added Value of Smart Farming Processes

If paradigm shifts from a linear to a circular model are happening and the processes of farming are changing, the tools to identify the reliability of technologies should also be appropriate for the time and the sector. In the past, the farming model referred to the industrial production model, the linear model of production [53]. Nowadays, starting or renewing a new business in agriculture needs an entrepreneurial approach and entrepreneurial tools that also consider social and environmental aspects. One of the main tools that can be used, when a farmer should approach its business, is the Business Model Canvas (BMC). As the creator, Osterwalder, says, "the Business Model describes the logic with which an organization creates, distributes and captures value." [54]. In smart farming this tool is not as well known as in other fields, even if it is strictly related to innovation. For this reason, the BMC is a necessary starting point for new businesses in agriculture and especially in smart farming. For instance, some studies refer to it being known that the current business models used by Climate Smart Agriculture (CSA) technological innovation providers are not optimized to meet current market demands and can, therefore, be seen as inhibiting the adoption and spread of CSA technological innovations [55]. Other studies focused on the Business Model innovation, concluding that many barriers exist when farmers take agricultural business model innovation into consideration. Some obstacles depend on human factors, such as behaviors, backgrounds, and beliefs of individuals. Other barriers are more specific in nature and related to the environment of a given industry or business. Besides, many other obstacles are more abstract, such as government laws, supply chain place, and environment [56]. Other authors point out that agricultural enterprise frameworks will protect their competitive advantages by developing and periodically renewing the business models canvases [57]. By going into the detail of the canvas, a BMC is composed of nine blocks—the central one is the "value proposition"; on the right side, there are four blocks focused on customer relationships, customer segments, channels, and revenue streams, on the left side four blocks focused on activities, resources, partners, and costs (Figure 3). In smart farming, BMC might be a tool to help enterprises to understand how to invest in PA to develop economically while also keeping an eye on social and environmental impact.

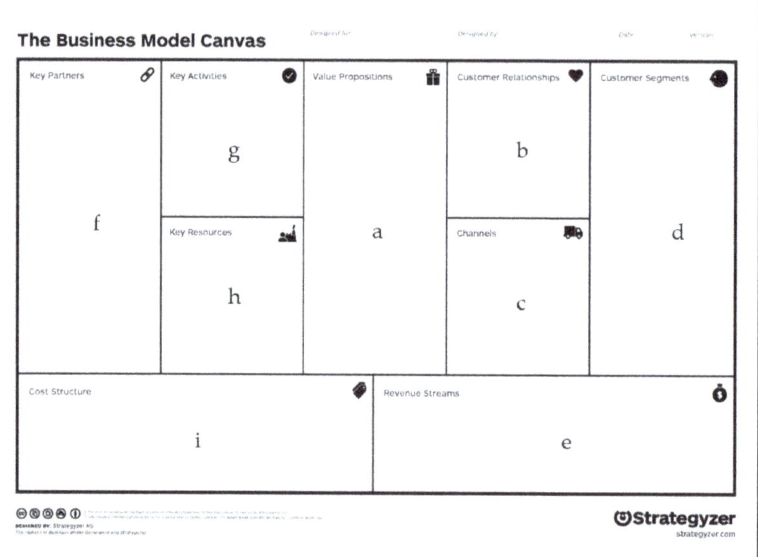

Figure 3. Business Model Canvas (BMC) Template: (**a**) value proposition; (**b**) customer relationships; (**c**) channels; (**d**) customer Segments; (**e**) revenue Streams; (**f**) key partners; (**g**) key activities; (**h**) key resources; (**i**) cost structure. Source: stategyzer.com.

Depending on the sector and the aim to achieve, BMC has been widened both in terms of impact for circular economy approach [58] and in terms of sustainability as in the case of the triple-layered Business Model Canvas [59]. In all developed versions, the core of the canvas remains the value proposition and how all the blocks are linked with and for implementing it, giving added value. The value proposition for a farm considers something that should be offered, some market options included, and customer needs satisfied. Thus, a clear and well-articulated value proposition helps companies with value-added do the right things and do them well. Above all, it allows them to concentrate on offering and providing high value to their customers to ensure that they can profitably acquire and stay competitive in the market [60]. Inside the smart farming approach, and in particular, taking as a practical example a winery, the value proposition is the link between what the farmer offers (grape and wine quality), what the customer needs (sustainability), what management needs in the light of climate change challenge (risk control), and market options (trust and technologies).

The value proposition, which contains the added value for which the company manages to maintain itself on the market, can be obtained with actions, practices, and technologies that must be assessed in terms of type, size, etc., as shown in Figure 4. In risk control and product quality there are important monitoring tools (e.g., proximal and remote sensing) such as Decision Support Systems (DSS) combined with variable rate applications (VRA) that respond to particular needs.

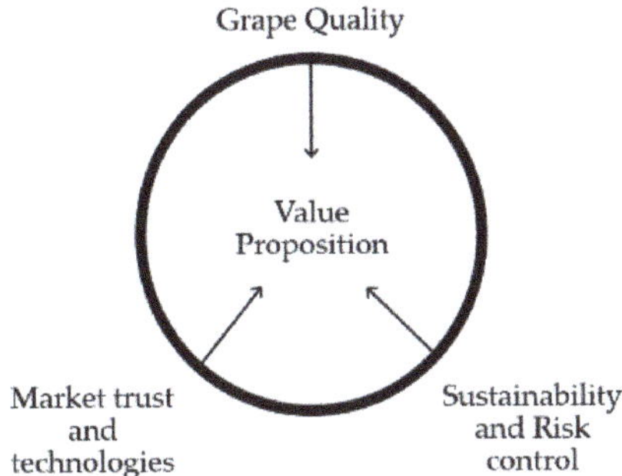

Figure 4. Value proposition structure in winery use case.

Achieving sustainability objectives in agriculture, and consequently in viticulture, first requires specific actions. For instance, in open field cultivation, automatic driving allows you to avoid overlaps with savings in terms of time and costs, but also CO_2 emissions. The sustainability obtained is therefore not only in environmental terms but in its triple form. which also includes impacts on work and on service costs [61].

In conclusion, BMC is a tool that could strongly help to highlight resources with added value, the right scale for technologies related to the business, activities, and key actors needed for upgrading a farm business. The introduction of the BMC tool enhances the evaluation of the farm business formula which has successfully implemented sustainable PA technologies. In addition, a new tool towards "smart farming" can be used to support companies that could invest in PA to expand economically, while at the same time reducing their environmental impact [62]. The "left" side of a BMC (key actors, key activities, key resources, and cost structure) is the crucial part for a farmer who already applies or wants to introduce smart farming in their business, that occurs to check and evaluate the feasibility of the choice made in terms of maturity and effectiveness of the chosen technology. The effectiveness of BMC in agriculture was tested during the SPARKLE Erasmus+ project (Table 1) combined with a "PA impact analysis" investigating drivers, barriers, benefits, and impact of PA technology adopted with the aim to drive consciously technological choices made by farmers. BMC was applied during the project on 20 farms (wine farms and others) in four different countries (Italy, Portugal, Spain, Greece). This application gave life to a training material [63] for farmers, students, and other targets.

5. Verifying the Reliability of New Technologies

In order to verify the reliability of new technologies, the Technology Readiness Levels (TRL) tool can be a valid instrument to evaluate the technology maturity. The concept of TRL was introduced by the National Aeronautics and Space Administration (NASA) in the mid-1970s [64]. This tool was designed to allow more effective assessment and communication regarding the maturity of new technologies, intended for use in a space mission (BSI Standards Publication Space Systems-Definition of the Technology Readiness Levels (TRLs) and Their Criteria of Assessment, 2013) [65]. In subsequent years, the TRL has spread to other productive sectors, such as chemical, fossil energy, electric mobility infrastructures, and also in the United States Department of Defense. Recently, TRLs were also introduced by the European Commission for the evaluation of research projects in the Horizon 2020 program.

The TRL tool assigns a scalar level from 1 to 9 to describe how mature a technology is. Figure 5 shows the technology maturity levels (adapted from the original NASA one), where level 1 is the lowest and level 9 the highest (BSI Standards Publication Space Systems—Definition of the Technology Readiness Levels (TRLs) and Their Criteria of Assessment) [65,66]. In particular, these levels describe, in a linear way, a whole production process, from the research phase to the development phase, to end with distribution.

Phase	TRL	Hardware	Software
Research	1	Basic principles	
Research	2	Concept and application formulation	
Research	3	Concept validation	
Development	4	Experimental pilot	
Development	5	Demonstration pilot	
Development	6	Industrial pilot	
Deployment	7	First Implementation	Industrialization detailed scope
Deployment	8	A few records of implementation	Release version
Deployment	9	Extensive implementation	

Figure 5. Technology Readiness Levels (TRL) table of technological products adapted by BSI Standards Publication Space Systems—Definition of the Technology Readiness Levels (TRLs) and Their Criteria of Assessment, 2013.

This tool, born as a single technology evaluation method, has expanded to complex solutions, i.e., the set of technologies that make up a product [66]. In this way, the TRL tool has spread to other communities, among which the agri-technology community has been recently added. Moreover, the TRL tool evaluates the technology maturity of a new product. Still, it does not take into account if this new product can turn into an innovation, defined as the whole of factors that make a technological solution adoptable in the production process [67]. In the technological evolution of agriculture, the tractor is an historical example of a new product that has turned into an innovation of the entire agricultural system. This evolution has been possible thanks to the creation of a reliable chain of actors, such as retailers, maintainers/repairers, tire services, etc. Only in this way has tractor use become profitable and become a system innovation.

In order to evaluate the reliability chain of actors and infrastructures, the Market Readiness Levels (MRL) tool has been introduced [68]. The MRL allows a technology readiness evaluation for commercialization and diffusion phases. This tool is based on a scale of 1 to 5, with 5 being the most diffused [69].

In a European study, the TRL and the MRL of precision agriculture technologies were evaluated. To summarize the ample supply of new products, the technologies were gathered in six major categories (nanotechnology, yield, soil mapping, drones, sensors, autonomy). The results showed that lower TRLs were recorded in nanotechnology and autonomy technologies, respectively 3.2 and 2.1. Instead, higher TRL were shown in soil mapping with 8.1 and yield with 7.6. In the middle, there were drones (6.6) and sensors (6.1).

Moreover, the MRL results for the same six categories reflected the TRL values, related to the shorter MRL scale. In fact, the MRL showed for autonomy a value of 1.4, for nanotechnology 1.5, for sensors 3.5, for drones 4.1, for yield 4.6, and for soil mapping 4.8. These results show that the most advanced technologies, such as nanotechnology or autonomous vehicles, although they are useful in agriculture, are not developed due to a low MRL. So, this means that the reliability chain of actors and infrastructure is not yet sufficient [69].

A recent market survey on Italian agriculture 4.0 conducted by the "Osservatorio Smart AgriFood" of the Polytecnic University of Milano has shown the Italian situation regarding precision agriculture and smart farming technologies. One of the most exciting survey sections showed the problems faced by the farms surveyed (≈300) in the introduction of new technologies in agricultural processes. Some of the most frequent answers, given by farmers, to this question, were "the malfunctioning of the solution", "insufficient technical support", and "lack of connectivity" in terms of lack of broadband connections [70].

The first evidence highlights that some technologies have been brought to commercial use with low values of TRL, which means they do not work correctly. This is a common problem that concerns new technologies. As a matter of this, companies launch their products without an appropriate test and verification to stay ahead of the competition and the result is that a new technology with low TRL will probably go to fill the "Valley of Death" of technologies [66].

The other two evidences underline how infrastructures around a new technology are not ready to welcome these new solutions. In this case, the MRL is too low to allow technologies to develop correctly. Therefore, if the infrastructure fails to grow in a short time, the new technology will quickly disappear. This risk is well described by Rogers in the "chasm phase" and more recently defined by Gartner Hype Cycle [6,71].

However, the major problem identified by farmers was the "lack of expertise". They stated that it is difficult to find appropriate skills to support these new technologies in the labor market. This common response by farmers shows that there are some things lacking in the educational system at any level, both in high schools/universities and in the repairs sector. Another issue to consider is that the MRL tool does not fully evaluate the system innovation, because it does not take into account the educational system. Therefore, it is necessary to establish the third evaluation tool, the Local Ecosystem Readiness Levels (LERL). This tool is likely to resolve the lack of an instrument that permits to evaluate the readiness of the actors chain around a new technology at local level.

LERL contains two important terms: On the one hand, the noun "ecosystem" (apart from the biological context), which can be interpreted as a complex network and interconnection between multiple entities. In the agricultural technology context, an ecosystem can be described as an aggregate of independent entities and interrelated factors to allow a system innovation in the whole sector [72]. Therefore, the ecosystem represents the linker between different actors and infrastructures of the technological chain. All actors must cooperate following the multi-actor approach to create a thriving and robust ecosystem [73]. Notably, the chain is composed of providers of high-tech systems, services for hardware and software, services for ICT (Information and Communication Technology), consultants, human capital, educational system, and governance. Only with the network between all these actors can a new product become a part of system innovation. However, this is a necessary condition but not sufficient. On the other hand, the adjective "local" strongly correlates to the ecosystem which needs to exist not only in the macro-area but also in the local area. In fact, the Local productive Ecosystem is the physical services and expertise network that supports innovation. The existence of a place in a local area where subjects can create a new product, using it or repairing it, is the necessary and inclusive condition to enable the use of a new product effectively, and lay the groundwork for a system innovation.

Therefore, the LERL is a tool that permits to evaluate the maturity level of the local ecosystem and to establish at which point of the transformation road, from new product to innovation, a new technology is located. Moreover, the LERL tool assigns a scalar level from 1 to 5 to describe how mature a local ecosystem is. Figure 6 shows the Local Ecosystem Readiness Levels, where level 1 is the lowest, and it represents the total absence of the ecosystem. Level 5 is the highest, and it establishes the complete saturation of it.

LERL	Description of presence and appropriateness of actors, services, infrastructures, and expertise.
1	Absence
2	Emergency
3	Development
4	Diffusion
5	Saturation

Figure 6. LERL: Local Ecosystem Readiness Levels. The maturity level of the local ecosystem supporting innovation.

6. Adjust Production Processes: Scheduling of Adoption Procedures

The scheduling of adoption procedures is the last activity in the farm. A proper scheduling requires the reliability verification of the technology that should be introduced (TRL), the compliance with the "added value" required, the appropriateness of the support services and infrastructures (LERL), the compatibility among the several factors of the BMC (i.e., resources, actions, actors), and the operational adoption checking in the production process.

All of this has to be thought of in light of the increasingly competitive emerging markets, where viticulture has to face the challenge of enhancing quality in the face of climate change. Current management of vine production has a considerable environmental and social burden which cannot be sustained indefinitely [36,74,75]. As a response to these critical questions, in other manufacturing sectors, new approaches such as design thinking are being adopted to rethink processes and products. The design thinking and the tools of "Lean Production" or "Lean Farming" may be endorsed as decision support tools also in the viticulture sector. The lean approaches, which are fundamentally anthropocentric, realistic, and firmly based on waste management, have been proven to be extremely compatible with companies sustainability policies and activities [76]. Several studies have investigated and established the capability of managerial practices to improve organizational success. [77]. Lean implementation has been noticeable across a diverse variety of economic sectors, contributing in many instances to increased economic performance and competitiveness [78]. However, this should go hand in hand with changes in legislation and regulations, as well as needs from the diverse stakeholders to develop greater environmental and social responsibility [79]. Farmers need to raise awareness of the environmental and social impact of their operations and to become more responsible for managing their businesses [80]. The integration between the lean philosophy and environmental sustainable production has been extensively documented [81–84]. Piercy and Rich stated that though lean methods and sustainability strategies emerged and continued independently, they have been shown to be very complementary mechanisms [85]. Recently, Reis posited the efficiency of lean and green systems in determining the level of maturity of lean and green integration, allowing benchmarking between organizations of the same industrial sector [86]. In the viticulture sector, studies on the use of lean and green systems are almost non-existent. The meta-analysis of the available literature highlighted five main categories regarding the driving factors that determine sustainable organizational efficiency through the implementation of lean methods:

1. Knowledge and training between workers and managers;
2. Awareness of the operative context;
3. Organizational structure;
4. Technology and decision support;

5. Implementing and cyclical enhancement of the adoption procedure.

6.1. Knowledge and Training between Workers and Manager

In the viticulture sector, the decision-making process in farm management is affected by several variables which take into account the production protocols, the orographical constraints, the administrative rules, the end-users, and supply chain stakeholders, as already stated by Pla [87]. The use of lean methods in smart farming cannot be separated by the overall knowledge in driving sustainable performance. Nevertheless, such knowledge, as stated by Wang Subramanian, must be collective for the subjects directly involved in the production process to allow an overall understanding of sustainability initiatives [88]. The main categories of knowledge identified as important include:

- knowledge on the environmental sustainability issues;
- knowledge on the innovations and practices that would improve sustainable performance;
- knowledge of the context in which the farm works;
- knowledge on the stakeholders' perceptions of value within the organization itself, between the subjects of the supply chain and by the customers.

In order to achieve a picture of knowledge level, we carried out a questionnaire on 26 wine farms located in central and northern Italy, taking into consideration business size class, orographic condition, and management model. The investigation was addressed to the worker and manager technicians who operate and manage the farms. The survey affected several focuses to achieve an overview of technical and economic issues and to understand what is lacking and needed. The general part of the questionnaire was devoted to the understanding of educational level. The results showed a low level among the workers and manager technicians which play active roles (Figure 7).

Workers Educational level distribution Vs Age

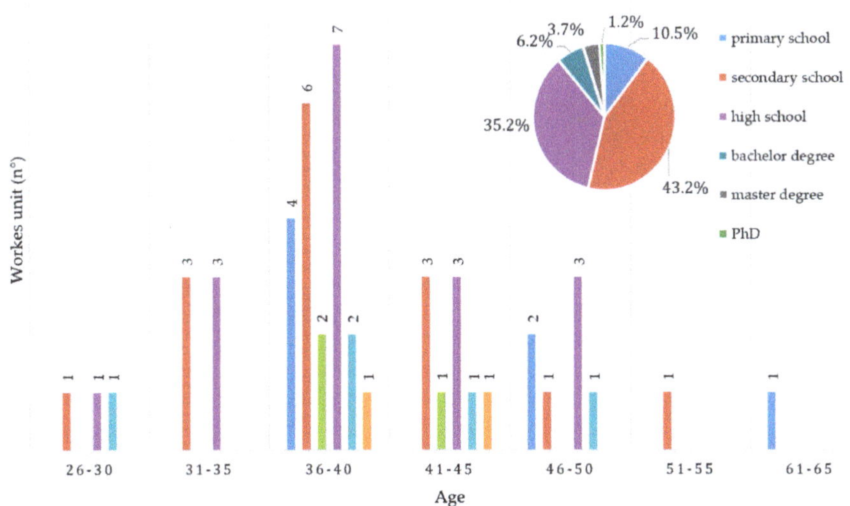

Figure 7. Workers' educational level distribution.

In particular, the highest level of qualification between workers was secondary school, with 43.2%, that is achieved at the age of 14. It represents the second cycle of studies of compulsory education in Italy. Secondary school lasts three years, usually from 11 to 14 years of age, as a continuation of primary school. The second most widespread qualification was high school, with 35.2%. It represents the education level following the attainment of the secondary school license. High school in Italy provides a great variety of addresses and includes various categories, which offer different preparations and professional outlets. Then followed the portion of operators with tertiary education, where the bachelor degree was 6.2%, the master's degree graduates constituted 3.7%, and PhD 1.2%. It follows that the workers' educational level is generally low or linked to experience accumulated over the years. With regard to the qualification of the technical managers, it was almost totally in agriculture sciences (Figure 8). Inside this, it was possible to observe a secondary qualification step in oenology masters and oeno-technicians. Around 30% had high school qualification, of which 18.5% were in agriculture. None of the interviewed people had a degree in engineering sciences. This scenario may represent both a limit in future operational farm management and the propensity to introduce technological innovations, as a consequence of the lack of skills and knowledge in this topic. It follows that a low education level is attributed to low knowledge that affects the implementation of new management protocols and the propensity to the innovation.

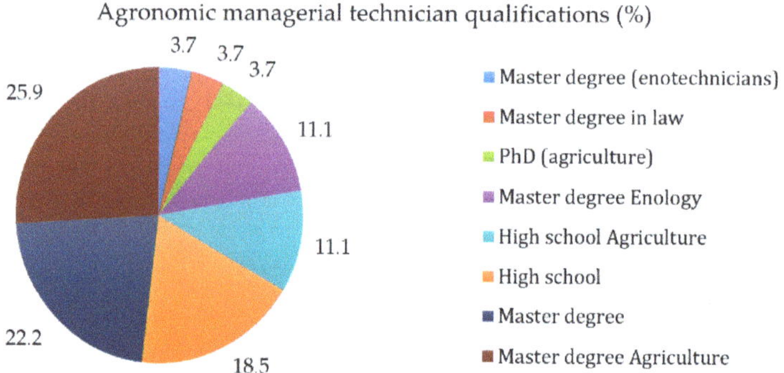

Figure 8. Technical managers' educational level distribution.

Currently, in the wine farms, the training is usually of three types: (a) safety, (b) systematic preventive approach to food safety with hazard analysis and critical control points (HACCP), and (c) technical subjects. Both (a) and (b) are mandatory in Italy. Therefore, all subjects operating in the company, albeit in different forms that vary according to the risks to which they are exposed, are subject to periodic training and updating. Technical training, which represents the most responsible part of the increase in knowledge, is unfortunately often absent in small (<5 ha) and medium-sized companies (<50 ha) and sporadic in large ones (>50 ha). Specific courses are organized in the latter, aimed at operators on the use of machinery, on specific practices such as pruning.

The increase in knowledge for technical managers on specific subjects is often delegated to their will and curiosity. It is most widespread in small- and medium-sized companies where technical managers are also owners and therefore have greater interest in farm growth. In small companies, technical training is carried out through cooperatives, trade associations, and manufacturers that organize events in collaboration with the producers of the raw materials or technologies which sponsor their products. This approach represents both a limit to the openness to knowledge and a risk of spreading standardized practices that are not replicable and efficient in each operating context. The training of key staff with extensive and multi-disciplinary knowledge which provides a broad view of the production processes and consequently a better ability in problem-solving may ensure that workers work independently and competently. Skilled personnel are strategic for the viticulture sector, where productions are generally extensive and decentralized. Nevertheless, as posited by Pearce, the diffusion of knowledge and introduction of innovations also require the assistance of internal or external experts to the farm [80]. The same approach should be followed in driving innovation and lean sustainability in grower operations. Training and development are key elements to the implementation of new practices and technology. Notwithstanding, as being supportive of lean and sustainable performance, these two activities are necessarily complementary, and the new technology cannot substitute good personnel [89].

6.2. Awareness of the Operative Context

Another critical point for a proper introduction of lean is the context where it is applied. Many researches have shown that lean methods are applicable in a wide range of contexts [90]. Anyway, the output outcomes of operational management activities are influenced by the context in which they are applied and must, therefore, be adjusted to the new scenario to achieve the desired results [91].

The viticulture sector and more in general, the agriculture sector, is a highly dynamic and complex (e.g., high orographical, soil, wheatear variability) context. Additionally, environmental and economical elements of viticulture are becoming increasingly complicated and dynamic over time (e.g., climate change and stakeholders demands such as organic or biodynamic wine), aligning with the claim of Azadegan [92] that manufacturers operate in ever more complex and volatile environments. A survey carried out by National Entity for Research and Training in Agriculture (ENAPRA) has highlighted for the Italian viticulture sector that complexities arise from the satisfaction of an ever more extensive and international demand from one country to another, with the maintenance of high-quality standards from year to year and secondly, as regards the high precision with which growers approach their grower operations [93]. Another critical point is the family transition, as 90% of Italian wineries are family-owned. In this sense, in 28% of cases, the most reported criticalities are the non-interest of the owners' heirs to continue the family entrepreneurial experience, in 27% the entrepreneur's reticence to pass the baton, in 26% of cases the organization's difficulties in accepting change, while 6% indicate the inability to leave management to external managers, and 5% the lack of entrepreneurial capacity of the emerging generation. The combination of complexity and dynamism factors of the wine sector makes the implementation of some lean practices more difficult. The changes in the production phases of the field and in the cellar transformation require continuous adaptations of the production protocols as a consequence of the annual variability that influences the quality of the harvest. Besides, in companies where innovations are introduced, learning times are required, which reconcile badly with the high intensity of some process phases, which need reduced execution times in order not to adversely affect quality. In light of the critical aspects of a sector in which the essential element of the production is a biological entity produced in a context subject to the variability of environmental factors, for a successful introduction of the lean method, a planning program in the medium term with concretely achievable objectives seems essential. This was confirmed by technical managers, who stated that on average, the technology adoption process requires at least three seasons to become fully operative.

6.3. Organisational Structure

The farm's size, layout, and resources available and at the same time, the network of developed suppliers and customers which constitute the organizational structure may act as determinants in a farm's ability to implement and sustain lean performance [94–96]. The organization needs to motivate workers, concentrating on more organized and responsible behavior, using more appropriate equipment, pattern sequence, and parallel working. All of this should be done during the effective production process on an ongoing basis. Longoni stated that organizational structures should be designed to develop capabilities and mindsets between the several subjects of the farm to effectively achieve lean [97].

Furthermore, another organizational aspect that must be considered for the proper introduction of the lean approach is the standardization of the geometries of the plants and management techniques. The regularity of the planting layout allows standardization of the settings of the machinery, a more efficient use of resources, and an optimization of the pathways by reducing downtime. Regular plants allow unvarying qualitative and quantitative yields and consequently a simplification of the transformation processes. The batches of grapes are qualitatively homogeneous, and the winemaking process is simpler to manage [98]. Moreover, uniform planting layout allows to optimize and standardize management stages, such as crop protection, which are those with the highest frequency, economic and environmental incidence, and directly responsible for production. Although the standardization of geometries and techniques represents an element in favor of lean management, this must also consider the agronomic and "green" aspects. Lean and green integration must foresee and consider the possibilities offered by sustainable precision agriculture, which deals with biological variability in a differentiated way. The interviews conducted with the farmers highlighted that the operational adjustment of the intervention between plots was almost negligible because it is time-consuming (e.g., manual or mechanical settings). Such behavior is in favor of the lean but not the green approach. For instance, farmers posited that the crop protection in the viticulture is carried out with the same application rate of 250 L ha^{-1} without taking into account the canopy growth stage. This technique is a simplification for analytical accountability and more generally from an organizational point of view, but the lack of adjustments between and within plots represent a "green" limit.

These assertions made by the growers and owners agree with the studies of Lapinski [99], which demonstrated that lean sustainable performance may be improved through the proper design of the layout of the operating area.

Among the available tools which can support the lean and green organizational adoption, there is the WebGIS digital platform. WebGIS together with satellite and telemetry monitoring of yards is an essential element for economic and environmental accounting [100]. In the short term, the WebGIS platforms allow managers to evaluate the progress of the execution of the works and directly or indirectly the process issues [101,102]. In the medium-long term it allows to obtain a database that can be used for the strategic planning of corrective managements actions, re-thinking production flow and investments.

6.4. Technology and Decision Support

Many studies stated that the process of lean adoption must be complementary to those that are technological. In this regard, Benner and Tushman indicated that lean process management integration within a company positively supports the organizational capacity for incremental technological innovation [103]. Furthermore, Ward and Zhou claimed that technological innovation might improve the efficiency advantages of lean organizational systems [104]. Regardless, the lean adoption has to face up to, on one hand, the compatibility of new technologies with existing systems and procedure, and on the other with the owners' expectations to get back the invested capital in a short time. Pears has highlighted two fundamental factors that contribute to the introduction and diffusion of technological innovations in business contexts in which lean management is implemented: the first is the role of pilot case studies that large companies have. In fact, where capital and size are greater, economies of scale can be achieved. The technology in these scenarios goes through a testing phase that determines its success. Success stories can then be replicated on smaller farm scales. The second that emerged and is connected to knowledge is the presence of qualified personnel who, as previously reported, are a fundamental prerequisite. The diffusion of technologies involves minimal changes from an operational point of view (the machines work intelligently supporting the management of the workers), but is relevant from an organizational point of view (changing in settings, maps elaboration, implements configuration). Variable-rate technologies, decision support tools such as evolution models of the primary pathogens, and parasites require sensors and mathematical models that generate data and therefore, information. The latter must be accompanied by effective organizational processes and procedures in order to exploit their potential to implement more sustainable practices.

6.5. Implementing and Cyclical Enhancement of the Adoption Procedure

The practical adoption of innovation through lean methodologies, as highlighted in the previous paragraphs, includes assessments and preparation of all the process elements: structures, plants, tools, services, staff training, operational protocols. The lean process is articulated into several phases and is characterized by a continuous adaptation process. Lean has a wide collection of tools and concepts. There are several types of lean that can be profitably used individually, which makes it much easier to get started. However, the simultaneous use of multiple instruments determines a synergistic effect which leads to a greater increase in benefits compared to the adoption of a single instrument.

The Australian Grape and Wine Authority has focused this process in the wine sector. The results have been published in the guide entitled, "Lean guide: a primer on lean production for the Australian wine industry" [92]. The guide highlights the usefulness of two methods, namely 5S workplace productivity and value stream mapping (VSM) [105,106]. The first eliminates waste that results from a poorly organized work area, while the VSM visually maps the flow of production. The VSM shows the current and future state of processes in a way that highlights opportunities for improvement. In particular, it exposes waste in the current processes and provides a roadmap for improvement through the future state. The tools presented indicate paths of continuous improvement of the production process through a cyclical sequence of stages: examine, identify, implement, assess, check of the feedbacks.

Dedicated to the optimization of the process in the cellar sector, the Australian lean guide represents a practical tool of great utility which, with the necessary adjustments, can also be transposed into farming, i.e., in the production of grapes.

Many of the phases which are performed in the vineyard are often repetitive (e.g., crop protection) and involve the use of resources such as men and machines like in the wineries. These factors inevitably determine a variability of operating outputs depending on workers' skills, the technical efficiency of the mechanical means, and depending on the managerial skills of the agronomical technical manager. Obviously, in farming, the complexity is higher because we operate in an open environment where we are conditioned by external uncertainties (i.e., climatic factors and market conditions), crop characteristics, and cultivating scenarios. Decision-making in this uncertainty is a central issue to be resolved. It is, therefore, necessary to define specific operational protocols capable of responding precisely to the manifestation of critical and emergency scenarios typical of viticulture. These protocols can support farms in planning and forecasting production, also considering the evolution of the market trends. In general, farming chains require systematic analysis for the development of optimization at all stages. Mathematical models, data, and digitalization are a promising direction for seeking solutions to the farming issues.

7. Conclusions

The smart farming methodology originated from the systematic literature review and a non-linear process analysis gained in research projects aimed at testing or introducing innovative technologies in farms to increase sustainability understood in its three environmental, economic, and social components and the fulfilment of the regulations imposed by the European community. The proposal was modeled on the vine sector's issues, which is facing a new management paradigm strongly affected by the use of digital technologies. The change taking place requires an overall re-thinking of the current management methods of wine farms and, more generally, agricultural ones. Greater awareness is needed in the adoption of innovations, as they require significant investments in capital and human resources so that they may bring an increase in competitiveness. At present, for our knowledge of the wine sector, there are no farms that have adopted the smart farming methodology in its entirety. In Italy, there are excellence cases where some of the above elements have been adopted mainly in order to solve critical issues or enhance a productive stage considered as a priority by individual companies, but not in an organic way on the entire business management process [23]. In adopting the smart farming methodology, data collection is an essential stage. Database collections are a basic tool during the decision-making process and the organization of operation and work in farms. However, driving data is a complex process that requires knowledge and competences in order to acquire and interconnect all the information present or provided in farms. The information collected in a farm can come in different formats and sources. Mainly, in smart farming, information comes from technologies and monitoring. For those reasons, to understand the strategic importance of any information, in this methodology, the layer concept is provided. In the layer concept, all the information and all the data are classified and stratified in hierarchical levels, with constraints and available resources at the bottom, and informatics/computing technologies at the top, that gives the agri-entrepreneur an overall picture of their farm and a way to strategically use all the data and the information in the decision-making process. BMC tool application could help to focus on the value proposition of the farm and then to point out the technologies needed to bring added values for the farm outcomes. In agriculture, the TRL tool permits evaluation of the maturity levels both of one technology and of a set of technologies. This tool indicates innovative products' (technology) readiness, but it does not indicate the development of the infrastructure related to the technology. That is why the MRL must integrate it. Indeed, this last tool can provide the degree of the technology readiness for commercialization and diffusion. However, even in this case, not all the variables that contribute to turning an innovative product in a system innovation are taken into account. It is in this perspective that the third evaluation tool (LERL) must be seen. This last indicator aims to evaluate the maturity level of the local ecosystem, i.e., the chain of infrastructures, actors, and formative system in each area or region. Moreover, the LERL may establish at which point of the transformation road, from innovative product to innovation, a new technology is located.

The increasing availability of agricultural technologies able to provide data requires an exact integration process. Tools such as the Business Model Canvas, the assessment of the TRL level and the restructuring of processes according to lean and, most recently, lean plus green methods, offer advantages that allow farms to acquire highly competitive margins.

The study carried out has made it possible to identify priority criteria that determine the success of smart farming, which can be summarized in the following points:

- The agricultural progress in smart farming could offer huge possibilities to enhance quality and profitability for the future agripreneurs;
- The enthusiasm for astonishing innovative products should be controlled, driving the whole entrepreneur process in a shared system of territorial rural innovation;
- Variability in type of farm, age of actors, and infrastructure (i.e., broadband) should be taken into consideration, scaling the introducing technology in an innovative systems design of new shared, connected services like the territorial digital platform to process data for all sizes of entrepreneurial farms of a productive community;
- It seems necessary to grow the diffused awareness of thinking in terms of added value, assess and allocate, prepare the change in the farming process and adopt, verify and tune-up through a lean-approach;
- More and more appropriate is the High Tech and ITC cluster networks participation to be aware and joined with knowledge at global and local level.

The analysis showed that farmers are too often viewed only as a source of knowledge to be used by researchers rather than as an active participant in rural territory development, management, and transformation choices in participatory processes.

In our experience, if projects are carried out with careful participatory processes, project objectives could change and implement during the process. The results of the "Oenosmart" project [18] are one of the sets of results confirming our experience. First of all, the system of European public funding, in this case an EAFRD, can economically stimulate the farmers to adopt PA technologies and "smart" solutions. By providing a trustworthy loan system, the EAFRD allows investment that, in other condition, are perceived by the farmers as not possible or not worthy. So, public funding is an essential base for PA and smart farming adoption. A second result concerns DSS use. As a matter of fact, in the project, the creation of a DSS system allowed the farms to verify themselves, permitting an optimization of the choice for PA adoption and to remodel the farm in a "smart" way, both in field operation and farm management. Indeed, the Oenosmart DSS provided the farmers with a comprehensive instrument for smart farming implementation. Another element deduced from the Oenosmart project is need and impulse to connectivity and sharing information between the farmer, other farmers, and other actors (providers, consultants, etc.). The Oenosmart project started as a weather data collection platform inside the farms of a defined territory (Montalcino, Tuscany, Italy). Beginning with this platform, the farmers understood the importance of data sharing. Little by little, Oenosmart led to the creation of a localized net of agricultural actors that share data, information, and experience on technologies. As an unexpected result, a platform born for weather forecasting evolved into a localized ecosystem, giving access to more data and real-time information to farmers and between farmers.

The findings suggest relevant implications, such as the need to raise awareness of precision agriculture and smart farming tools and encourage the dissemination of information aimed at reducing the degree of perceived complexity, in light of the Common Agricultural Policy Reform post 2020. Based on the conclusion and statements of this study and the opportunity of framework implementation in real wine farm cases, the logical and natural evolution of this research aims to the on-farm validation of the multidisciplinary approach. The next step will be the evaluation of the effectiveness of the method, which may be quantified with economic, environmental, and social impact indicators. Finally, this will open the question of how and which technologies are the most profitable for a given production scenario, because technologies are individually adaptable and cannot be standardized.

Author Contributions: Conceptualization, M.V., D.S., S.L., A.P., C.P.; methodology, D.S., S.L.; investigation D.S., S.L., A.P., C.P. V.D.P., R.L., M.R., G.C., M.V.; data curation, D.S., S.L., A.P., C.P.; writing—original draft preparation, D.S., S.L., A.P., C.P.; writing—review and editing, all authors; supervision, D.S., M.V. All authors have read and agreed to the published version of the manuscript.

Funding: This research received no external funding.

Acknowledgments: The authors would like to thank all the farms and the subjects who interacted with the acquisition of the elements for the development of the methodology.

Conflicts of Interest: The authors declare no conflict of interest.

References

1. Bochtis, D.; Sørensen, C.A.G.; Kateris, D. 1-Agricultural Production Through Technological Evolution. In *Operations Management in Agriculture*; Bochtis, D., Sørensen, C.A.G., Kateris, D.B.T.-O.M., Eds.; Academic Press: Cambridge, MA, USA, 2019; pp. 1–18, ISBN 978-0-12-809786-1.

2. Van Der Ploeg, J.D.; Renting, H.; Brunori, G.; Knickel, K.; Mannion, J.; Marsden, T.; De Roest, K.; Sevilla-Guzmán, E.; Ventura, F. Rural development: From practices and policies towards theory. *Sociol. Rural.* **2000**, *40*, 391–408. [CrossRef]

3. Oliver, M.A.; Bishop, T.F.A.; Marchant, B.P. *Precision Agriculture for Sustainability and Environmental Protection*; Routledge: Abingdon, UK, 2013.

4. Komninos, N. The age of intelligent cities. *Age Intell. Cities* **2014**. [CrossRef]

5. Lombardo, S.; Sarri, D.; Vieri, M.; Baracco, G. Proposal for spaces of agrotechnology co-generation in marginal areas. *Atti Della Soc. Toscana Sci. Nat. Mem. Ser. B* **2018**, *125*, 19–24. [CrossRef]

6. Rogers, E.M.; Singhal, A.; Quinlan, M.M. *Diffusion of Innovations*; Simon & Schuster International: New York, NY, USA, 2019; ISBN 9781351358712.

7. Cera, M. *Meccanica Agraria vol.1 Meccanizzazione Agricola*; Pàtron: Bologna, Italy, 1979; ISBN 8855513702.

8. Open Data Portal for the European Structural Investment Funds—European Commission |Data| European Structural and Investment Funds. Available online: https://cohesiondata.ec.europa.eu/funds/eafrd#top (accessed on 27 August 2020).

9. Agriculture, Forestry and Fishery Statistics—Statistics Explained. Available online: https://ec.europa.eu/eurostat/statistics-explained/index.php/Agriculture,_forestry_and_fishery_statistics (accessed on 24 August 2020).

10. Wine Market Observatory|European Commission. Available online: https://ec.europa.eu/info/food-farming-fisheries/farming/facts-and-figures/markets/overviews/market-observatories/wine (accessed on 24 August 2020).

11. Di Pasquale, V.; Franciosi, C.; Iannone, R.; Malfettone, I.; Miranda, S. Human error in industrial maintenance: A systematic literature review. In Proceedings of the XXII Summer School "Francesco Turco", Palermo, Italy, 13–15 September 2017; pp. 164–170.

12. Valero, C.; Krus, A.; Barreiro, P.; Diezma, B.; Hernández, N.; Garrido, M.; Moya, A.; Ramírez, J.J.; Moreda, G.; Muñoz, B.; et al. Mejorando la formación en nuevas tecnologías para una agricultura digital: Proyecto Sparkle. *Congr. Ibérico Agroingeniería* **2019**. [CrossRef]

13. TINIA Project. Agricoltura di Precisione per le PMI Cerealicole. Available online: https://www.agrismartlab.unifi.it/p118.html (accessed on 26 June 2020).

14. INTRACERT Project. Innovazione e Tradizione Nella Filiera dei Cereali Tipici. Available online: https://www.dagri.unifi.it/p714.html (accessed on 26 June 2020).

15. KATTIVO Project. Piano Strategico per lo Sviluppo di un Kit per la Modifica di Atomizzatori in Grado di eseguire Trattamenti con Tecnologia Innovativa a dose Variabile Ottimizzata in Funzione della Chioma e Ridurre il Rilascio di Sostanze Inquinanti e Fitofarmaci. Available online: https://www.kattivo.it/ (accessed on 26 August 2020).

16. SUSTAIN BIO Project. Viticoltura di Precisione e Innovazione Digitale Della Ottimizzazione dei Trattamenti del Comprensorio Viticolo. 2020. Available online: https://www.dagri.unifi.it/p718.html (accessed on 26 August 2020).

17. SMASH Project. Smart Machine for Agricultural Solutions Hightech. 2020. Available online: https://www.dagri.unifi.it/p441.html (accessed on 26 August 2020).

18. OENOSMART Project. Piattaforma Territoriale di Servizi per L'agricoltura Sostenibile di Precisione. 2018. Available online: https://brunello.ciatoscana.eu/ (accessed on 26 August 2020).

19. SPARKLE Project. Sustainable Precision Agriculture: Research and Knowledge for Learning How to be an Agri-Entrepreneur. 2020. Available online: http://sparkle-project.eu/ (accessed on 26 August 2020).

20. Marte + Meccanizzazione Project. Mare, Ruralità e Terra: Potenziare L'unitarietà Strategica. 2020. Available online: https://www.agrismartlab.unifi.it/vp-101-download.html (accessed on 26 August 2020).

21. VELTHA Project. Vite e Vino, Eccellenza del Territorio, dell'Habitus e dell'Ambiente. 2015. Available online: http://www.velthapif2015.it/ (accessed on 26 August 2020).

22. SEMIA Project. Indirizzi di Sanità, Sostenibilità ed Eccellenza per la Futura Olivicoltura Mediterranea. Available online: https://www.semiadss.it/home/index.html (accessed on 26 June 2020).

23. Vecchio, Y.; Agnusdei, G.P.; Miglietta, P.P.; Capitanio, F. Adoption of precision farming tools: The case of italian farmers. *Int. J. Environ. Res. Public Health* **2020**, *17*, 869. [CrossRef]

24. van der Burg, S.; Bogaardt, M.J.; Wolfert, S. Ethics of smart farming: Current questions and directions for responsible innovation towards the future. *NJAS Wagening. J. Life Sci.* **2019**, *90–91*, 100289. [CrossRef]

25. Lal, R. Soil carbon sequestration impacts on global climate change and food security. *Science* **2004**, *304*, 1623–1627. [CrossRef]

26. Lal, R. Enhancing eco-efficiency in agro-ecosystems through soil carbon sequestration. *Crop Sci.* **2010**, *50*, S120–S131. [CrossRef]

27. Montgomery, D.R. Soil erosion and agricultural sustainability. *Proc. Natl. Acad. Sci. USA* **2007**, *104*, 13268–13272. [CrossRef]

28. Quinton, J.N.; Govers, G.; Van Oost, K.; Bardgett, R.D. The impact of agricultural soil erosion on biogeochemical cycling. *Nat. Geosci.* **2010**, *3*, 311–314. [CrossRef]

29. Robertson, G.P.; Vitousek, P.M. Nitrogen in agriculture: Balancing the cost of an essential resource. *Annu. Rev. Environ. Resour.* **2009**, *34*, 97–125. [CrossRef]

30. Morison, J.I.L.; Baker, N.R.; Mullineaux, P.M.; Davies, W.J. Improving water use in crop production. *Philos. Trans. R. Soc. B Biol. Sci.* **2008**, *363*, 639–658. [CrossRef] [PubMed]

31. Stoate, C.; Báldi, A.; Beja, P.; Boatman, N.D.; Herzon, I.; van Doorn, A.; de Snoo, G.R.; Rakosy, L.; Ramwell, C. Ecological impacts of early 21st century agricultural change in Europe—A review. *J. Environ. Manag.* **2009**, *91*, 22–46. [CrossRef] [PubMed]

32. Hernández, A.F.; Parrón, T.; Tsatsakis, A.M.; Requena, M.; Alarcón, R.; López-Guarnido, O. Toxic effects of pesticide mixtures at a molecular level: Their relevance to human health. *Toxicology* **2013**. [CrossRef] [PubMed]

33. Montzka, S.A.; Dlugokencky, E.J.; Butler, J.H. Non-CO$_2$ greenhouse gases and climate change. *Nature* **2011**, *476*, 43–50. [CrossRef] [PubMed]

34. Macleod, M.; Eory, V.; Gruère, G.; Lankoski, J. Cost-effectiveness of greenhouse gas mitigation measures for agriculture: A literature review. *OECD Food Agric. Fish. Pap.* **2015**, *89*. [CrossRef]

35. Litskas, V.D.; Irakleous, T.; Tzortzakis, N.; Stavrinides, M.C. Determining the carbon footprint of indigenous and introduced grape varieties through life cycle assessment using the island of Cyprus as a case study. *J. Clean. Prod.* **2017**. [CrossRef]

36. Recchia, L.; Sarri, D.; Rimediotti, M.; Boncinelli, P.; Cini, E.; Vieri, M. Towards the environmental sustainability assessment for the viticulture. *J. Agric. Eng.* **2018**, *49*, 19–28. [CrossRef]

37. Aguilera, E.; Guzmán, G.; Alonso, A. Greenhouse gas emissions from conventional and organic cropping systems in Spain. II. Fruit tree orchards. *Agron. Sustain. Dev.* **2015**, *35*, 725–737. [CrossRef]

38. Bramley, R.G.V.; Pearse, B.; Chamberlain, P. Being profitable precisely—A case study of precision viticulture from Margaret River. *Aust. N. Z. Grapegrow. Winemak.* **2003**, *473a*, 84–87.

39. Kitchen, N.R.; Sudduth, K.A.; Myers, D.B.; Drummond, S.T.; Hong, S.Y. Delineating productivity zones on claypan soil fields using apparent soil electrical conductivity. *Comput. Electron. Agric.* **2005**. [CrossRef]

40. FAO. Citrus Fruit fresh and processed. *Ccp.Ci/St/2016 FAO* **2016**, *47*. [CrossRef]

41. Angers, D.A.; Eriksen-Hamel, N.S. Full-inversion tillage and organic carbon distribution in soil profiles: A meta-analysis. *Soil Sci. Soc. Am. J.* **2008**, *72*, 1370–1374. [CrossRef]

42. Khan, S.A.; Mulvaney, R.L.; Ellsworth, T.R.; Boast, C.W. The myth of nitrogen fertilization for soil carbon sequestration. *J. Environ. Qual.* **2007**, *36*, 1821–1832. [CrossRef] [PubMed]

43. Waldrop, M.P.; Zak, D.R.; Sinsabaugh, R.L.; Gallo, M.; Lauber, C. Nitrogen deposition modifies soil carbon storage through changes in microbial enzymatic activity. *Ecol. Appl.* **2004**, *14*, 1172–1177. [CrossRef]

44. Plant, R.E.; Pettygrove, G.S.; Reinert, W.R. Precision agriculture can increase profits and limit environmental impacts. *Calif. Agric.* **2000**, *54*, 66–71. [CrossRef]

45. Balafoutis, A.T.; Koundouras, S.; Anastasiou, E.; Fountas, S.; Arvanitis, K. Life cycle assessment of two vineyards after the application of precision viticulture techniques: A case study. *Sustainability* **2017**, *9*, 1997. [CrossRef]

46. Bekmezci, M. Companies' profitable way of fulfilling duties towards humanity and environment by sustainable innovation. *Procedia Soc. Behav. Sci.* **2015**. [CrossRef]

47. van Evert, F.K.; Gaitán-Cremaschi, D.; Fountas, S.; Kempenaar, C. Can precision agriculture increase the profitability and sustainability of the production of potatoes and olives? *Sustainability* **2017**, *9*, 1863. [CrossRef]

48. Crop Calendar. Available online: http://www.fao.org/agriculture/seed/cropcalendar/welcome.do (accessed on 25 June 2020).

49. Wolfert, S.; Ge, L.; Verdouw, C.; Bogaardt, M.J. Big data in smart farming—A review. *Agric. Syst.* **2017**, *153*, 69–80. [CrossRef]

50. Hira, S.; Deshpande, P.S. Data analysis using multidimensional modeling, statistical analysis and data mining on agriculture parameters. *Procedia Comput. Sci.* **2015**, *54*, 431–439. [CrossRef]

51. Sixth Assessment Report—IPCC. Available online: https://www.ipcc.ch/assessment-report/ar6/ (accessed on 29 June 2020).

52. ICT (Information and Communication Technologies) in "Dizionario di Economia e Finanza". Available online: http://www.treccani.it/enciclopedia/ict_%28Dizionario-di-Economia-e-Finanza%29/ (accessed on 25 June 2020).

53. Porter, M.E. *Competitive Advantage: Creating and Sustaining Superior Performance*; The Free Press: New York, NY, USA, 1985; ISBN 9780029250907.

54. Osterwalder, A.; Pigneur, Y.; Bernarda, G.; Smith, A. *Value Proposition Design: How to Create Products and Services Customers Want*; John Wiley & Sons: Hoboken, NJ, USA, 2014.

55. Long, T.B.; Blok, V.; Poldner, K. Business models for maximising the diffusion of technological innovations for climate-smart agriculture. *Int. Food Agribus. Manag. Rev.* **2016**, *20*, 5–23. [CrossRef]

56. Sivertsson, O.; Tell, J. Barriers to business model innovation in swedish agriculture. *Sustainability* **2015**, *7*, 1957–1969. [CrossRef]

57. Dudin, M.N.; Lyasnikov, N.V.; LeontEva, L.S.; Reshetov, K.; Sidorenko, V.N. Business model canvas as a basis for the competitive advantage of enterprise structures in the industrial agriculture. *Biosci. Biotechnol. Res. Asia* **2015**, *12*, 887–894. [CrossRef]

58. Circulab Circular Canvas. Available online: https://circulab.fr/wp-content/uploads/2015/09/circulab-board2.pdf (accessed on 8 June 2020).
59. Joyce, A.; Paquin, R.L. The triple layered business model canvas: A tool to design more sustainable business models. *J. Clean. Prod.* **2016**, *135*, 1474–1486. [CrossRef]
60. Amanor-boadu, V.; Verni, R. *Value Proposition Definition for an Agricultural Value-Added Business*; Department of Agricultural Economics. Kansas State University: Manhattan, KS, USA, 2004; pp. 1–4.
61. Soto, A.I.; Barnes, A.; Balafoutis, A.; Beck, B.; Vangeyte, J. *The Contribution of Precision Agriculture Technologies to Farm Productivity and the Mitigation of Greenhouse Gas Emissions in the EU*; Joint Research Centre: Ispra, Italy, 2019; ISBN 9789279928345.
62. Maria, P.; Aikaterini, P.; Marco, V.; Lombardo, S. A Tool for Business Model Canvas Analysis: Effective Implementation of Sustainable Precision Agriculture. EBEEC Conf. Proc. 2020. Available online: http://ebeec.ihu.gr/documents/oldConferences/EBEEC2020_abstracts.pdf (accessed on 9 June 2020).
63. Handbook for Business Model Canvas in the Field of Sustainable Precision Agriculture Sustainable Precision Agriculture: Research and Knowledge for Learning How to be an Agri-Entrepreneur. Available online: http://tiny.cc/quam6y (accessed on 27 August 2020).
64. Mankins, J.C. Technology readiness assessments: A retrospective. *Acta Astronaut.* **2009**, *65*, 1216–1223. [CrossRef]
65. BSI. *BSI Standards Publication Space Systems-Definition of the Technology Readiness Levels (TRLs) and Their Criteria of Assessment*; BSI: London, UK, 2013.
66. Earto. *The TRL Scale as a Research & Innovation Policy Tool, Earto Recommendations*; Earto: New York, NY, USA, 2014; pp. 1–17.
67. Pérez-Luño, A.; Wiklund, J.; Cabrera, R.V. The dual nature of innovative activity: How entrepreneurial orientation influences innovation generation and adoption. *J. Bus. Ventur.* **2011**, *26*, 555–571. [CrossRef]
68. Aasrud, A.; Baron, R.; Karousakis, K. Market Readiness: Building Blocks for Market. Approaches. 2010. Available online: https://www.oecd-ilibrary.org/content/paper/5k45165zm8f8-en?crawler=true (accessed on 11 June 2020).
69. Maciejczak, M. Assessing readiness levels of production technologies for sustainable intensification of agriculture. *APSTRACT Appl. Stud. Agribus. Commer.* **2018**, *12*, 47–52. [CrossRef]
70. Osservatori.net digital innovation. IL DIGITALE È SERVITO! Dal Campo Allo Scaffale, La Filiera Agroalimentare è Sempre Più Smart. 2019. Available online: https://www.osservatori.net/it_it/convegno-risultati-ricerca-osservatorio-smart-agrifood-2020 (accessed on 10 July 2020).
71. Hype Cycle Research Methodology. Available online: https://www.gartner.com/en/research/methodologies/gartner-hype-cycle (accessed on 29 June 2020).
72. Oosterwaal, L.; Stam, E.; Van den Toren, J.P. *Openbaar Bestuur in Regionale Ecosystemen Voor Ondernemerschap*; Utrecht University Centre for Entrepreneurship: Utrecht, The Netherlands, 2017.
73. Rogge, E.; Dessein, J.; Verhoeve, A. The organisation of complexity: A set of five components to organise the social interface of rural policy making. *Land Use Policy* **2013**, *35*, 329–340. [CrossRef]
74. Foley, J.A.; Ramankutty, N.; Brauman, K.A.; Cassidy, E.S.; Gerber, J.S.; Johnston, M.; Mueller, N.D.; O'Connell, C.; Ray, D.K.; West, P.C.; et al. Solutions for a cultivated planet. *Nature* **2011**, *478*, 337–342. [CrossRef]
75. Power, A.G. Ecosystem services and agriculture: Tradeoffs and synergies. *Philos. Trans. R. Soc. B Biol. Sci.* **2010**, *365*, 2959–2971. [CrossRef] [PubMed]
76. Fliedner, G.; Majeske, K. Sustainability: The new lean frontier. *Prod. Invent. Manag. J.* **2010**, *46*, 6–13.
77. Chiarini, A. Environmental policies for evaluating suppliers' performance based on GRI indicators. *Bus. Strategy Environ.* **2017**, *26*, 98–111. [CrossRef]
78. Martínez-Jurado, P.J.; Moyano-Fuentes, J. Lean management, supply chain management and sustainability: A literature review. *J. Clean. Prod.* **2014**, *85*, 134–150. [CrossRef]
79. Gordon, P.J. *Lean and Green—PROFIT for Your Workplace and the Environment*; Berrett-Koehler Publishers: San Francisco, CA, USA, 2001.
80. Pearce, D.; Dora, M.; Wesana, J.; Gellynck, X. Determining factors driving sustainable performance through the application of lean management practices in horticultural primary production. *J. Clean. Prod.* **2018**. [CrossRef]
81. Dhingra, R.; Kress, R.; Upreti, G. Does lean mean green? *J. Clean. Prod.* **2014**, *85*, 1–7. [CrossRef]

82. Fercoq, A.; Lamouri, S.; Carbone, V. Lean/green integration focused on waste reduction techniques. *J. Clean. Prod.* **2016**, *137*, 567–578. [CrossRef]

83. Rolo, A.; Pires, A.R.; Saraiva, M. Supply Chain as a Collaborative Virtual Network Based on LARG Strategy. In *Advances in Intelligent Systems and Computing, Proceedings of the Eighth International Conference on Management Science and Engineering Management, Lisbon, Portugal, 25–27 July 2014*; Xu, J., Cruz-Machado, V.A., Lev, B., Nickel, S., Eds.; Springer: Berlin/Heidelberg, Germany, 2014; pp. 701–711.

84. Verrier, B.; Rose, B.; Caillaud, E.; Remita, H. Combining organizational performance with sustainable development issues: The lean and green project benchmarking repository. *J. Clean. Prod.* **2014**, *85*, 83–93. [CrossRef]

85. Piercy, N.; Rich, N. The relationship between lean operations and sustainable operations. *Int. J. Oper. Prod. Manag.* **2015**, *35*, 282–315. [CrossRef]

86. Reis, L.V.; Kipper, L.M.; Giraldo Velásquez, F.D.; Hofmann, N.; Frozza, R.; Ocampo, S.A.; Taborda Hernandez, C.A. A model for lean and green integration and monitoring for the coffee sector. *Comput. Electron. Agric.* **2018**, *150*, 62–73. [CrossRef]

87. Plà, L.M.; Sandars, D.L.; Higgins, A.J. A perspective on operational research prospects for agriculture. *J. Oper. Res. Soc.* **2014**, *65*, 1078–1089. [CrossRef]

88. Wang, Z.; Subramanian, N.; Gunasekaran, A.; Abdulrahman, M.D.; Liu, C. Composite sustainable manufacturing practice and performance framework: Chinese auto-parts suppliers' perspective. *Int. J. Prod. Econ.* **2015**, *170*, 219–233. [CrossRef]

89. Thoumy, M.; Vachon, S. Environmental projects and financial performance: Exploring the impact of project characteristics. *Int. J. Prod. Econ.* **2012**, *140*, 28–34. [CrossRef]

90. Holweg, M. The genealogy of lean production. *J. Oper. Manag.* **2007**, *25*, 420–437. [CrossRef]

91. Sousa, R.U.I.; Voss, C.A. Quality management: Universal or context dependent? *Prod. Oper. Manag.* **2001**, *10*, 383–404. [CrossRef]

92. Azadegan, A.; Patel, P.C.; Zangoueinezhad, A.; Linderman, K. The effect of environmental complexity and environmental dynamism on lean practices. *J. Oper. Manag.* **2013**, *31*, 193–212. [CrossRef]

93. Enapra. La Formazione Per La Competitività Delle Imprese Vitivinicole Italiane. In Proceedings of the La Formazione Per la Competitività Delle Imprese Vitivinicole Italiane, Verona, Italy, 17 April 2018.

94. Longoni, A.; Pagell, M.; Johnston, D.; Veltri, A. When does lean hurt?—An exploration of lean practices and worker health and safety outcomes. *Int. J. Prod. Res.* **2013**, *51*, 3300–3320. [CrossRef]

95. Verrier, B.; Rose, B.; Caillaud, E. Lean and green strategy: The lean and green house and maturity deployment model. *J. Clean. Prod.* **2016**, *116*, 150–156. [CrossRef]

96. Wu, P.; Low, S.P.; Jin, X. Identification of non-value adding (NVA) activities in precast concrete installation sites to achieve low-carbon installation. *Resour. Conserv. Recycl.* **2013**, *81*, 60–70. [CrossRef]

97. Longoni, A.; Golini, R.; Cagliano, R. The role of new forms of work organization in developing sustainability strategies in operations. *Int. J. Prod. Econ.* **2014**, *147*, 147–160. [CrossRef]

98. Gatti, M.; Squeri, C.; Garavani, A.; Frioni, T.; Dosso, P.; Diti, I.; Poni, S. Effects of variable rate nitrogen application on cv. Barbera performance: Yield and grape composition. *Am. J. Enol. Vitic.* **2019**, *70*, 188–200. [CrossRef]

99. Lapinski, A.R.; Horman, M.; Riley, D.R. Lean processes for sustainable project delivery. *J. Constr. Eng. Manag.* **2006**, *132*, 1083–1091. [CrossRef]

100. Sarri, D.; Lombardo, S.; Pagliai, A.; Zammarchi, L.; Lisci, R.; Vieri, M. A technical-economic analysis of telemetry as a monitoring tool for crop protection in viticulture. *J. Agric. Eng.* **2020**, *51*, 91–99. [CrossRef]

101. Koerkamp, P.W.G.G.; Lokhorst, C.; Ipema, A.H.; Kempenaar, C.; Groenestein, C.M.; van Oostrum, C.G.; Ros, N.J. New Engineering Concepts for a Valued Agriculture. In Proceedings of the European Conference on Agricultural Engineering AgEng2018, Wageningen, The Netherlands, 8–12 July 2018; Wageningen University & Research: Wageningen, The Netherlands, 2018.

102. Sarri, D.; Martelloni, L.; Vieri, M. Development of a prototype of telemetry system for monitoring the spraying operation in vineyards. *Comput. Electron. Agric.* **2017**, *142*, 248–259. [CrossRef]

103. Benner, M.J.; Tushman, M. Process management and technological innovation: A longitudinal study of the photography and paint industries. *Adm. Sci. Q.* **2002**, *47*, 676–769. [CrossRef]

104. Ward, P.; Zhou, H. Impact of information technology integration and lean/just-in-time practices on lead-time performance*. *Decis. Sci.* **2006**, *37*, 177–203. [CrossRef]

105. Ohno, T. *Toyota Production System: Beyond Large-Scale Production*; CRC Press: Portland, OR, USA, 1988.
106. Rother, M.; Shook, J.; Institute, L.E. *Learning to See: Value Stream Mapping to Add Value and Eliminate Muda*; Taylor & Francis: Oxfordshire, UK, 2003.

Article

Sustainable Circular Mobility: User-Integrated Innovation and Specifics of Electric Vehicle Owners

Simone Wurster [1,*], **Philipp Heß** [1], **Michael Nauruschat** [1] and **Malte Jütting** [2]

1   Department of Innovation Economics, Technische Universität Berlin (TU Berlin), 10587 Berlin, Germany;
philipp.hess@tu-berlin.de (P.H.); michael.nauruschat@tu-berlin.de (M.N.)

2   Fraunhofer IAO, Fraunhofer Institute for Industrial Engineering, Center for Responsible Research and
Innovation (CeRRI), 10623 Berlin, Germany; malte.juetting@iao.fraunhofer.de

*   Correspondence: simone.wurster@tu-berlin.de

Received: 15 July 2020; Accepted: 19 September 2020; Published: 24 September 2020

Abstract: The circular economy (CE) represents an environmentally and sustainability-focused economic paradigm that has gained momentum in recent years. Innovation ecosystems are the evolving interconnected sets of actors, activities, artefacts, and institutions who are vital to the innovative performances of single actors or actor groups consisting largely of firms in the products and services sector. To develop sustainable CE ecosystems, participating firms need to involve the consumers and users in their innovation processes. The automotive industry is to a large extent an industry in which incorporating customer requirements in product development is critical to success. In addition, growing expectations and growing awareness of environmental issues drive the industry to develop environmentally friendly products. However, CE solutions and, specifically, sustainable tyres have not yet been given due consideration. Likewise, the specific preferences of the end-users of sustainability-focused cars such as electric vehicles (EVs) and users of biofuels are unknown in the CE context so far. Based on the current state of research, this article addresses an important, unexplored topic of product circularity. Being the first article on consumer interests and active contributions to CE automotive products, it also extends the first articles on CE software products. A survey of 168 traditional car owners (no EV/biofuels users), 29 users of biofuels, and 40 EV affine consumers was conducted in Germany to create an empirical foundation for the specification of CE configuration software for sustainable automotive products, particularly sustainable tyres. The results show different preferences among these user groups, but also the importance of other characteristics not captured by the distinction by car ownership. In particular, the perception of climate change and the use of test reports or rating portals were variables that had significant influence on configuration preferences.

Keywords: circular economy; sustainability; user integration; innovation ecosystems; cars; electric vehicles; biofuels

---

1. Introduction

1.1. CE Innovation Ecosystems and User Integration in the Automotive Industry

The United Nation's (UN's) Agenda for Sustainable Development defined 17 sustainable development goals (SDGs), which are intended to be reached by 2030. 'Sustainability' is defined by the UN Brundtland Commission as 'meeting the needs of the present without compromising the ability of future generations to meet their own needs' [1]. Sustainability topics include, for example, responsible consumption and construction (goal 12), climate action (goal 13), as well as economic growth, employment, decent work for all, and social protection in goal 8 [2]. The circular economy (CE) plays an essential role in the pursuit of the UN's sustainability goals [3]. It represents an environmentally and sustainability-focused economic paradigm that has gained momentum in recent years [4]. A CE is an economic system with the following characteristics: 'The value of products and materials is maintained for as long as possible; waste and resource use are minimized, and resources are kept within the economy when a product has reached the end of its life, to be used again and again to create further value' [5].

This article refers to the development of a CE ecosystem in the automotive sector, which relies on the principles of open innovation and the contributions of different user groups to specify CE configuration software for sustainable tyres and appropriate tyre characteristics. In this context, the attribute 'sustainable' does not only refer to CE aspects but also to the use of renewable resources and social factors, for example, appropriate working conditions in the product life cycle. More information on material-related aspects of sustainable tyres is, for example, provided in Section 1 of [6].

Innovation plays an essential role in the pursuit of economic growth as mentioned in goals 8.2 and 8.3 of [2]. Considered in SDG 9, fostering innovation is even an individual goal of the SDGs [2].

The concept of open innovation has generated 'an avalanche of interest' since its discovery by [7,8] Its fundamental goal is to help firms 'to span their boundaries in both the creation and commercialization of innovations' and 'to shift the dominant logic of R&D away from the internal discovery toward external engagement' [8].

Scholars also observed the development of the open innovation research direction beyond the dyadic interaction between two firms towards collaborations with external networks, ecosystems, and communities [8]. Likewise, Ref. [9] highlighted that, from a conceptual point of view, the innovation perspective, originally focused on product/service and business model innovation, has recently widened to include ecosystems.

Innovation ecosystems are defined as 'the evolving set(s) of actors, activities, and artifacts, and the institutions and relations, including complementary and substitute relations, that are important for the innovative performance of an actor or a population of actors' [10]. They require the creation of economically successful alliances of loosely coupled organizations to interact with each other to achieve a collective outcome and the alliances' continuous development to achieve lasting success [9] further developed this idea by also considering the creation of the ecosystem.

This article applies the innovation ecosystem approach to the CE concept. On that basis, we define a circular economy (CE) innovation ecosystem as the evolving set(s) of actors, activities, and artefacts, and the institutions and relations that are important for the innovative performance within a circular economy.

Research on the CE has various facets. From a linguistic point of view, a specific kind of wording can be observed. Frequently, the single term 'circular' replaces the composed name 'circular economy', for example, 'circular business model' and 'circular design' [11] 'circular goods' [12] and 'circular innovation' [9]. Likewise, this paper uses the term 'circular mobility' for CE solutions for the mobility sector.

Ref. [11] specified the CE strategies and business model archetypes 'narrow', 'slow', and 'close', complemented by [9] by the strategies 'regenerate' and 'inform'. 'Regenerate' refers to the minimised use of toxic substances and the need for an increase of renewable materials and energy in a circular economy. 'Inform' is a support strategy based on the importance of information technology in enabling a CE. An example for the 'inform' function is provided by material database ecosystems describing the characteristics of materials and components in products so that products can be more easily reused and their materials recovered [9].

CE ecosystems have to be innovative and user-centred (adopted from [13] which refers to 'human-centred'), to be sustainable. Linked with the user-centricity, another new view on open innovation has emerged in addition to the ecosystem perspective. Nowadays, the perception of innovation, something initially understood as a firm-centric activity, has shifted towards a 'joint accomplishment', where it also considers the users' and customers' abilities to contribute to value creation [14].

User integration into the product development process means the cooperative development of new products and services focusing on high customer benefits. By involving the customer in the development process at an early stage, specific demand and market conditions can be taken into account during the product development phase, thus reducing the risk of failures, see [15].

According to [16] the importance of user information and user participation for seeking business opportunities has been widely acknowledged in a variety of industries.

In the CE context, the recognition of the users as valuable contributors in shaping business strategies [14] has to be applied to ecosystems. Since users are one of the most significant stakeholder groups of economic activities [16], the supply side of the ecosystem needs to understand users' preferences and reflect these needs. It is critical to incorporate their needs, ideas, and feedback in the system's innovation management for its sustainable growth (see [16] in this context concerning sustainable business models in general).

Particular topics of innovation research on the user side thus far include

- Studies on different integration levels such as user behaviours, users as a source of innovation, and the user role in innovations;
- Comparisons of user innovation with supplier-driven innovation;
- Interactions between users and manufacturers and suitable forms of governance for user innovation; and
- User communities and crowdsourcing for innovation (see [14] for an overview).

User communities and crowdsourcing for innovation belong to the most recent research topics [14]. This article extends the user integration research in the context of CE innovation ecosystems.

Ref. [17] emphasized the importance of considering customer-perceived values in establishing selection criteria in innovation management decisions. Likewise, they stressed that the automotive industry is substantially a business-to-consumer sector, in which incorporating customer requirements is critical to innovation. It is also shaped by a growing awareness of environmental issues, driving the industry to strive to develop environmentally friendly vehicles. In addition, customer requirements are becoming more sophisticated [17]. Consumer values play a crucial role in this context. Nevertheless, little effort has been made to reflect customer-perceived value in establishing criteria to select innovation projects. This applies also, as [17] stated, to the reflection of their interest in innovative, environmentally friendly product features. In the CE context, specific preferences and the role of the end-users of more sustainable cars such as electric vehicles (EVs) and biofuels are unknown so far.

This article is structured as follows:

- Section 1: Introduction,
- Section 2: Materials and methods of this study,
- Section 3: Results,
- Sections 4 and 5: Discussion and conclusions.

1.2. Users as CE Innovation Partners in Literature

Various research gaps shape the user aspect in the CE. As an example, 'the consumers' purchase intention of recycled circular goods' is an 'important' *'unexplored'* topic of 'product circularity' [12]. While [12] addressed this gap with a survey on CE fashion products, our article is the first one on consumer views and contributions regarding CE automotive and mobility products and, in particular, on related software products. In addition, consumers are part of the innovation process, while [12] considered mainly their opinion only.

A key research stream on CE software is dedicated to software for industrial symbioses, 'interconnected network(s) which strives to mimic the functioning of ecological systems, within which energy and materials cycle continually with no waste products produced' [18]. An overview was provided by [18]. Industrial symbioses refer mainly to production waste, while the focus of the present article is on consumption waste. Furthermore, the innovation processes and the consumer focus are underexplored in the industrial symbioses articles in [18] overview.

Specific newer research work on CE software was provided, for example, by [9,19,20]. Ref. [19] focused on a design tool for architects based on the research question 'How can architects, non-expert to the CE, be stimulated and systematically guided towards circular design?' It refers to long-life products and considers consumers only in the definitions and in the annex. The authors of [20] which was published one year later, referred to 37 existing building design tools and aimed at identifying 'specific needs of building designers and advising engineers ... for design support tools (and their features) for circular building The users of the building are not explicitly considered. Ref. [9]'s focus was on a tool to create CE systems. A tool for frequent interactions with consumers within a CE was not a priority. Ref. [9]'s tool creation relied on workshops with participants from incumbent organizations and start-ups, also not including the consumers. The present article provides specific input on consumer contributions.

1.3. Sustainable Mobility, Electric Vehicles, and Sustainable Car Components

1.3.1. Sustainable Mobility: Electric Vehicles and Biofuels

This sub-section describes the general characteristics of EV owners and consumers who are inclined to use biofuels.

The topic of electric mobility and EVs is currently a large field of research, especially when it comes to the potential environmental effects of sustainable transport, energy transition, smart grids, as well as charging infrastructure and policy initiatives to promote the diffusion of electric mobility. The characteristics and motivations of early plug-in Electric Vehicle (PEV) buyers and user behaviour in the context of charging patterns and mobility habits are frequently analysed. Less is known, however, when it comes to the broader sustainability perception and behaviour of current EV owners and potential EV buyers.

According to [21], EV owners, in particular owners of PEVs, have normally the following characteristics:

- Demographics: higher income, middle age, male, higher education;
- Travel patterns (context): longer commutes;
- Motivations—identity, personality: agreeableness, pro-environmental identity;
- Motivations—priorities, beliefs: environmental impacts, low costs.

The use of EVs is linked with symbolic behaviour and social signalling. Scholars also found a range of ample examples of positive, but also negative and uncertain, perceptions of societal–functional impacts, such as ultimate impacts on CO_2 emissions, air pollution, noise, and safety [21].

In general, consumers interested in EVs are often interested in the car's sustainability. Inaccurate sustainability information may be a barrier to the adoption of EVs ('Show me they are truly sustainable.') [22]. However, [22] found, nevertheless, that sustainability has low importance in an EV purchase compared to cost and performance, which applies in particular to driving range or charging time and infrastructure.

The adoption of EVs can be improved by advancing these performance characteristics and addressing consumers' trust by appropriate measures. The development of credible rating systems for the cars' safety, in particular, regarding electric safety, and working with user profiles appear to be useful [21]. Ref. [23] also found that digitalisation is having a major impact on the transport and automotive industry, especially on EVs: owners of EVs are more interested in digital driving solutions than other car owners.

To improve the adoption of sustainable EVs, improving the charging infrastructure is a fundamental factor. In addition, Ref. [21] highlighted regarding the interrelations with the consumers that

- More pro-environmental users can be reached by improving the certainty of environmental benefits (e.g., by comparisons with traditional solutions);

- Marketing and framing should move beyond just 'pro-environmental framing' to include pro-technology, practicality, and other motives;
- A variety of innovation options should be offered (e.g., styles, models, membership packages) to reach a wide variety of user groups.

Likewise, as mentioned above, addressing consumers' trust by the development of credible systems for ratings and user profiles appears to be useful. At present, the current state of research refers to safety ratings mainly [21].

Regarding the characteristics of the users of biofuels, the state of research does not provide information of comparable depth. Significant findings were, however, provided by [24] who found that socio-demographics do not represent suitable predictors of drivers' willingness to pay for biofuels. On the other hand, consumer characteristics, which might be used for a target group segmentation, are usually not based on simple indicators and are not subject to demographic statistics. Ref. [25] identified two consumer characteristics that have a positive impact on the use of biofuels: (a) preferences for buying organic food, and (b) preferences for electric and hybrid vehicles. Based on these findings, buyers of organic food deserve further consideration in user-centric sustainable mobility research.

The findings above show three specific aspects:

- First hints exist that consumers interested in biofuels may also be interested in other sustainable car products,
- Consumers interested in EVs are often interested in sustainability information,
- Consumers interested in EVs have also a specific interest in digital solutions.

With regard to the price, it is essential to note that an EV requires a higher investment.

1.3.2. Sustainable Car Components, Tyres, and Research Questions

Environmental car characteristics go far beyond the type of energy used and the amount of energy consumption. The end-of-life stage is an additional essential aspect in this regard, both of the whole car and its individual components, as well. Components of the car requiring regular replacement, tyres specifically being a top concern, need far more careful consideration. According to [26], 5 million additional used tyres have to be managed daily worldwide. Solutions are available, but it is important to identify appropriate user segments. 'While consumers become more and more sensitive to sustainability and climate issues in various topics, e.g., regarding plastic bags, food packaging and mobility in general, car tyres are still not in the focus of the conscious consumer' [6]. For this reason, involving consumers in CE innovation processes in the context of car tyres requires at first an appropriate awareness-raising to make them aware of the tyre sustainability issue and the end-of-life tyre problem.

Various mixes of materials, functional and sustainability characteristics, and prices are possible in order to provide sustainable tyres, while specifying appropriate configurations is an essential task. CE configuration software, fed with suitable user input, may provide significant support in this context.

On this basis, an important research question is:

1. Which input can be gained by conscious consumers to specify

    (a) sustainable tyres,
    (b) innovative CE software to configure such tyres

after making them aware of the sustainability problems of tyres?

Users of EVs and biofuels proved to have a specific environmental orientation, which requires further exploration.

Sustainability-conscious drivers who use biofuels are also interested in other sustainable car features. Ref. [24] showed the positive relation with regard to the interest in biofuels and EVs. Therefore, it is important to learn whether car owners interested in environmentally friendly energy are also interested in additional environmentally friendly car characteristics concerning tyres. Specifically, it is important to answer the question:

2a.   Are sustainability-conscious drivers who use biofuels also more interested in sustainable tyres than other car drivers?

The particular interests of EV owners require more in-depth analyses, as well. Tyres for EVs must have specific characteristics. Many EVs are, for example, heavier than comparable non-EVs, which means more strain for the tyres. For this reason, EV tyre sidewalls have to be stronger. The heavy battery also increases the weight. The tyre cavity's shape must ensure that the tread remain in contact with the road. An additional issue is the car's grip and rolling resistance. Proper grip and rolling resistance characteristics make EVs, which are in general quieter than other cars, even more quiet. Individual block and groove sizes play an essential role in this context. A few tyre manufacturers produce EV-specific tyres already, for example, Michelin or Continental with its Conti.eContact™. Alternatively, high-performance tyres can be used for EVs, as well [27].

Environmentally friendly tyres provide few opportunities for social signalling. This aspect raises the question of whether EV affine consumers have a specific interest in these tyres nevertheless. The next question is therefore:

2b.   Are sustainability-conscious owners of EVs also more interested in sustainable tyres than other car drivers?

The next question refers to the intended innovative CE configuration software, its specification, and target groups, also considering the specific market segment for EV tyres. EV owners showed a specific interest in digital car solutions. The next question is therefore:

3   Are sustainability-conscious owners of EVs, due to their particular interest in digital solutions, more interested in tyre configuration software than others?

Our approach to address these question is explained in detail in Section 2.

The analysis of consumers within a CE innovation ecosystem in this article forms part of a larger, currently evolving stream of literature. Whereas the innovation ecosystem construct has been used to describe the joint development and commercialization of technologies, products, and services for several years now [28], it has only recently been applied to the sustainability context [29]. Exploring the phenomenon of CE innovation ecosystems, aiming at accelerating the transition towards a CE, is of particular interest in this regard, as the works of [9,30–33] illustrated. Interestingly, previous studies primarily focussed on triple helix structures, thus the cooperation between politics, science, and industry, as well as on inter-firm collaboration. Alternatively, an explicit user focus, as adopted in this article, sheds light on an ecosystem actor who has received little attention in the circular ecosystem context so far.

2. Materials and Methods

As described in Section 1, this article refers to the development of software to support an innovative CE system specifically focused on customised automotive solutions and sustainable tyres, whose features are defined by user integration. The intended ecosystem consists of four groups of actors. Suppliers (producers, retailers, etc.) (1), consumers (2), and (public or private) recyclers (3) are directly involved in the material flow of the recycling objects as well as indirect participants, actors who influence the material flow indirectly (e.g., the state and associations) (4).

In the automotive industry, the group of suppliers consists mainly of car manufacturers and dealers. Users or vehicle owners include private households and also organisations with cars or car fleets. The group of collectors, recyclers, shredders and recyclers belong to the group of disposers (see Figure 1). Landfills are not pictured, as landfilling tyres is forbidden in Europe.

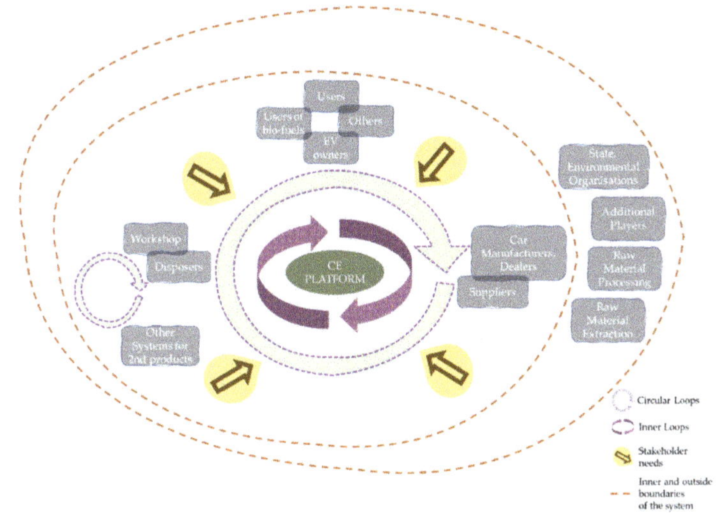

Figure 1. The ecosystem and stakeholder contributions to innovation.

The provision of new tyres, from day one during manufacturing, to the continued replacement of new tyres, is a supply service largely covered by tyre dealerships, workshops, or the internet.

A specific feature of the intended software system (refer to Figure 2) is a configuration software, which provides consumers with the opportunity to specify tyres according to their preferences, for example to purchase products from suppliers who guarantee sustainable sourcing throughout the whole value chain.

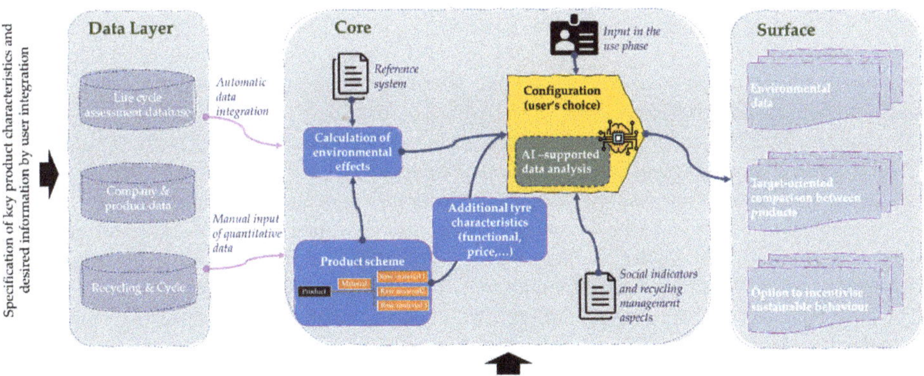

Figure 2. Intended circular economy (CE) software system and user integration in the development. Source: ConCirMy (modified).

The configuration shown in orange in the middle of Figure 2 relies on information on general characteristics of possible tyres as well as environmental information and social indicators, for example, regarding the treatment of workers in the value chain. To specify the tyre characteristics as part of the system's selection options, user integration methods were applied as a means to stimulate innovation (see frames of the figure). Ref. [16] distinguished four types of user innovation models: the workshop-based, the consortium-based, the crowdsourcing-based, and the platform-based models.

The workshop-based kind of user innovation refers to specific interaction with the stakeholders to specify early results of user interaction. Instead of workshops, surveys can also be used as a means of communication. Inspired by [16] the scientific approach of the present research is pictured in Figure 3.

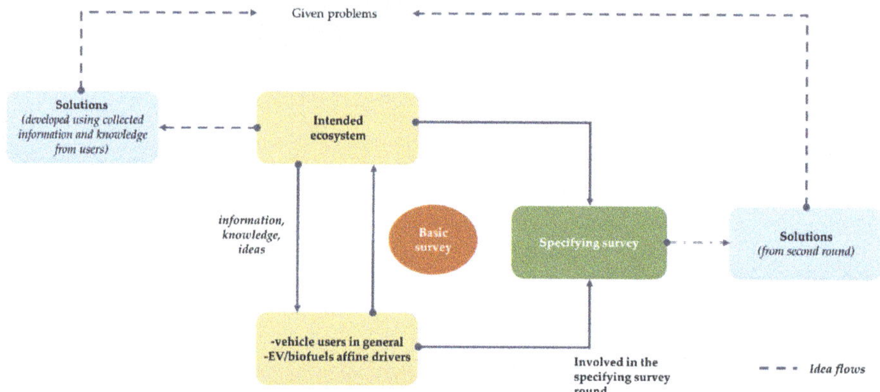

Figure 3. User integration concept of this research.

Illustrated in Figure 3, the user integration concept developed for this research has four key elements: vehicle users, the supply side of the intended ecosystem, a *basic* survey, and a *specifying* survey. This article presents the results of the basic survey.

Ref. [34] identified three possible consumer groups as target groups of CE business models: quality-oriented consumers, cost-oriented consumers and, in particular, 'green' consumers who are interested in environmentally friendly products. The survey relied on volunteer sampling among sustainability-conscious consumers, all of whom already purchase, high quality organic food. These participants responded to a nationwide advert in Germany among all newsletter subscribers of an online platform for food from regional organic farms that focuses on conscious meat consumption and transparency along the entire value chain.

In total, 5773 consumers were contacted. The survey was conducted from 16 January to 13 February 2020 and attracted 451 participants. 311 completed the survey, and 140 partially completed the survey. On this basis, the response rate was 7.8% (of total consumers contacted).

To sensitise consumers to the tyre sustainability issue and the problem of end-of-life tyres, the survey started with information on these aspects. The high volume of end-of-life tyres and their recycling as a global challenge were highlighted. Specific topics of the survey are shown in Table 1.

They included: 1. Demographic aspects and other fundamental user characteristics, and 2. Questions on user preferences. The general user characteristics included also, for example, the users' perceptions of climate change with three answer options: 'How do you estimate the impact of climate change for you personally and your family?'—'I regard the climate change for me and my family . . . as an existential danger/ . . . as a medium danger/ . . . not as a danger'. Related to the consumers' use of information in the tyre purchase process, the survey also included, for example, the fundamental question 'Do you use test reports or rating portals to get information before buying a tyre?'—'Yes'/'No'.

The section on driving habits included filtering questions about whether the participant drives a car frequently, whether they own a car, etc. To identify users of EVs and biofuels, the following questions were used:

- Have you bought or leased a car privately in the past or were you involved in the selection decision?
- Do you or a member of your household currently own a car? If yes, is it an EV (fully electric or hybrid vehicle)?

- Do you plan to buy or lease a car in the next two years? If yes, an EV (fully electric or hybrid)?
- As a car driver, do you have any experience in the use of biofuel? If yes, could you imagine using this fuel in the future?

Table 1. Overview of the survey topics.

Topic Area	Survey Questions
Demographic aspects and other fundamental user characteristics	•Age, education, income •Attitudes towards sustainability in general •Driving habits (car ownership, brand, tyre purchase)
Questions on user preferences	•Preferred tyre brand •Place of tyre purchase •Use of information for tyre purchase •Importance of general tyre characteristics •Familiarity with EU tyre labelling •Importance of this label when buying tyres •Importance of information on environmental and social aspects in the purchasing of tyres •Importance of specific environmental and social aspects •Willingness to buy sustainable tyres in the future and relevant information in this context •Potential benefits of a sustainable tyre label to support purchasing decisions •Interest in innovative recycling management approaches •Innovative collaboration and software for sustainable tyres •Relevant functions of this software

Another filter question asked whether the participant had been involved in tyre purchase. Forty-three percent of the participants gave a non-confirmative answer or did not answer the question. These participants were not asked further questions on tyre purchasing and their preferences. Figure 4 illustrates the screening process.

Figure 5 shows the profiles of the participants with regard to the use of environmentally oriented energy sources in detail.

According to Figure 5, 66% (the clear majority) of the remaining participants had no relation to any of the sustainable mobility options. However, just over a quarter (26%) demonstrated sustainable interest and/or behaviour. These participants had an EV, planned to buy one, or confirmed an interest in the future use of biofuels. One percent of that group even applied multiple types of these sustainable practices. The research presented relied on the contribution of 237 persons (168 traditional car users, 29 users of biofuels, and 40 EV affine consumers (16 EV owners and 24 car owners who planned to buy an EV in the next two years)). Eight percent of the participants did not answer the question.

Concerning responses to general tyre characteristics, up to 98% of the EV affiliated participants selected performance characteristics on a four-part scale from very important to not important at all: driving behaviour of the car—92%, fuel consumption—88%, wet braking behaviour—98%, grip in ice and snow—98%, and mileage/life expectancy—90%. The rates of the whole group were a little lower (with driving behaviour of the car—74%, fuel consumption—54%, wet braking behaviour—81%, grip in ice and snow—82%, and mileage/life expectancy—58%). These figures suggest a clear correlation between the consumers who look to purchase an EV or HV, and a broader regard and interest in a holistic approach to sustainability when looking at the automotive industry as a whole.

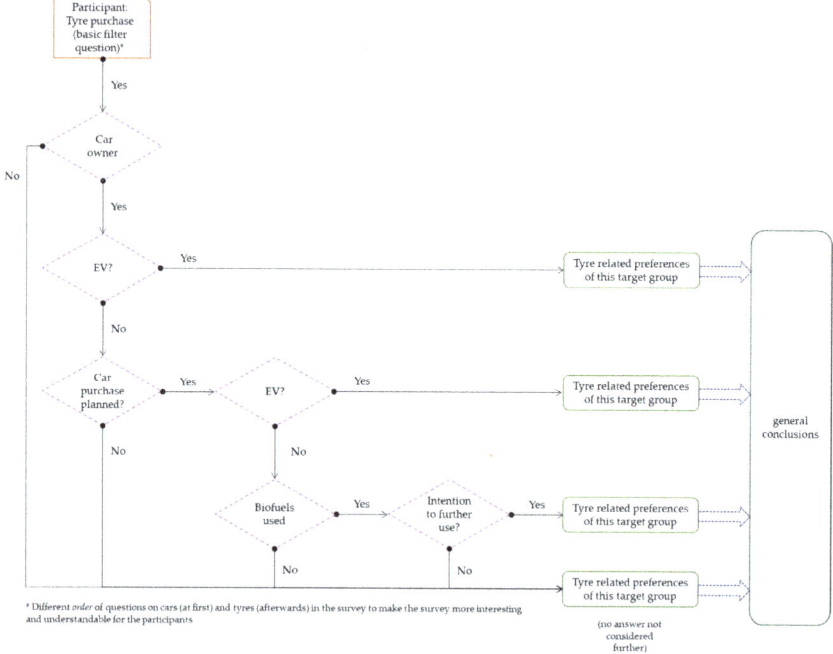

Figure 4. Identification of sustainability-oriented drivers in this study.

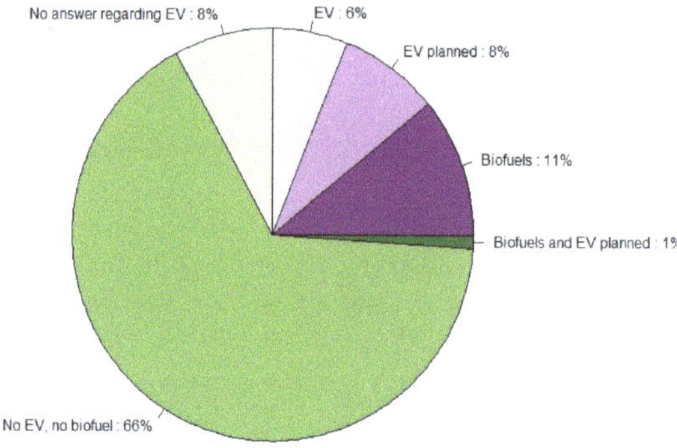

Figure 5. Distribution of car owners in the survey (N = 256).

3. Results

3.1. Importance of Sustainability Information to the Target Groups

Based on the survey, data on the following topics was explored to specify the innovative CE configuration software and characteristics of sustainable tyres:

- Interest of different user groups in various types of sustainability aspects in the tyre purchase,
- Interest in specific sustainable tyres,
- Interest in individual tyre configuration and configuration software.

Of particular interest was the consumers' interest in sustainable tyre characteristics and specific types of sustainable tyres. A first question considered the importance of environmental information in the tyre purchase. Among the consumers who regarded environmental information as very important in their buying decisions, EV owners were, according to Figure 6, clearly leading with 69%, followed by biofuels users with 56% of the participants. Among consumers of the group 'no EV, no biofuel', the share was the lowest with 48%.

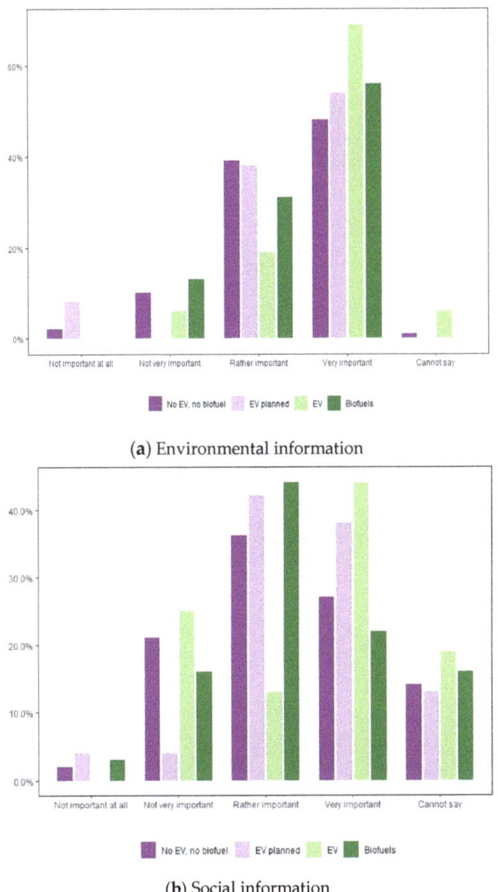

(**a**) Environmental information

(**b**) Social information

Figure 6. Importance of environmental and social information in tyre purchases for different tyre buyers (N = 236).

Regarding social issues, owners of EVs were once again the user group in which the most participants regarded information as very important for a tyre purchasing decision, according to Figure 6. A total of 44% of this user group's members gave this evaluation. However, compared to the results on the importance of environmental information, the results on social aspects were more varying. With regard to specific environmental aspects, resource-related information, and social issues, 16 items were discussed, from, for example, the content of recycled material in a tyre to fair play and working conditions. These items were only shown to consumers who expressed an interest in environmental and/or social information in the tyre purchase in general.

As shown by all segments of Figure 7, among all consumers with this general interest, the owners of EVs or biofuels users again selected the proposed specific criteria most often.

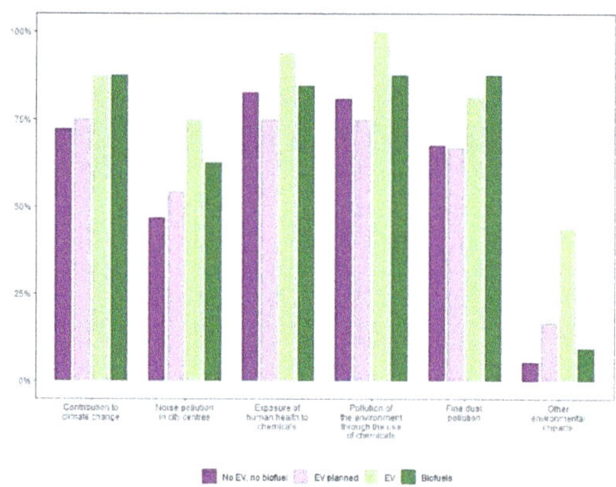

(a) Specific environmental information

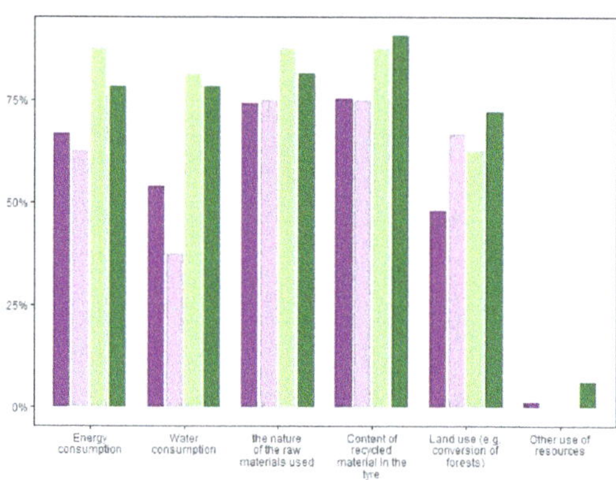

(b) Specific resource-related information

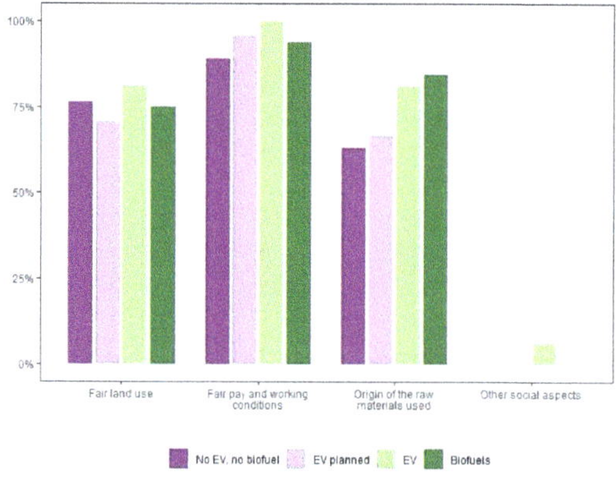

(c) Specific information on social issues

Figure 7. Importance of specific sustainability aspects for tyre purchases (N = 233).

3.2. Interest in Sustainable Tyres and Software Tools

A particular question referred to the interest in sustainable tyres concerning three specific tyre types: tyres with a high proportion of materials from renewable resources, tyres with a high ratio of recycled content, and retreaded tyres.

The interest of biofuels users led for all three types of sustainable tyres, according to Figure 8, followed by the users of EVs.

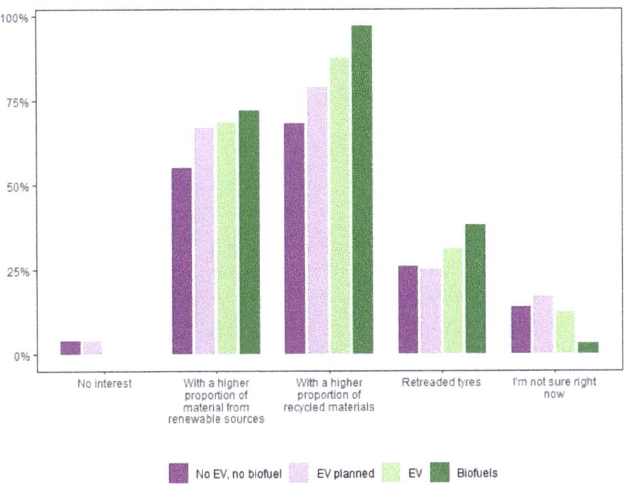

Figure 8. Interest in sustainable tyres per user group (N = 229).

With regard to an increase in interest in sustainable tyres due to options to select and configure tyre characteristics, EV users were the most affirmative with 100% positive answers to the question: 'In the ConCirMy project, software is being developed that suggests tyres based on preferences when buying a car. Would it be attractive for you to determine the characteristics of your tyres with this software or to configure them yourself during the ordering process?' (see Figure 9).

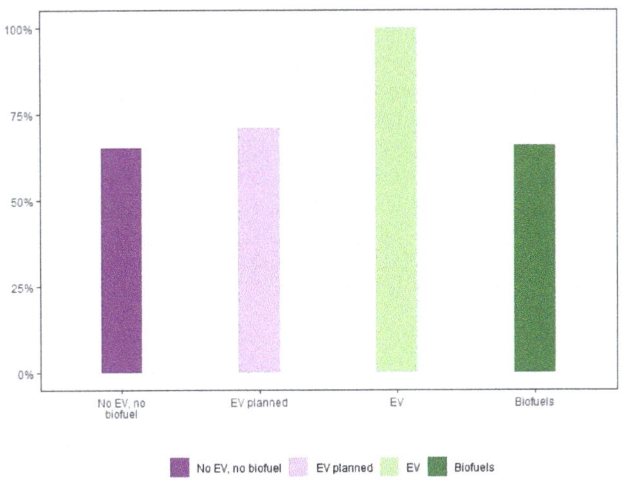

Figure 9. Interest in sustainable tyre configuration software per driver group (N = 217).

The question was introduced by a filtering question: 'Can your willingness to purchase environmentally friendly tyres be increased by an opportunity to select and configure tyre characteristics according to your preferences?'

Regarding specific selection features of the configuration software, the previously described interest in specific tyre characteristics was repeated with minor variations. Key preferences were observed, highlighting, for example, the importance of 'Wet breaking behaviour', 'Contribution to climate change', 'Pollution of the environment through the use of chemicals', 'Exposure of human health to chemicals', 'Recyclability', and 'Fair pay and working conditions'.

In summary, the results of the first statistical analyses suggested that car users are interested in innovative sustainable tyres and specific sustainability information for their purchasing decisions. In addition, the results showed that there is indeed interest in an innovative CE software system to specify tyres regarding individual sustainability preferences in combination with functional aspects. Section 3.3 analyses the data in more detail.

3.3. Multivariate Analyses

Following the analysis of descriptive survey statistics, a multivariate regression analysis was carried out in a second step to further investigate the variation across participant groups in their assessments of (1) the usage of configuration software, (2) the general interest in configuring tyres, and in tyres with a higher proportion of (3) recycled or (4) renewable materials. As the dependent variables were dichotomous, a logistic regression model was selected. The model controlled for demographic aspects and potential confounders and included the following variables:

- Age,
- Income,
- Perception of climate change,
- Use of tyre rating portals,
- The relevance of social and environmental aspects when buying tyres,
- Ownership of EVs or the willingness to purchase EVs in the future.

These variables were selected to represent the two main concepts that were theorised to be the underlying motivators for a more profound interest in environmentally friendly tyres:

(a) The perception of the relevance of environmental topics, and
(b) A general willingness to engage with the topic of tyres.

To avoid problems related to multicollinearity or a disproportion between the number of regressors and sample size, we selected specific variables (such as the perception of climate change) according to their correlation with similar variables and the extent to which they captured the relevant concept. Age and income were included in the model as continuous variables, based on the assumption of equidistant categories. Perceptions of climate change and EV ownership were modelled on three levels (no danger, medium danger, existential danger; no EV, EV planned, EV owned). The usage of tyre rating portals, and the relevance of social and, respectively, environmental aspects when buying tyres were represented by binary variables. Table 2 shows parameter estimates and associated standard errors for the logistic regression models (1) to (4).

Table 2. Result from logistic regression.

	Interested in	Using Configuration Software	Configuring Tyres	Tyres with a Higher Proportion of Recycled Materials	Tyres with a Higher Proportion of Materials from Renewable Resources
Age		0.0748 (0.1732)	0.0751 (0.1614)	−0.5671 (0.1994) **	−0.2031 (0.1602)
Income		−0.1961 (0.184)	0.0073 (0.1673)	0.1444 (0.1921)	0.4367 (0.1713) *
Perception of climate change	medium danger	0.8394 (0.5762)	0.0132 (0.5772)	−0.3587 (0.7431)	1.0331 (0.5845)
	existential danger	1.2765 (0.621) *	0.4711 (0.612)	−0.5873 (0.7682)	0.9122 (0.6033)
Use of tyre rating portals		0.8257 (0.4095) *	0.7828 (0.3791) *	0.6543 (0.4406)	0.1717 (0.3849)
Relevance when buying tyres	social aspects	−0.0101 (0.2841)	−0.4651 (0.2805)	0.0326 (0.3283)	0.2771 (0.2887)
	environmental aspects	−0.0127 (0.3137)	0.2556 (0.2996)	0.1841 (0.3611)	−0.2357 (0.3278)
EV	owns EV	16.4175 (1155.3287)	1.5169 (1.0779)	0.02 (0.8442)	0.3346 (0.7282)
	plans to buy EV	0.0528 (0.6117)	−0.2773 (0.5631)	0.4387 (0.7154)	0.4572 (0.6387)
Const		0.8772 (1.5659)	1.0306 (1.4596)	3.4838 (1.7638) *	−0.8766 (1.4826)
Observations		155	154	151	151
Pseudo R^2		0.007	0.022	0.114	0.015
AIC		181	202	165	202

Logistic regression estimates with standard errors in parentheses. ** $p \leq 0.01$, * $p \leq 0.05$.

The interest to generally configure tyres was positively associated with a prior interest in tyres (expressed by the usage of tyre rating portals). This association was confirmed for the use of configuration software for environmentally friendly tyres, which was furthermore positively associated with the participants' perceptions of climate change as an existential danger. Interest in tyres with a higher proportion of recycled materials was negatively associated with age, while the interest in tyres with a higher proportion of materials from renewable sources was positively associated with income. Even though these relationships were significant on 0.95 confidence levels, the models have to be interpreted cautiously and do not necessarily represent conclusive robust results, as particularly EV ownership included very small sample sizes, and overall, most models did not manage to greatly outperform baseline models. However, the presented results are intuitive and can be considered input for subsequent studies.

Among the survey participants who perceived climate change as an existential threat to themselves and their families, a positive relationship was observed with regard to the willingness to use tyre configuration software.

The relation between age and the interest in tyres with a higher proportion of recycled materials was negative, which means that older people seem to have a lower willingness to buy these tyres. Survey participants using tyre rating portals had a higher interest in configuration software.

Furthermore, the relation between income and the interest in tyres with a higher proportion of material from renewable resources was positive, which means that people with a higher income seem to have a greater willingness to purchase tyres with a higher proportion of material from renewable resources.

While simple survey statistics showed that EV owners and users of biofuels were the most interested user groups, estimations from the regression model that controlled for demographic aspects rather suggested that the perception of climate change (as proxy for general environmental awareness) and the prior propensity to become informed about tyres were main influencing factors that determined interest in configuration software. The implications from these findings are discussed in the next section.

4. Discussion

User integration is an important method to develop and market products that meet consumer interest. Creating attractive, environmentally friendly products and marketing them successfully is of particular importance in this context.

The present research, which focused on an innovative CE automotive ecosystem and sustainable tyres, helped to specify product characteristics and target group characteristics, as well. Specifically, it aimed to provide answers to three questions:

1. Which input can be gained by conscious consumers to specify

    (a) sustainable tyres
    (b) innovative CE software to configure such tyres

    after making them aware of the sustainability problems of tyres?

2. Are sustainability-conscious

    (a) drivers who use biofuels
    (b) owners of EVs

    also more interested in sustainable tyres than other car drivers?

3. Are sustainability-conscious owners of EVs more interested in tyre configuration software than others?

With regard to Question 1, the majority of the conscious users showed interest in sustainable tyres with a higher proportion of materials from renewable resources and recycled materials, and in innovative configuration software for tyres. Their input to specify tyre characteristics focussed in particular on environmental properties. Even 48% of the consumers with no specific sustainability preferences regarding EVs or the use of biofuels so far regarded environmental information as very important for the purchase of tyres. However, regarding social aspects and information, the share was much lower. It ranged between 22% and 44%, depending on the type of car drivers.

Most specific environmental and resource-related aspects analysed by the survey were regarded as important by the majority of the participants. Examples included 'contribution to climate change' and 'pollution of the environment through the use of chemicals', chosen by more than 70%, which will encourage tyre producers and software developers to customise their products and product information accordingly.

Regarding Question 2, we did indeed observe a specific interest in sustainable tyres among car users who had a general interest in sustainable car characteristics. For the innovative ecosystem, their input was therefore of particular value.

Concerning Question 3, our descriptive survey statistics suggested a positive answer as well with regard to the conscious consumers of the present study. The results reflected the findings of (9,23) that consumers interested in EV show a specific interest in digital solutions and are also interested in sustainability information, requiring proof of true sustainability. In addition, the logistic regression showed the importance of the variables *perception of climate change* and *the use of tyre rating portals*.

As described in Section 2, all participants of the survey were quality-oriented consumers interested in environmentally friendly products, a target group of particular interest for CE products, according to [34]. The characteristics of EV owners were embedded in this context. Beyond this, they differed only slightly regarding additional attributes from the peers, for example, regarding the fact that the ownership of EVs, which are more expensive than cars in general, was often linked with a higher income. In this context, the logistic regression suggests also that people with a higher income have a greater willingness to purchase tyres with a higher proportion of material from renewable resources.

Because of the small sample size, especially in the case of EV owners, regression results might not sufficiently capture additional differences between EV, biofuel, and conventional fuel groups. For this reason, an additional study of 1000 average consumers with a driver's licence is prepared to launch. This may also provide tyre producers with additional hints as to what extent a stronger focus on producing more sustainable tyres suitable for EVs might be attractive. More studies should follow in the core phase of the majority stage of EV adoption.

5. Conclusions

5.1. Research on Sustainable Mobility

The present research used consumers as a valuable source of information to specify features of the output of an innovative CE ecosystem. Nevertheless, there are three limitations:

1. The results are based on an online survey, not on real behaviour and real buying decisions. The real behaviour will also depend on the marketing and perception of the configuration software.
2. The present study includes a relatively small sample of consumers with a specific sustainability focus in Germany only. Inspired by [24], the survey addressed consumers who showed a specific kind of sustainability behaviour by buying sustainable food to analyse additional mobility-related sustainability characteristics. Based on [34] s classification, they represented the specific characteristics of not only one but two CE target group(s): sustainability and quality-oriented customers. EV ownership was a frequent characteristic of the participants with higher income but was also embedded in the specific context of pro-environmental, quality-seeking behaviour.
3. The survey was conducted before the beginning of the Coronavirus crisis and therefore before current discussions to support the automotive industry during the crisis by providing incentives to buy new cars, and EVs, in particular. In future research, it will also be important to distinguish between early EV adopters and mainstream adopters buying EVs as a reaction to these stimuli. This will give more in-depth insight into what extent early and mainstream EV adopters are interesting target groups for additional sustainable car products.

In summary, more research will be necessary to update our results in the given context. A broader study with 1000 consumers with no specific sustainability-oriented habits is already prepared.

5.2. Research on Innovation Ecosystems

According to Section 1, an explicit user focus, as adopted in this article, sheds light on an ecosystem actor who has received little attention in the circular ecosystem context so far, despite being particularly interesting. On the one hand, such a perspective encompasses direct users, whose customer demands do not only need to be addressed for commercial reasons but also depict a valuable source of additional innovation impulses (see Section 5.3). On the other hand, the approach can easily be expanded to include non-users and the broader society, whose expectations should be taken into account, as broad societal acceptance (beyond immediate users) is indispensable for a successful transition towards the circular economy. Whereas the former perspective is at the core of this article, the latter opens up promising avenues for further inquiry. Future research may thus focus on questions of how to best integrate both users and societal stakeholders in (CE) innovation ecosystems, at what point in the innovation process such involvement should take place and which formats are best suited to enable fruitful collaboration. Only if both policy makers as well as innovation practitioners adopt a broad ecosystem understanding and commit to involving stakeholders beyond their immediate value chain will they be able to leverage the full potential of their ecosystem for a circular economy transition.

5.3. Research on User Innovation

Innovation processes with users can be differentiated according to different levels of user involvement and user contribution. Processes with a strong user involvement are those in which the users themselves are the innovators. Those users are also called user–innovators [35]. In the given context, the users gave input to shape the future ecosystem and its products, specifying important ecosystem and product characteristics. Nevertheless, the users in the given case were not involved in the decision-making to determine the final system and product features.

In contrast to this, various examples exist in which the users had a bigger influence on the innovation process, while innovation processes were also very successful, for example, in online user communities and with regard to smart energy innovations [35–37].

In the CE context, information on such processes so far barely exists. To exploit knowledge on successful user innovation, stimulating innovation with a higher level of user involvement to support the CE is suggested as well as an analysis of the factors that make such processes successful.

Author Contributions: Conceptualization, S.W.; Data curation, S.W.; Formal analysis, S.W., P.H. and M.N.; Funding acquisition, S.W.; Investigation, S.W., P.H. and M.N.; Methodology, S.W. and P.H.; Resources, S.W., M.N.; Supervision, S.W.; Software: P.H.: Validation, S.W., P.H.; Visualization, S.W. and P.H.; Writing–original draft, S.W., M.N. and M.J.; Writing–review & editing, S.W., P.H., M.N. and M.J. All authors have read and agreed to the published version of the manuscript.

Funding: We acknowledge the support by the German Federal Ministry of Education and Research (BMBF) (ConCirMy project, measure ReziProK, funding No. 033R236E), the German Research Foundation and the Open Access Publication Fund of TU Berlin.

References

1. United Nations. Report of the World Commission on Environment and Development Our Common Future. 1987. Available online: http://www.un-documents.net/wced-ocf.htm (accessed on 4 August 2020).

2. United Nations. Transforming Our World: The 2030 Agenda for Sustainable Development: Resolution Adopted by the General Assembly on 25 September 2015. Available online: https://www.un.org/en/development/desa/population/migration/generalassembly/docs/globalcompact/A_RES_70_1_E.pdf (accessed on 4 August 2020).

3. Schroeder, P.; Anggraeni, K.; Weber, U. The Relevance of Circular Economy Practices to the Sustainable Development Goals. *J. Ind. Ecol.* **2019**, *23*, 77–95. [CrossRef]

4. Lacy, P.; Rutqvist, J. The Roots of the Circular Economy. In *Waste to Wealth: The Circular Economy Advantage*; Lacy, P., Rutqvist, J., Eds.; Palgrave Macmillan: Basingstoke, UK, 2015; pp. 19–23.

5. European Commission. Circular Economy. 2015. Available online: https://ec.europa.eu/growth/industry/sustainability/circular-economy_en (accessed on 4 August 2020).

6. Wurster, S.; Schulze, R. Consumers' Acceptance of a Bio-circular Automotive Economy: Explanatory Model and Influence Factors. *Sustainability* **2020**, *12*, 2186. [CrossRef]

7. Chesbrough, H.W. *Open Innovation: The New Imperative for Creating and Profiting from Technology*; Harvard Business School Press: Boston, MA, USA, 2003.

8. West, J.; Salter, A.; Vanhaverbeke, W.; Chesbrough, H. Open innovation: The next decade. *Res. Policy* **2014**, *43*, 805–811. [CrossRef]

9. Konietzko, J.; Bocken, N.; Hultink, E.J. A Tool to Analyze, Ideate and Develop Circular Innovation Ecosystems. *Sustainability* **2020**, *12*, 417. [CrossRef]

10. Granstrand, O.; Holgersson, M. Innovation ecosystems: A conceptual review and a new definition. *Technovation* **2020**, *90*, 102098. [CrossRef]

11. Bocken, N.M.P.; Pauw, I.; Bakker, C.; Grinten, B.V.D. Product design and business model strategies for a circular economy. *J. Ind. Prod. Eng.* **2016**, *33*, 308–320. [CrossRef]

12. Calvo-Porral, C.; Lévy-Mangin, J.-P. The Circular Economy Business Model: Examining Consumers' Acceptance of Recycled Goods. *Adm. Sci.* **2020**, *10*, 28. [CrossRef]

13. Lofthouse, V.; Prendeville, S. Human-Centred Design of Products And Services for the Circular Economy—A Review. *Des. J.* **2018**, *21*, 451–476. [CrossRef]

14. Ketonen-Oksi, S.; Valkokari, K. Innovation Ecosystems as Structures for Value Co-Creation. *Tim Rev.* **2019**, *9*, 25–35. [CrossRef]

15. Amelingmeyer, J. (Ed.) *Technologiemanagement & Marketing: Herausforderungen Eines Integrierten Innovationsmanagements*, 1st ed.; Gabler Edition Wissenschaft; Dt. Univ.-Verl.: Wiesbaden, Germany, 2005.

16. Cho, C.; Lee, S. How Firms Can Get Ideas from Users for Sustainable Business Innovation. *Sustainability* **2015**, *7*, 16039–16059. [CrossRef]

17. Lee, S.; Cho, C.; Choi, J.; Yoon, B. R&D Project Selection Incorporating Customer-Perceived Value and Technology Potential: The Case of the Automobile Industry. *Sustainability* **2017**, *9*, 1918. [CrossRef]
18. Maqbool, A.; Mendez Alva, F.; van Eetvelde, G. An Assessment of European Information Technology Tools to Support Industrial Symbiosis. *Sustainability* **2019**, *11*, 131. [CrossRef]
19. Amory, J. *Research and Development Directions for Design Support Tools for Circular Building*; Faculty of Architecture & the Built Environment, Delft University of Technology: Delft, The Netherlands, 2019.
20. Cambier, C.; Galle, W.; de Temmerman, N. Research and Development Directions for Design Support Tools for Circular Building. *Buildings* **2020**, *10*, 142. [CrossRef]
21. Axsen, J.; Sovacool, B.K. The roles of users in electric, shared and automated mobility transitions. *Transp. Res. Part D Transp. Environ.* **2019**, *71*, 1–21. [CrossRef]
22. Egbue, O.; Long, S. Barriers to widespread adoption of electric vehicles: An analysis of consumer attitudes and perceptions. *Energy Policy* **2012**, *48*, 717–729. [CrossRef]
23. Mihet-Popa, L.; Saponara, S. Toward Green Vehicles Digitalization for the Next Generation of Connected and Electrified Transport Systems. *Energies* **2018**, *11*, 3124. [CrossRef]
24. Lanzini, P.; Testa, F.; Iraldo, F. Factors affecting drivers' willingness to pay for biofuels: The case of Italy. *J. Clean. Prod.* **2016**, *112*, 2684–2692. [CrossRef]
25. Li, T.; McCluskey, J.J. Consumer preferences for second-generation bioethanol. *Energy Econ.* **2017**, *61*, 1–7. [CrossRef]
26. GENAN. Nachhaltiges Altreifenrecycling. 2019. Available online: https://www.genan.de/ (accessed on 1 July 2020).
27. Dillard, S. Hybrids & Electrics. Electric Car Tires: Do You Need to Buy a Special Tire for an EV? 2019. Available online: https://www.motorbiscuit.com/electric-car-tires-do-you-need-to-buy-a-special-tire-for-an-ev/ (accessed on 1 July 2020).
28. Gomes, L.A.V.; Facin, A.L.F.; Salerno, M.S.; Ikenami, R.K. Unpacking the innovation ecosystem construct: Evolution, gaps and trends. *Technol. Forecast. Soc. Chang.* **2018**, *136*, 30–48. [CrossRef]
29. Liu, S. Exploring Innovation Ecosystem from the Perspective of Sustainability: Towards a Conceptual Framework. *JOItmC* **2019**, *5*, 48. [CrossRef]
30. Brown, P.; Bocken, N.; Balkenende, R. Why Do Companies Pursue Collaborative Circular Oriented Innovation? *Sustainability* **2019**, *11*, 635. [CrossRef]
31. Brown, P.; Bocken, N.; Balkenende, R. How Do Companies Collaborate for Circular Oriented Innovation? *Sustainability* **2020**, *12*, 1648. [CrossRef]
32. Parida, V.; Burström, T.; Visnjic, I.; Wincent, J. Orchestrating industrial ecosystem in circular economy: A two-stage transformation model for large manufacturing companies. *J. Bus. Res.* **2019**, *101*, 715–725. [CrossRef]
33. Konietzko, J.; Bocken, N.; Hultink, E.J. Circular ecosystem innovation: An initial set of principles. *J. Clean. Prod.* **2020**, *253*, 119942. [CrossRef]
34. Lüdeke-Freund, F.; Gold, S.; Bocken, N.M.P. A Review and Typology of Circular Economy Business Model Patterns. *J. Ind. Ecol.* **2019**, *23*, 36–61. [CrossRef]
35. Grosse, M. How User-Innovators Pave the Way for a Sustainable Energy Future: A Study among German Energy Enthusiasts. *Sustainability* **2018**, *10*, 4836. [CrossRef]

36. Grosse, M.; Pohlisch, J.; Korbel, J. Triggers of Collaborative Innovation in Online User Communities. *JOItmC* **2018**, *4*, 59. [CrossRef]
37. Grosse, M.; Send, H.; Loitz, T. Smart Energy in Deutschland: Wie Nutzerinnovationen Die Energiewende Voranbringen (Smart Energy in Germany: The Benefits of User Innovation). *SSRN J.* **2018**. [CrossRef]

Article

Open Innovation 4.0 as an Enhancer of Sustainable Innovation Ecosystems

Joana Costa [1,2,*] and João C.O. Matias [1,2]

[1] DEGEIT—Department of Economics, Management, Industrial Engineering and Tourism, University of Aveiro, Campus Universitário de Santiago, 3810-193 Aveiro, Portugal; jmatias@ua.pt
[2] GOVCOPP—Research Unit on Governance, Competitiveness and Public Policies, Campus Universitário de Santiago, 3810-193 Aveiro, Portugal
* Correspondence: joanacosta@ua.pt

Received: 15 July 2020; Accepted: 29 September 2020; Published: 1 October 2020

Abstract: Innovation matters. Business success increasingly depends upon sustainable innovation. Observing recent innovation best practices, the emergence of a new paradigm is traceable. Creating an innovative ecosystem has a multilayer effect: It contributes to regional digitalization, technological start-up emergence, open innovation promotion, and new policy enhancement retro-feeding the system. Public policy must create open innovation environments accordingly with the quintuple helix harmonizing the ecosystem to internalize emerging spillovers. The public sector should enhance the process, providing accurate legal framework, procurement of innovation, and shared risks in R&D. Opening the locks that confine the trunks of community, academic, industry, and government innovation will harness each dimension exploiting collective and collaborative potential of individuals towards a brighter sustainable future. In this sense, the aim of this study is to present how open innovation can enhance sustainable innovation ecosystems and boost the digital transition. For that, firstly, a diachronic perspective of the sustainable innovation ecosystem is traced, its connection to open innovation, and identification of the university linkages. Secondly, database exploration and econometric estimations are performed. Then, we will ascertain how far open innovation frameworks and in particular the knowledge flows unveiled by the university promote smart and responsible innovation cycles. Lastly, we will propose a policy package towards green governance, empowering the university in governance distributed ecosystem, embedded in the community, self-sustained with shared gains, and a meaningful sense of identity.

Keywords: open innovation; innovation ecosystems; sustainability; logit models

---

1. Introduction

Challenges faced worldwide are too large to tackle in isolation; the creation of new shared value through innovation is urgent. Convergent globalized features likewise digital transformation, universal interaction, and sustainability provided momentum to exponential growth in innovation led shared value. Approaching the new innovation paradigms requires integrated collaboration, co-creation, value sharing, hosting ecosystems, and fast adoption; joint research will speed up the process and raise the standards of the outcomes. Dissimilarly from previous industrial shifts, Industry 4.0 relies on pairing and amalgamation of subjects and expertise [1].

Creating shared value, sustainable growth, and development, relying on innovation, will overflow the economic and social sphere as under the collaborative paradigm firms will move from the maximization of short-term financial performance to long-term economic and social responsibility [2].

To present standards, assessing innovative performance based on patent count or R&D expenditures seems to be insufficient; co-creation emerging from knowledge network connections seems to be more robust, based upon the interactive contact between universities providing cutting edge knowledge and competitive firms [3]. Therefore, the development of innovation demands a particular ecosystem in which they will emerge as a result of the collaboration and co-creation among different players. The ecosystem approach emphasizes the position and roles of local and public actors in developing innovative activities, and the public policy challenge is to provide the means and instruments to transform traditional environments in innovative milieu [4].

The purpose of this research is twofold: At first, identifying the relevant players in the promotion of sustainable innovation ecosystems as the underlying foundation for the digital transformation and followed by investigating if their role does hold for the different innovation types. The second is to further explore the importance of relating open innovation to the academia and the user community to establish self-sustained networks and propose a policy package to support these ecosystems along the digital transformation. This topic is central in the international policy agenda as sustainable communities will better accommodate the challenges of the future putting humans at the core of the process thus enhancing prosperity and welfare.

2. Literature Review

2.1. Sustainable Innovation Ecosystems

Rapidly obsolescent assets and constant shifts in demand in a globally competitive environment are putting pressure on both the industry and national innovation systems. Trading blocks, nations, regions, and clusters worldwide face ongoing structural changes trying to appraise global innovation trends and technological shifts. Accommodating this hectic pace demands efficient and reinforced innovative ecosystems.

The concept of innovation ecosystem gained increased popularity over the last decade, due to its particular link to open innovation. The term was firstly coined by Tansley [5], to name one ecological element embedding the living creatures and their environment. Moore [6] revived the concept to describe a framework of players in coopetition, highlighting the geographic dimension of knowledge spillover sharing. This increased popularity put the debate on its relevance and definition. Presently, the innovation ecosystem comprises a multilayer framework in which institutions interconnect to develop and share information and knowledge required for the development of new innovation processes [4]. It evidences the co-creation and sharing of firms to provide a coherent solution to meet the challenges of the demand.

Innovation is strongly connected to problem solving, and presently, the challenges relate to complex problems demanding structural changes in individual and collective living such as sustainable development. According to the World Commission on Environment and Development (1987) [7], this state is a dynamic process of change, allowing the present exploitation of resources, the completion of investments, and the path of technological and institutional change, combining the welfare maximization of present and future generations.

Sustainable innovations will work as catalysts for cleaner production, meeting societal challenges in both the short and long run, encompassing economic and environmental targets in local and global dimensions. Sustainable development practices provide background for any context in which humans and the environment are found. These innovations will underpin sustainable development relying upon the networks, local communities, and corporate sustainability as think tanks developing benign solutions to societal challenges [8,9].

As a consequence, sustainability-driven innovation consists of developing goods and services that raise present welfare while efficiently allocating the endowments of resources for both the present and future generations [10,11].

The innovation ecosystem is a network of relationships combining actors and objects that establish connections, both complementary and substitute reinforcing the importance of the institutions and the environment, providing information and knowledge flows through systems of value co-creation, enhancing sustainability [9]. Sustainable innovation relies on sustainable development, encompassing ethical, social, economic, and environmental principles [4,8].

Sustainable innovation deems sustainable welfare an intrinsic value and income generation an instrumental value. Focusing on both local and global communities and the *innovative milieu* is the ecosystem and not the national system of innovation [8].

The ecosystem will consist of a dynamic, interactive network embedded in an innovation *mindset*, an interactive set-up focused in knowledge creation and diffusion. These ecosystems might be virtual due to the digital transformation we are facing globally; however, they need some grounded hub as members need to physically meet to interact and co-create, to develop new ideas benefiting from their multidisciplinary skills and competences [10].

A vigorous innovation ecosystem will provide firms an innovation environment of "tropical rain forests" where they can share value with a community with shared interests; this process will include governments, the value chain, and the user community, which communicate and promote innovation in order to create valuable new products [11]. It will be reinforced by openness and flexibility, enlarging participation to unusual partners to grasp the knowledge arising from the quintuple helix. Innovative activities are not developed inside the firm borders anymore; they are part of broader interaction with the environment, involving various players embedded within an interdependent innovation ecosystem [12]. These frameworks present a straightforward agenda to bring together human resources boosting entrepreneurial initiatives in a bidirectional way, therefore becoming collective intelligence catalysts. The ecosystem will be revived when fed by external knowledge and contributions, which will spawn an innovative *mindset* [13,14].

Traditional ecosystems tend to centralize in one entity, which benefits the most from the added value; hence, this concentration should be avoided, placing the entire community at the epicenter of the ecosystem. Establishing organized interactions will favor the continuity of the ecosystem, which should be settled on trust, sharing, and a meaningful sense of identity that will consolidate the network based on shared values, which will enhance sustainable practices [12,13].

Sustainability does not come itself; it requires enough resources and capabilities; moreover, present environmental problems call for more environmentally benign technology. The best way to survive market volatility and survive the long run is throughout innovation management and technological innovation to enhance sustainability [15].

Recent theoretical developments such as Reynolds and Uygun [16] argue that inside modern ecosystems there will be high level of interaction between key players such as universities, the value chain, and the user community to create innovative capabilities. This is further reinforced by Song [17], underlining the importance of external ties with suppliers, competitors, and user community within a centralized interaction model.

Shifting from value chains to ecosystems is more prone to organizations adopting industry 4.0 frameworks, service, or customer orientations as they are emerged in networked ecosystems; still, this movement calls for changes in the business model and increasing enrolment with stakeholders [18, 19]. The existence of solid community networks with different roles and interests will generate mutual challenges requiring sustainable practices to uphold the ecosystem.

The consistent emergence of innovations requires a dynamic and sustainable ecosystem encompassing universities and research agencies, financial endowments, sufficient demand, human capital, specialized knowledge, and willingness to collaborate in a global perspective [4,16]. Sustainable innovations add new features to conventional innovations, linked to market-desirable attributes such as durability, locality, resource and energetic efficiency, and reduction of environmental burdens [8]. Performing innovation inside the firm walls is no longer possible given the agile requests of the environment.

Endowments of intellectual capital feed the collective knowledge (explicit or tacit), serving both firms and society to amplify the ability to generate income of other productive factors, reinforcing competitive advantages. Its existence reinforces the firms' absorptive capacity, embedded in processes and capabilities inside the firm domain, which will also enrich the ecosystem [20].

Innovation tends to cluster among certain sectors or geographies with faster levels of growth and imply structural changes [21]. Indeed, regional development is happening in large clusters, cities, and metropolis. Still, R&D activities, patenting, and value creation occur in globalized innovation hubs. In that sense, smaller regions must complete the effort to identify innovative potential to fully exploit this framework. This process goes along four stages: Inception, implementation, consolidation, and renewal, hence the final stage is not observed in many regions [17].

As a consequence, it is imperative for firms to shift innovation strategy from organization-centered to ecosystem co-creation. This framework will approach organizations and improve sustainable and smart product development based on co-creation, leveraging institutional integrations, and improving the allocation of knowledge and assets inside the ecosystem [22].

Persistent growth in the extraction and use of resources in absolute terms, due to the magnitude of production growth and overconsumption, has led to waste overwhelming resource endowments. Civil society, governance, and private institutions demand for sustainability-oriented innovation systems to increasingly rationalize consumption [23].

Increasing awareness about environmental depletion has pushed innovation towards sustainability in both technological and consumption domains, resulting among others in eco-innovations with positive societal multi-level impacts. The development of these actions relies upon knowledge inflows and outflows exchanged with other agents outside the firm to speed up internal innovation and enlarge the market for innovations with increased value for the environment and society [24]. Innovation has been seen as an important tool for achieving sustainability [25], forming a key binomial in the pursuit of environmental, economic, and social development [26].

Interfirm collaboration is therefore central for sustainability purposes; however, its effect will differ according to both firm and product characteristics such as size, innovation type, elasticity, market share, and production costs [27].

2.2. Open Innovation 4.0

Open innovation is an innovation model that relies on the purposeful use of inflows and outflows of knowledge to leverage internal innovation processes reaching new paths to market, as the firms look to advance their technologies. The organization's boundaries become more flexible, permitting the combination of the internal resources with the external co-operators [18]. This model of innovation was firstly proposed in 2003, redirecting the flows of knowledge and the innovation strategies to boost collaboration among firms and other agents inside and outside the value chain, shifting towards a co-innovation paradigm in which the firm speeds up the innovation pace and the organization changes the business model buying and selling knowledge as needed [18,28,29].

Opening the innovation strategy plays a key role towards effective strategic sustainable management. In doing so, firms can leverage knowledge production and management promoting sustainable innovations that retro-feed organizational sustainability. Efforts will be put into knowledge management and the incoming ideas from the external stakeholders, such as research centers, universities, suppliers, and customers. If there is a breakdown of values, in which knowledge arises through partners, the network will acquire relevant skills to manage knowledge and innovation as complements [30].

Blurring the boundaries between the firm and its environment will enable transferring innovations to different marketplaces, with bidirectional knowledge flows circulating outside the organizational borders, highlighting the increased benefit of knowledge sharing throughout partnerships and networks. It implies leveraging external sources of knowledge such as other firms, consumer community, and the ecosystem. In doing so, organizations will combine internal and external know-how, extending the collaboration with the rest of the ecosystem, mostly the Academia and the user community, thus accelerating the innovative process [18,28,29].

Regardless of the centrality of the user community in driving socio-technical transitions, its role within sustainable innovation remains largely overlooked by policymakers. Empirical evidence proves that these agents can no longer be neglected; still, policymakers remain apprehensive about the potential of the user community in this process [31].

Researchers, practitioners, and policy makers quickly understood the importance of the shifting paradigm, and OI was granted important acceptance and diffusion, due to its adherence to reality. Nearly a decade later, the framework was updated arguing that success of the process depends on the knowledge flows and that they should be carefully managed inside and outside the firm boundaries with straightforward mechanisms providing already-established solutions, accordingly to the business model for all kinds of knowledge flows [29].

Despite some skeptical considerations [32], the concept was awarded the trust of the community and remained in solid position, being refined by several authors e.g., [33–36]. Most of the criticisms relied on the need for strong clarification about the agents needing to be involved in the process and their role in the development of the actions [37,38]; however, decentralization in governance is the major challenge put forward by this framework [39]. Innovation is a complex and uncertain process with natural hindering factors; however, open innovation will naturally speed up the pace innovation outputs arise.

Ten years after the concept proposal, Open Innovation 2.0 was reshaped, connecting to the quadruple helix, adding the civil society to the usual players (government, university, and firms), and as a consequence, adding the structural changes driven by user-oriented innovation models; in these frameworks, the speed of the innovation process is accelerated as the different phases co-exist and are set a real world context [3]. The second version of the framework underlines new foundations enhancing the importance of networks and collaborations, promoting interdependencies, relying on corporate entrepreneurship, promoting R&D, and specific intellectual property management, which combined with the accelerated exchange of ideas will boost innovation success, powered by synergies and complementarities [40,41]. The establishment of trusted relations in aligned communities, networks, and stakeholders will be integrated in the surrounding communities thus creating an ecosystem. Innovation 3.0 was proposed in 2010 as conceptual approach, as *"Embedded Innovation"*; the framework encompasses the digital transformation. SMEs (Small and Medium Sized firms) that emerged in a digital and dynamic environment should rely on combined knowledge as it is the most important source of innovation, being essential for survival and growth [42].

This framework captures how companies survive and the way they embed with the other players, focusing on the idiosyncrasies of each. The embedding of the different organisms requires the promotion of the "innovation ecosystems" and business models for innovation to generate sustainable ecosystems. Given the dynamic nature of the innovative process, the organizational process needs to encompass the exploration/exploitation binomial to survive the demanding environment [42,43].

When fed with innovation, embeddedness is a self-sustained process in which the firms along with its stakeholders interact in a certain environment, coexisting and stressing for survival; the process will shape the environment. Mutual influences are exerted, and the innovation process is intertwined with the environment along the innovation life cycle [43].

Embedded innovation practices require the consumer involvement, and the perception of long-term ties between agents creating value and critical development of complementarities among them, seeding the sustainable economies of the future. As a consequence, the framework cannot be considered a substitute but a complement to the conventional models. Structural innovations, despite their importance, are bounded to firms' internal resources, limiting the ability to stay competitive in the long run and beat the competition [42,44].

The improvement of already existing products also generates value to the consumer and the civil society due to the engagement with smart techniques and responsibility. This model will allow for the maintenance of the demanding pace of the innovation locomotive. Additionally, valuing consumer habits will address the environmental responsibility issues, which are at the core in present policy actions [44]. Embedded innovation therefore builds upon structural innovation.

The creation of sustainable sources of growth will rely on expanding from structural innovation to the consumer-community, tying in with the society. These enlarged communities will encompass diverse people, with different backgrounds, working together towards the creation of an interdependent existence [45]. This paradigm requires potential focus, relationship-based value, and transformational stakeholder engagement to create a sustainable competitive advantage connected to the business model [44]. When moving to the ecosystem level, sustainable innovative practices include the co-creation of knowledge, the engagement of stakeholders, and the value chain to promote improvements in products, technology, and environment [46]. The ideas that feed the innovation process come from people; as a consequence, they need to be stimulated to generate, discuss, and share them. This atmosphere will leverage firm performance and accelerate the innovation cycle, so training people to acquire this *mindset*, creating and recognizing relevant knowledge, will enhance innovation [45]; notwithstanding, larger firms and firms operating in broad marks are more prone to adopt the framework as they have increased awareness about the importance of the environment in their competitiveness [47].The implementation of open innovation in smaller firms is dependent on their persistent managing control over complexity. Due to their versatility, they can be more effective in combining alternative practices, introducing new products and new markets, creating virtuous innovation cycles based upon systematic emergence and partially on complexity control; this will keep the openness of the culture and the business model alive, rising altruism and promoting trust-based collaborations [48].

The integration of the value chain (vertical and horizontal) and the interoperability will break down firm boundaries into a network; this dynamic process will change and create the existing roles of agents. Creating and benefiting from the value emerging from the ecosystem goes beyond the individual value chains, forcing business models to change, shifting to industry 4.0 or alike [49,50]. In this environment, firms will develop new capabilities and meet the consumer-community by means of digital tools gathering and analyzing data to support evidence-based strategies being part of a multi-sided dynamic ecosystem rather than a linear value chain [51].

Industry 4.0 was firstly proposed by the German government in 2011 as a milestone to the fourth industrial revolution. It encompasses multiple advances in digitization, automation, and robotizations. It will transform the value chain to global business model, based on the construction of systems and relationships between machines and machines and humans [1].

This model promotes an accelerated innovation cycle, which accommodates fast-changing consumer expectations, switching to automatization, digitization, and digital security. Consequently, at its core is fostering cooperation and networking among businesses and other entities in the ecosystem to tackle challenges but also to develop new innovations, ideas, or even new businesses [52]. Regardless of the individual characteristics, the open innovation framework will facilitate the creation and understanding of different linkages to be established with players to raise the efficiency of the innovative process, as well as the exchange of resources and knowledge with other entities.

The evident need for transition towards the digital requires a technological push from firms, and with the acceleration of the innovation processes, firms must be able to quickly identify the value of the open innovation processes [53], and leverage their competences to speed up the transition to digital. In the presence of global business models, in which there is constant communication, networking is essential among similar and different players. This exchange of information and knowledge transfer at a macro scale will make the best practices available to everyone, embedding all members with front-edge technology, which will optimize the use of resources, respect the environment, and include community values, being a consequence of sustainable ecosystems.

The literature highlights the existence of an N-shaped curve called the open innovation paradox, which has to be fold back. With the digital transformation, the dynamics of open innovation is increasing rapidly overcoming the inflection point, decreasing the costs of these dynamics, which is in "rocket-shooting stage" for firms, institutions, and all other players in the ecosystem [54]. The university seems to play a major role in this process, demanding mutual adjustments [55,56]; still, the transition demands for the establishment of alliances [57] and the exploitation of the academic research in industrial innovations [58], as the knowledge produced in universities is increasingly valuable to firms [59].

Developing open innovation strategies intertwines external sources of knowledge with the internal; this dynamic process combines knowledge, human and financial resources, and all other players in the collaborative ecosystem [60]. Specific external supports are therefore required mainly in the case of smaller firms; a good practice in this field is partnership development, multi-dimensional clustering with legal support responding to market changes with prompt sustainable innovations [47].

Open innovation will help turn the immense challenge Europe is facing into an opportunity by investing in the future. European Green Deal the digital transformation initiatives will boost employment throughout innovative and inclusive growth, fostering the resilience of societies environmental sustainability [61].

2.3. The Loose Links in OI and the Ecosystem: University-Industry-Collaborations

Knowledge is the masterpiece of the whole open innovation framework, in particular, the external (tacit or explicit), as it relies on the flows; the process requires the coordination of research and development within and integrated horizontal concept to minimize the costs, by means of outsourcing research results developed by the company [30].

Open innovation has been addressed in different perspectives and the conceptual framework has evolved, however the analysis of the patterns of linkage and the associated gains with external collaborations is still overlooked [62], along with the drivers of collaboration and why and with which external entities to collaborate [56,57].

According to Shin et al. [63], six knowledge capacities are required to build an open innovation framework: (1) Transformative capacity; (2) connective capacity; (3) inventive capacity; (4) absorptive capacity; (5) innovative capacity; and (6) desorptive capacity. These competences will allow the retention of internal and external knowledge, its exploration, and exploitation. The connective capacity refers to the ability to establish links with other agents to access external knowledge bases, through inter-organizational relationships such as strategic alliances.

Sustainability is undoubtfully essential these days; nevertheless, going through its path is approached differently among communities. Innovation has recently emerged as a means to achieve sustainability. An integrative capability-based framework (including exploration, retention, and exploitation phases of innovation) based on the classic evolutionary model must be implemented, as sustainability requires more diverse and particular sustainable partners. The effectiveness of the innovation process depends on the context, enabling the combination of open innovation capabilities with the specific context of sustainable development [64].

Universities are at the center of the innovation process given their role in educating students as agents of innovation, promoting and inspiring their critical spirit [65], and transferring their knowledge to promote organizational aptitudes and development tools, which are extremely valuable assets inside the organizations [66,67].

The analysis of the economic performance along with the innovative strategies of developed economies reinforces the importance of knowledge production and diffusion in different technological regimes as a booster of competitive advantages [68]. According to Xie and Wang [14], there are six principal modes by which firms engage in open innovation ecosystems: (a) Firm–university–institute cooperation, (b) interfirm cooperation, (c) firm intermediary cooperation, (d) firm-user cooperation, (e) asset divestiture, and (f) technology transfer; here, the focus will be put on the first.

As a consequence, university firm collaborations play a determinant role in the promotion of those flows, promoting enlarged exchanges in seminal and specific domains [56–58]. The challenges of the future and the empowered role of the Academia in the ecosystem demand institutional and bureaucratic adjustments. The active enrolment in research projects, which will shift the priorities and the financing incentives of the institutions, will force them to abandon the ivory tower fostering closer relations with the entrepreneurial sphere [55]. University–industry collaborations encompass a multi-layer framework of collaborative research, scientific consulting, or research contracts intertwining the theoretical and real-world dimensions [69,70].

The interconnectedness of universities and firms will backup policy frameworks such as the Research and Innovation Strategy for Smart Specialization (RIS3) harmonizing regional and national innovation ecosystems with the helix [71], with multiple influences among institutions with constant and bidirectional influences sustaining the entrepreneurial ecosystem with the richness of a reef [72]. At present, the dissemination of the digital transformation is eroding the traditional boundaries of the industry, requiring the redesign of the existing Business Models [44].

The Fourth Industrial Revolution is considered as the most powerful innovation driver at present, causing an entire innovative wave; features such as real-time capability, interoperability, and the horizontal and vertical integration of production systems through ICT systems will soon become a reality [52]. Meeting this transformation is urgent for firms to survive a globalized competition and the ever-changing demand, which shortens innovation and product lifecycles demanding for a more interactive environment [50].

The shape and intensity of the collaborations will be strongly influenced by the university research resources along with the competences of the research teams and the institutional orientation for the commercialization of science [67,69,73].

More than ever, universities and firms will collaborate based on the belief that their complementarity will reinforce the strength and the gains concerning they partnership, given that the absence of competition will prevent opportunistic behaviors, reinforcing trust [57,69] and combining different endowments of physical and human resources with expertise [69,74]. Very often, these collaborations pushed universities to move from traditional to entrepreneurial organizations [75,76].

Traditionally, universities connect to knowledge transfer by means of education. Skilling the labor force will bridge the knowledge from the inside to the outside world. Through research, they become engines of knowledge creation and diffusion, producing new technologies and tools useful for the entrepreneurial agents, making fruitful connections. These R&D dynamos bridge scientific knowledge to applied knowledge by means of scientific networks, which will help the firms in speeding up their innovation cycles and decreasing innovation costs [71,72,76]. The knowledge transfer should incorporate the regional needs consequently creating positive *spillovers* to the local community. Nowadays, the university cannot be detached from its ecosystem, as it must commit to the community needs in the generation of sustainable actions and welfare. Moreover, most universities draw upon public funding, which means that they shall give back to tax payers, and firms may use their facilities and technology to reduce their costs, tethering internal R&D with the external knowledge. The last decades have shown an exponential increment of this concern, and presently, the third mission is the touchstone with the innovative ecosystem multiplying the extent of the public policies [59,65,71].

The role of Universities in the promotion of open innovation strategies goes in line with its missions as fostering more efficient innovation cycles including public R&D, in this line mutual enhancement, expectably the innovation ecosystems will raise. The metamorphosis of the university into an entrepreneurial institution is inevitable [51], and the new innovation policy frameworks are now assessed in terms of social coordination rather than market oriented [57,62]. However, public research is unique due to its independence and the distancing with commercial leitmotifs having the ability to radically change the landscape of the industry.

In addition to knowledge generation, universities will train their students, supplying the industry with graduated individuals [55,66]. Hence, the university also benefits from its involvement with the industry gaining awareness of the real-world concerns and technological trends, and finding occupations for their students [57,58]. The role of the universities in regional and national systems of innovation is widely accepted, thus the policy actions and instruments to tackle this issue is fuzzy.

Recently, policy makers have concentrated, in a semi-myopic way, on the university's third mission, neglecting to some extent, the others. These institutions are asked to teach research and contribute to the civil society through knowledge creation and diffusion [59,65].

Open innovation helps in framing the entrepreneurial mission of universities as it promotes networking between the heterogeneous players to transfer and commercialize knowledge [42,46]. Loaded with enlarged missions, universities are seen as the holy grail for the development of open innovative ecosystems [42]; with the increased demand for technological progress and disruptive innovations, firms operating in globalized markets have an increased incentive for companies to exploit collaborations with universities [51], given their value as sources of information for innovation [53].

Policy makers consider universities as knowledge crafters, at the center of the ecosystem catalyzing connections and flows of valuable information that will enhance the regional ability to generate income [20]. Multilateral connections appear, in the sense of university–industry–government collaborations, to promote the skilling of the workforce, knowledge diffusion, and sharing with the involvement of the civil society [56].

The promotion of sustainable innovation ecosystems is a co-creative process in which players must contribute and benefit from knowledge creation relying upon absorptive capacities and improvements [76]. In this symbiotic process, the role of each agent overlaps mutually benefiting from different competences that will complement each other.

The university's third mission is leveraged by facilities such as science parks, industrial clusters knowledge transfer offices, incubators, and other infrastructures, which are placed nearby to simplify the link between the actors [72]. Consequently, shifting from conventional structures to networked structures has been the priority of many institutions [73,76]. At present, research projects are often co-subsidized by public and private institutions, with private funding assisting government-funded projects and vice versa [58].

However, the emerging entrepreneurial universities are highly focused on university–industry collaboration and may find themselves relatively limited to the type of research carried out at the institution. Ongoing research may be subject to the requests of the private agents with a solution-driven logic, which will serve the purposes of the ecosystem in the short term, consequently restricting curiosity-driven experimentation and random research performed by scientists [75].

This transformation possibly endangers basic research and, with no investments in long-term incremental R&D, biased profitable initiatives will emerge, as no incentive schemes are created for non-commercial research. Additionally, the paid interest of industry is the existing research that releases the government from supporting public research, as universities become less dependent on public funding, arguing in favor of university funding closely linked to its economic impact [58,65]. The situation will worsen as the enrolled private agents, either as sponsors or as free-riders, become successful in appropriating the knowledge produced by the public sector, with the contribution of tax payers, patenting the findings and internalizing monopoly profits on the one hand and preventing the general spread of the progress on the other [62,65].

The implementation of generalized open innovation strategies in the ecosystem towards the digital transformation process will approach different players, placing them in convergent technological paths. The reinforcement of this paradigm by means of public policy, with the active intervention of all agents in the ecosystem, is fundamental to achieve long-term sustainability, guaranteeing the continuity of regions.

3. Methodology

To measure the role of open innovation strategies in the promotion of sustainable innovative ecosystems, five logit models were run. In doing so, empirical evidence will be gathered to discuss the impact of the open innovation strategy along with reliance on public funds and the user community, as well as firm structural characteristics that impact the propensity to perform the different innovation types, and address the role of the innovation sources, such as the universities, as enhancers of the effect.

3.1. Database Description

Empirical analysis will rely on data from the Community Innovation Survey (CIS) 2016, from Portugal, covering the biennia of 2014–2016. The database includes 6775 firms operating in Portugal with heterogeneous structural characteristics and innovative profiles. This survey is the most comprehensive concerning innovation-related issues providing enlarged evidence for the relevant variables in use.

Innovative strategies take time to produce effects, and innovation cycles are long. The efforts performed by this period will hopefully produce the desired effects in the future. Addressing the case of Portugal is of particular interest in this period as it is amongst the regions that have made the greatest evolution in the recent years, being classified in 2019 and 2020 as a "strong innovator". To the European Innovation Scoreboard, transition towards the leading stage was reached only in a group of seven countries. In Portugal, these years were the greatest "leaps" after a persistent classification as "moderate innovator"; additionally, its leading capacity was highlighted in domains such as "innovation in SMEs" [61].

3.2. Exploratory Analysis

To provide a detailed description of the variables in use, Table 1 addresses a description of each variable and its scale of measurement. In most cases, the measurement follows the CIS original scale; in others, simple mathematical conversions were performed.

Table 1. Variable description.

VARIABLES	DESCRIPTION	MEASUREMENT
TEC_REG (1)	Technological regime of the firm (according to Boliacino and Pianta) [77]	1 = supplier dominated; 2 = scale intensive; 3 = specialized supplier; 4 = science based
SIZE (2)	Firm dimension	1–3 degree
EXP_PROP (3)	Proportion of the turnover exported	Decimal
EMPUD (4)	Human Capital intensity	1–6 degree
OPEN (5)	Performing inbound and outbound innovation	Binary
FUNDS (6)	Beneficiary of funds	1 to 4 count
SUNI (7)	Relying upon Universities as source of innovation	Binary
USER_COMM (8)	Relying upon User Community as source of innovation	Binary
HSI_LEG (9)	Concerned about Hygiene and Security legislation	Binary
ENV_LEG (10)	Concerned about Environmental legislation	Binary
INT_ASSET (11)	Having registered copyrights or others	Binary
PROD_INNOV (12)	Having performed product innovation	Binary
PROC_INNOV (13)	Having performed process innovation	Binary
ORG_INNOV (14)	Having performed organizational innovation	Binary
MKT_INNOV (15)	Having performed marketing innovation	Binary
TECH_INNOV	Having performed product or process innovation	Binary

Concerning technological regimes, firms were divided into four categories, according to technological intensities. Firm dimension encompasses small, medium, and large, following the European Commission methodology and the European Innovation Scoreboard [61]. Human capital was divided into intensities, following the CIS methodology and the EIS [61]. Innovation types were appraised as binary variables, depending on the firm response. The proxies in use follow the survey methodology and are further described in Table 1, and were similarly used in [78].

Table 2 provides the descriptive statistics and the correlations among the variables in use. Regarding correlation, moderate intensity is found, which highlights the accuracy of the set used and guarantees the inexistence of multicollinearity. This was further reinforced with VIF (variance inflation factor) tests, which pointed in the same direction.

Table 2. Descriptive statistics.

VARIABLES	Mean	S.D.	Min	Max	(1)	(2)	(3)	(4)	(5)	(6)	(7)	(8)	(9)	(10)	(11)	(12)	(13)	(14)	(15)
TEC_REG (1)	1.753	1.100	1	4	1.00														
SIZE (2)	1.351	0.572	1	3	0.05	1.00													
EXP_PROP (3)	0.217	0.319	0	1	0.06	0.24	1.00												
EMPUD (4)	2.413	1.830	0	6	0.44	0.15	0.04	1.00											
OPEN (5)	0.103	0.304	0	1	0.11	0.22	0.14	0.20	1.00										
FUNDS (6)	0.295	0.700	0	4	0.13	0.18	0.19	0.17	0.34	1.00									
SUNI (7)	0.476	0.499	0	1	0.14	0.20	0.09	0.23	0.23	0.21	1.00								
USER_COMM (8)	2.363	2.383	0	6	0.10	0.13	0.11	0.21	0.23	0.27	0.41	1.00							
HSI_LEG (9)	0.718	0.483	0	5	0.08	0.12	0.10	0.20	0.21	0.26	0.03	0.62	1.00						
ENV_LEG (10)	0.722	0.489	0	5	0.07	0.12	0.10	0.19	0.22	0.27	0.06	0.61	0.96	1.00					
INT_ASSET (11)	0.289	0.705	0	6	0.13	0.11	0.15	0.20	0.26	0.28	0.18	0.24	0.20	0.19	1.00				
PROD_INNOV (12)	0.335	0.472	0	1	0.04	0.10	0.17	0.08	0.26	0.29	0.08	0.40	0.45	0.45	0.23	1.00			
PROC_INNOV (13)	0.495	0.500	0	1	0.03	0.14	0.12	0.10	0.25	0.29	0.03	0.48	0.62	0.61	0.16	0.39	1.00		
ORG_INNOV (14)	0.358	0.480	0	1	0.10	0.12	0.11	0.22	0.24	0.23	0.15	0.43	0.47	0.47	0.23	0.27	0.40	1.00	
MKT_INNOV (15)	0.396	0.489	0	1	0.05	0.04	0.01	0.20	0.20	0.21	0.12	0.49	0.50	0.50	0.26	0.35	0.37	0.42	1.00

4. Econometric Analysis

4.1. Econometric Estimations

In order to appraise the determinants of the innovative performance for each innovation type, five logit models were run, which are presented in the following Table 3. Models 1 and 2 refer to the conventional types of innovation: Product and process, respectively. Model 3 encompasses having performed any of the types of technological innovation (product or process). Non-technological types of innovation such as marketing or organizational appear in Model 4 and Model 5, respectively. Open innovation studies focus on product and process innovations; however, given the emergent importance of the other types, analysis was further extended.

Table 3. Econometric estimations—marginal effects after logit estimation.

VARIABLES	PROD_INNOV	PROC_INNOV	TECH_INNOV	ORG_INNOV	MKT_INNOV
TEC_REG	−0.053 *	−0.109 ***	−0.184 ***	−0.035	−0.137 ***
	(0.032)	(0.037)	(0.053)	(0.032)	(0.033)
SIZE	−0.110 *	0.216 ***	0.051	0.061	−0.218 ***
	(0.057)	(0.072)	(0.103)	(0.058)	(0.059)
EXP_PROP	0.726 ***	0.242 *	0.592 ***	0.182 *	−0.520 ***
	(0.105)	(0.129)	(0.205)	(0.104)	(0.107)
EMPUD	−0.082 ***	−0.101 ***	−0.181 ***	0.145 ***	0.144 ***
	(0.021)	(0.024)	(0.035)	(0.021)	(0.021)
OPEN	*0.536 ****	*0.430 ****	*0.422 ***	*0.554 ****	*0.375 ****
	(0.098)	(0.125)	(0.190)	(0.098)	(0.101)
FUNDS	0.188 ***	0.181 ***	0.180 **	0.106 **	0.091 **
	(0.044)	(0.055)	(0.085)	(0.043)	(0.045)
SUNI	*−0.028*	*−0.049*	*0.101*	*0.004*	*−0.141 **
	(0.074)	(0.089)	(0.130)	(0.074)	(0.076)
USER_COMM	*0.114 ****	*0.094 ****	*0.080 ****	*0.161 ****	*0.254 ****
	(0.016)	(0.020)	(0.029)	(0.017)	(0.017)
HSI_LEG	−0.050	0.560	0.665	0.072	0.003
	(0.241)	(0.349)	(0.501)	(0.243)	(0.247)
ENV_LEG	0.365	−0.282	−0.640 *	0.091	0.103
	(0.229)	(0.267)	(0.345)	(0.219)	(0.227)
INT_ASSET	0.346 ***	−0.033	0.134	0.300 ***	0.527 ***
	(0.049)	(0.052)	(0.088)	(0.047)	(0.054)
Constant	−0.551 **	0.780 **	2.606 ***	−1.432 ***	−0.685 ***
	(0.243)	(0.332)	(0.455)	(0.239)	(0.242)
Observations	4229	4229	4229	4229	4229

Standard errors in parentheses: *** $p < 0.01$, ** $p < 0.05$, * $p < 0.1$.

Given the binary nature of the dependent variable (whether or not having performed innovation), a binary count model (logit) was implemented. Notwithstanding, most of the empirical evidence presents different frameworks in terms of the dependent variables or establishes international comparisons, which precludes direct comparisons.

Logit estimations were omitted due to the impossibility of interpreting the coefficients; Table 3 presents the marginal effects that quantify the impact on the propensity to innovate caused by changes in the independent variables.

Among the five models, the dependent variable does change to address the impacts of the exogenous variables in each type of innovation. Explanatory variables and controls are the same among the five models in use, allowing for inter-model comparisons.

4.2. Results and Discussion

Firms belonging to more knowledge intensive technological regimes are naturally more prone to develop innovative activities irrespective of the innovation type [77,79]. Empirical evidence proves to be contrary in this case, with firms in more knowledge intensive segmentations being less prone to innovate, and the impact being higher in technological innovations. This result is somehow deceptive as the policy makers tend to favor these areas.

There is a positive correlation between firm size and innovation; larger organizations look for appropriating and binding all the relevant ideas that lie beyond their boundaries [80,81]. Contrarily from expected, size increments reduce the propensity to perform product or marketing innovation; conversely, concerning process innovation (Model 2), size raises the innovative propensity. These results evidence that policy makers cannot generalize policy requests based upon size.

Export propensity plays a positive effect on the propensity to innovate regardless the innovation type, with the exception of marketing innovation. Firms operating in external markets are more prone to have a dynamic innovative strategy; as a consequence, incentives promoting the access to foreign markers will reinforce innovative behaviors.

Concerning human capital intensity, the effect on technological innovations (product and process innovations) contradicts previous expectations [45], as enlarged stocks of employees with an undergraduate or more will deter innovations. On the contrary, non-technological innovations such as organizational or marketing will be enhanced by human capital intensity. Results evidence the fact that technological innovations are complementary enhanced physical and human capital.

Public funding raises the probability to perform innovation, regardless of the innovation type [58]; this result reinforces the importance of the conventional policy instruments in the promotion of innovation cycles.

Pursuing open innovation strategies, and their role promoting innovative ecosystems supporting the digital transformation [53], are the core of the empirical analysis. These strategies were proxied by the combination of inbound and outbound R&D flows; and appear with a strong positive impact in innovative propensity for all innovation types. In detail, firms performing open innovation strategies are 42.2 percentage points more prone to develop technological innovations than their non-innovative counterparts. Even for non-technological innovation, being open is still a strong booster raising the probability to innovate by 55.4 pp and 37.5 pp in organization and marketing domains.

Relying upon the university as a source of information for innovation is, according to the literature, an enhancer of the innovative performance [9,70], however the variable fails to be statistically significant notwithstanding the innovation type. The lack of significance of this variable needs to be further appraised; however, these results go along with previous research for the Portuguese case [78]. Despite the positive evolution of the innovation performance as a whole [61], more needs to be done to promote University industry collaborations.

The effect of the augmented helix in innovative performance demands user-community involvement and the civil society along with the adoption of responsible practices in innovative strategies [19]. The influence of the user-community on the innovation initiatives was measured by a binary variable, which takes the value 1 if the firm indicates having considered the user opinion and further contributions to develop innovations. In all dimensions of analysis, the variable appears with a positive impact in innovation. This result points towards the existence of co-creation dynamics encompassing the knowledge flows emerging from the ecosystem, retro-feeding innovation practices.

Social responsibility dimensions were appraised by means of legal aspects such as the hygiene and security along with the environment [73]. In general, these vectors are not yet relevant as innovation determinants. These results should get the attention of policy makers reinforcing the urgency for the redesign of these regulations. Sustainability strongly relies on public policy and its accuracy in generating desirable behaviors; as a consequence, the promotion of corporate responsibility cannot be neglected.

5. Concluding Remarks

5.1. Theoretical and Empirical Implications

Innovation is the main driving force for sustainable development and the promotion of growth; as a consequence, the topic gained centrality in the agenda of policymakers, practitioners, and researchers. So, innovation and sustainability are considered, to some extent, two dimensions of the same reality [82,83]. Regarding international regulations in the field of environmental protection, the focus was put on the preservation of natural resources [10].

Drawing on a cross-sectional sample of firms operating indifferent sectors and running several logistic models to identify the determinants of the different innovation types, empirical evidence proves that open innovation, operation out of boundaries, public funding, and the user community promote innovation cycles. Moreover, technological regimes and dimension produce mismatched effects. This implies that economic and sustainability innovation goals can be attained at once.

As industrial innovations are becoming increasingly more open, it is important to understand where open innovation will add value to knowledge intensive processes. The digital transformation demands for connections, networks, and high speed in the innovation cycles. The present paper contributes to the identification of the importance of this new player recently identified and overlooked in the extant literature [37,83–85].

As we journey, further efforts need to be made to systematically include the user community in the ecosystem, given its increasing importance [86]. The empirical findings underline that having the awareness and the proximity to these agents will avoid mistakes being made, as they are the entry gate to marketplace acceptance.

Additionally, the article theoretically and empirically identifies open innovation strategies as innovation boosters regardless of the innovation type, therefore grounding innovation ecosystems. Given the importance of adopting permeable organizational boundaries, managers should focus on this strategy, transforming their business models to meet the fast-changing requirements of the consumer-user community, combining internal endowments with the external, accelerating the innovation cycle, and reducing its costs.

Universities play a key role in creating human capital and in generating new knowledge through research. The article addresses the three missions of the university, and, contrarily to the previous, in empirical terms, the first mission seems to be properly established, as the human capital intensity influences innovative propensity.

Conversely, the connection with the university as a source of relevant knowledge to innovation fails to affect the innovation strategies. This unsettling fact deserves further attention and the identification of the adjustments to put universities at the center of knowledge networks. Nowadays, universities play a determinant role in the advance of basic research while in the 1980s, that mission was led by firms' R&D departments. Business will drain all possible appliances for their knowledge. Still, the firm focuses on their own business model whilst society seizes competition between different ideas. So, new R&D strategies are emerging and need further analysis.

Digital transformation will multiply interactions and sources in unprecedented ways, and open innovation will provide the opportunity to ground these connections, empowering users, optimizing the match between demand and supply, and minimizing waste. Competition pressure will come from the value chain and the empowered user [86], and this shift demands accurate open innovation strategies. In accordance to Chesbrough [82], the empirical evidence proves the need for addressing the intellectual property issues under the open innovation *mindset*, to meet the singularities of the innovation types and the changes brought by the cooperative innovation processes.

5.2. Limitations and Future Research

The importance of the value chain along with the Academia in the promotion of the innovation ecosystems has grasped the attention of academics, practitioners, and policy makers since the 1980s with the introduction of Etzkowitz's helix. As society evolves along with the technology, new players should be considered. The present article shed some light on the role of the user community, however the next generation of studies in the field needs to further explore the role of the digital revolution in the reshaping of the knowledge creation and diffusion processes and mostly the emergence of a central player: The user community.

Nowadays, consumers claim for sustainability requirements; traditional price competitive models are insufficient and the active involvement in environmental protection and social responsibility is demanded by stakeholders. In this vein, sustainability-oriented innovations are the new core of entrepreneurial innovative strategies [10,82]. The present research does not measure the impacts on the firm performance, which deserves further attention.

Despite the numerous studies on university–industry collaborations, there is still insufficient evidence, concerning the upstream, midstream, and downstream aspects of this connection. It is also important to empirically address the role of open innovation as facilitator of this process. This loose link is probably deterring the development of more efficient innovation ecosystems [20,87,88] in the Portuguese case [78]. The question that remains unanswered is: Why does the university fail to impact innovative propensity? This finding should be further analyzed to shed some light in what is hampering the establishment of solid connections between the Academia and the Industry. Open innovation is undermined if this pillar keeps lacking. All in all, practitioners and managers should be aware of the importance of this source of knowledge for their processes.

A more detailed analysis is required to understand the positive effect on non-technological innovation compared to the unexpected negative effect on technological innovations. The variable in use is perhaps broad, encompassing general degrees rather than specific competences being a limitation. Apparently, skilling is biased towards non-engineering competences. Separating the number of engineers from the grand total is impossible in the CIS database, but it could provide a finer conclusion. Despite the robustness of the respondent sample, present results emerge from a sectional analysis and they may represent an exceptional coordination rather than a long-term trend. Running the same empirical analysis in a diachronic perspective would reinforce the findings, which is an open avenue of research for future works.

Regardless of their structural characteristics, firms increasingly rely upon external sources of knowledge rather than closing themselves off by being confined to the exploitation of internal resources. However, perhaps due to uncertainty and complexity of the legal framework, small and medium-sized firms are less confident in open innovation networks, anticipating appropriability problems more often than their larger counterparts. As these organizations are the backbone of the industrial fabric in most countries, policy makers should address their weaknesses with a more effective legal system protecting property rights. Policy action should help SMEs in finding more modern business models, promoting the digital transition. Sharing best practices in open innovation networks underlying the extant gains will certainly encourage the adhesion to the ecosystem.

5.3. Policy Recommendations

In a globalized era with proliferating open innovation practices, policy makers must ensure a democratic access to knowledge and technology. Policy changes will ascertain firms and entrepreneurs towards co-creation. Open innovation is the key for supporting networks rather than individual firms in the promotion of market competition.

The new innovation policy must abandon the centrality of large companies and consider the roles of human capital, competition, financing, intellectual property, and public data in promoting an open innovation ecosystem encompassing smaller organizations.

In most cases, current policy instruments still rely upon the closed innovation paradigm, focusing upon large markets, traditional sectors; protect national companies; and subsidize the larger organizations.

Empirical findings sustain that large firms are potentially starting to outsource innovation and entrepreneurship projects in environments that are considered more open and more agile, as their size contributes to their lower propensity to innovate. It is worth mentioning that the Portuguese achievements in terms of SME dynamism towards innovation were highlighted in the European Innovation Scoreboard 2020 [65]. Again, policy makers' attention is deserved as often large firms are positively discriminated towards public funding.

The insignificant results of the environmental and hygiene legislation in and security domains should be further analyzed, as they are strongly tied to sustainable practices and responsible attitudes towards resource use. Designing demanding policy measures will discipline industrial practices and standardize desirable behaviors.

The positive effect of funding demand more sophisticated actions such as policy mixes reallocating spending, as technology intensive sectors are not the leading innovative group. Small and medium-sized firms deserve specific supports given their high propensity to innovate. Incentives and subsidization should rely upon technology maturity. Start-ups encompassing immature technologies should be subsidized, and more mature organizations should perhaps benefit from lighter incentives such as fiscal benefits, tax credits, or loan guarantees.

The transition from closed to open innovation requires new funding frameworks combining the strengths and the weaknesses of the ecosystem throughout smarter policy packages such as mandatory consortia to reach public grants. Managers have certainly understood the importance of the public support so they will link to the University.

Natural environment preservation has gained momentum and the international community presented concepts such as "green economy, green growth, and green development". Open innovation effectively deals with market failure correction such as externalities minimizing economic distortions related to economic value and green value. Implementing green governance will foster resource preservation and expectably provide next generations a promising future. Still, this new governance paradigm should avoid the "governance failure" caused by "collective action dilemma" [89].

Developing sustainability-oriented open innovation framework requires a broader stakeholder approach [90]. Governments must promote innovative strategies building upon latent capabilities and focusing on regionally relevant problems, promoting societal mind-changing road maps. Major weaknesses will be surpassed, and the innovation requirements will be identified. Vertically integrated labs are shifting to disintegrated networks of innovation tying agents into ecosystems centralized in the private sector, public action needs to adapt.

Facing the present challenges of the uprising economic crisis with unprecedented consequences, given the simultaneous cut in both aggregate demand and supply, reinforces the need of a knowledge-based model of development [88]. Enhancing regional capabilities will increase cohesion and create self-sustained ecosystems, which will pace up the speed of recovery through inclusive growth [20]. Concerning scientific decision-making processes and the long-term equilibria between man and nature, governance should address sustainability and the concept of green governance should be implemented in a timely manner [89].

Nurturing open innovation ecosystems is vital for the acceleration of recovery, boosting sustainable and responsible practices along with the respect for local communities. This innovation strategy contrasts with conventional models as, for the first time, it places the human at the center of disruptive innovation with different roles being played.

Author Contributions: Conceptualization, J.C.; methodology and empirical analysis, J.C.; formal analysis, J.C.O.M. and J.C.; writing—original draft preparation J.C. and J.C.O.M., writing—review J.C. and J.C.O.M. All authors have read and agreed to the published version of the manuscript.

Funding: This research received no external funding.

Conflicts of Interest: The authors declare no conflict of interest.

References

1. Schwab, K. *The Fourth Industrial Revolution*; Crown Business: New York, NY, USA, 2017; ISBN 9781524758868.
2. Porter, M.E.; Kramer, M.R. *Creating Shared Value*; Harvard Business Review: Cambridge, MA, USA, 2011.
3. Curley, M.; Slamelin, B. *Open Innovation 2.0: A New Paradigm*; European Commission: Brussels, Belgium, 2013.
4. Granstrand, O.; Holgersson, M. Innovation ecosystems: A conceptual review and a new definition. *Technovation* **2020**, *90*, 102098. [CrossRef]
5. Tansley, A.G. The use and abuse of vegetational concepts and terms. *Ecology* **1935**, *16*, 284–307. [CrossRef]
6. Moore, J. Predators and prey:A new ecology of competition. *Harv. Bus. Rev.* **1993**, *71*, 75–86. [PubMed]
7. WCED. *World Commission on Environment and Development. Our Common Future: Report of the World Commission on Environment and Development*; Oxford University Press: Oxford, UK, 1987.
8. Oksanen, K.; Hautamäki, A. Sustainable innovation: A competitive advantage for innovation ecosystems. *Technol. Innov. Manag. Rev.* **2015**, *5*, 24–30. [CrossRef]
9. Adner, R. Match your innovation strategy to your innovation ecosystem. *Harv. Bus. Rev.* **2006**, *84*, 98–107.
10. Maier, D.; Maier, A.; Aschilean, I.; Anastasiu, L.; Gavris, O. The relationship between innovation and sustainability: A bibliometric review of the literature. *Sustainability* **2020**, *12*, 4083. [CrossRef]
11. Boons, F.; Luedeke-Freund, F. Business models for sustainable innovation: State-of-the-art and steps towards a research agenda. *J. Clean. Prod.* **2013**, *45*, 9–19. [CrossRef]
12. Zeng, D.; Hu, J.; Ouyang, T. Managing innovation paradox in the sustainable innovation ecosystem: A case study of ambidextrous capability in a focal firm. *Sustainability* **2017**, *9*, 2091. [CrossRef]
13. Ding, L.; Wu, J. Innovation ecosystem of CNG vehicles: A case study of its cultivation and characteristics in Sichuan, China. *Sustainability* **2018**, *10*, 39. [CrossRef]
14. Xie, X.; Wang, H. How can open innovation ecosystem modes push product innovation forward? An fsQCA analysis. *J. Bus. Res.* **2020**, *108*, 29–41. [CrossRef]
15. Zhang, Y.; Khan, U.; Lee, S.; Salik, M. The influence of management innovation and technological innovation on organization performance. A mediating role of sustainability. *Sustainability* **2019**, *11*, 495. [CrossRef]
16. Reynolds, E.; Uygun, Y. Strengthening advanced manufacturing innovation ecosystems: The case of Massachusetts. *Technol. Forecast. Soc. Chang.* **2018**, *136*, 178–191. [CrossRef]
17. Song, J. Innovation ecosystem: Impact of interactive patterns, member location and member heterogeneity on cooperative innovation performance. *Innovation* **2016**, *18*, 13–29. [CrossRef]
18. Chesbrough, H.; Schwartz, K. Innovating business models with co-development partnerships. *Res. Technol. Manag.* **2007**, *50*, 55–59. [CrossRef]
19. Liu, Z.; Stephens, V. Exploring innovation ecosystem from the perspective of sustainability: Towards a conceptual framework. *J. Open Innov. Technol. Market Complex.* **2019**, *5*, 48. [CrossRef]
20. Etzkowitz, H.; Klofsten, M. The innovating region: Toward a theory of knowledge-based regional development. *RD Manag.* **2005**, *35*, 243–255. [CrossRef]
21. Fagerberg, J. What do we know about innovation and socio-economic change? Lessons from the TEARI project. In *National Innovation, Indicators and Policy*; Elgar, E., Earl, L., Gault, F., Eds.; Elgar: Cheltenham, UK, 2006.
22. Yin, D.; Ming, X.; Zhang, X. Sustainable and smart product innovation ecosystem: An integrative status review and future perspectives. *J. Clean. Prod.* **2020**, *274*, 123005. [CrossRef]
23. Kuzma, E.; Padilha, L.; Sehnem, S.; Julkovski, D.; Roman, D. The relationship between innovation and sustainability: A meta-analytic study. *J. Clean. Prod.* **2020**, *259*, 120745. [CrossRef]
24. Garcia, R.; Wigger, K.; Hermann, R. Challenges of creating and capturing value in open eco-innovation: Evidence from the maritime industry in Denmark. *J. Clean. Prod.* **2019**, *220*, 642–654. [CrossRef]

25. Adams, R.; Jeanrenaud, S.; Bessant, J.; Denyer, D.; Overy, P. Sustainability oriented innovation: A systematic review. *Int. J. Manag. Rev.* **2016**, *18*, 180–205. [CrossRef]
26. Michelino, F.; Cammarano, A.; Celone, A.; Caputo, M. The linkage between sustainability and innovation performance in IT hardware sector. *Sustainability* **2019**, *11*, 4275. [CrossRef]
27. Aldieri, L.; Vinci, C. Firm size and sustainable innovation: A theoretical and empirical analysis. *Sustainability* **2019**, *11*, 2775. [CrossRef]
28. Chesbrough, H. The era of open innovation. MIT Sloan. *Manag. Rev.* **2003**, *127*, 34–41.
29. Chesbrough, H.; Bogers, M. Explicating open innovation: Clarifying an emerging paradigm for understanding innovation. In *New Frontiers in Open Innovation*; Chesbrough, H., Vanhaverbeke, W., West, J., Eds.; Oxford University Press: Oxford, UK, 2014; pp. 3–28.
30. Lopes, C.; Scavarda, A.; Hofmeister, F.; Thome, A.; Vaccaro, G. An analysis of the interplay between organizational sustainability, knowledge management, and open innovation. *J. Clean. Prod.* **2017**, *142*, 476–488. [CrossRef]
31. Nielsen, K.R. Policymakers' views on sustainable end-user innovation: Implications for sustainable innovation. *J. Clean. Prod.* **2020**, *254*, 120030. [CrossRef]
32. Dahlander, L.; Gann, D.M. How open is innovation? *Res. Pol.* **2010**, *39*, 699–709. [CrossRef]
33. Laursen, K.; Salter, A. Open for innovation: The role of openness in explaining innovation performance among U.K. manufacturing firms. *Strateg. Manag. J.* **2006**, *27*, 131–150. [CrossRef]
34. Lichtenthaler, U. Open innovation in practice: An analysis of strategic approaches to technology transactions. *IEEE Trans. Eng. Manag.* **2008**, *55*, 148–157. [CrossRef]
35. Gassmann, O.; Enkel, E.; Chesbrough, H. The future of open innovation. *RD Manag.* **2010**, *40*, 213–221. [CrossRef]
36. West, J.; Salter, A.; Vanhaverbeke, W.; Chesbrough, H. Open innovation: The next decade. *Res. Policy* **2014**, *43*, 805–811. [CrossRef]
37. Huizingh, E.K.R.E. Open innovation: State of the art and future perspectives. *Technovation* **2011**, *31*, 2–9. [CrossRef]
38. Brettel, M.; Cleven, N.J. Innovation culture, collaboration with external partners and NPD performance. *Creat. Innov. Manag.* **2011**, *20*, 253–272. [CrossRef]
39. Manzini, R.; Lazzarotti, V.; Pellegrini, L. How to remain as closed as possible in the open innovation era: The case of lindt & sprüngli. *Long Range Plann.* **2017**, *50*, 260–281.
40. Curley, M.; Salmelin, B. *Open Innovation 2.0—A New Paradigm*; EU Open Innovation and Strategy Policy Group: Brussels, Belgium, 2013.
41. Curley, M. Twelve principles for open innovation 2.0. *Nature* **2016**, *533*, 314–316. [CrossRef]
42. Hafkesbrink, J.; Schroll, M. Innovation 3.0: Embedding into community knowledge-collaborative organizational learning beyond open innovation. *J. Innov. Econ. Manag.* **2011**, *1*, 55–92. [CrossRef]
43. Simanis, E.; Hart, S. *Innovation from the Inside Out*; MIT Sloan: Cambridge, MA, USA, 2009.
44. Gerlitz, L. Design management as a domain of smart and sustainable enterprise: Business modelling for innovation and smart growth in industry 4.0. *Entrepreneurship Sustain. Issues* **2016**, *3*, 244–268. [CrossRef]
45. Sartori, R.; Ceschi, A.; Costantini, A. *Open Innovation: Unveiling The Power Of The Human Element*; World Scientific: Singapore, 2017.
46. Enkel, E.; Gassmann, O.; Chesbrough, H. Open R&D and open innovation: Exploring the phenomenon. *RD Manag.* **2009**, *39*, 311–316.
47. Stanisławski, R. Open innovation as a value chain for small and medium-sized enterprises: Determinants of the use of open innovation. *Sustainability* **2020**, *12*, 3290. [CrossRef]
48. Yun, J.; Zhao, X.; Park, K.; Shi, L. Sustainability condition of open innovation: Dynamic growth of alibaba from SME to large enterprise. *Sustainability* **2020**, *12*, 4379. [CrossRef]
49. Iivari, P.; Komi, M.; Tihinen, M.; Valtanen, K. toward ecosystemic business models in the context of industrial internet. *J. Bus. Models* **2016**, *4*, 42–59.
50. Arnold, C.; Kiel, D.; Voigt, K.I. Innovative business models for the industrial internet of things. *BHM Berg Hüttenmänn. Mon.* **2017**, *169*, 371–381. [CrossRef]
51. Bauer, M.; Schlund, S.; Vocke, C. Transforming to a hyper-connected society and economy–towards an "Industry 4.0". *Proc. Manuf.* **2015**, *3*, 417–424. [CrossRef]

52. Kaggermann, H. Change through digitazation-value creation in the age of the industry 4.0. *Manag. Perm. Chang.* **2015**, *22*, 23–45.

53. Herskovits, R.; Grijalbo, M.; Tafur, J. Understanding the main drivers of value creation in an open innovation program. *Int. Entrepreneurship Manag. J.* **2013**, *9*, 631–640. [CrossRef]

54. Yun, J.; Zhao, X.; Jung, K.; Yigitcanlar, T. The culture for open innovation dynamics. *Sustainability* **2020**, *12*, 5076. [CrossRef]

55. Etzkowitz, H.; Webster, A.; Gebhardt, C.; Regina, B.; Terra, C. The future of the university and the university of the future: Evolution of ivory tower to entrepreneurial paradigm. *Res. Policy* **2000**, *29*, 313–330. [CrossRef]

56. Maria, I.; Freitas, B.; Geuna, A.; Rossi, F. Finding the right partners: Institutional and personal modes of governance of university—Industry interactions. *Res. Policy* **2013**, *42*, 50–62.

57. Bercovitz, J.; Feldman, M. Fishing upstream: Firm innovation strategy and university research alliances. *Res. Policy* **2007**, *36*, 930–948. [CrossRef]

58. Mansfield, E. Academic research underlying industrial innovations: Sources, characteristics, and financing. *Rev. Econ. Statist.* **2014**, *77*, 55–65. [CrossRef]

59. Etzkowitz, H. *MIT and the Rise of Entrepreneurial Science*; Routledge: London, UK, 2020.

60. Gao, H.; Ding, X.H.; Suming, W. Exploring the domain of open innovation: Bibliometric and content analyses. *J. Clean. Prod.* **2020**, *275*, 122580. [CrossRef]

61. *European Innovation Scoreboard*; European Commission: Brussels, Belgium, 2020.

62. Stefan, I.L.; Bengtsson, L. Unravelling appropriability mechanisms and openness depth effects on firm performance across stages in the innovation process. *Technol. Forecas. Soc. Chang.* **2017**, *120*, 252–260. [CrossRef]

63. Shin, K.; Kim, E.; Jeong, E. structural relationship and influence between open innovation capacities and performances. *Sustainability* **2018**, *10*, 2787. [CrossRef]

64. Behnam, S.; Cagliano, R.; Grijalvo, M. How should firms reconcile their open innovation capabilities for incorporating external actors in innovations aimed at sustainable development? *J. Clean. Prod.* **2017**, *170*, 950–965. [CrossRef]

65. Ranga, M.; Etzkowitz, H. Triple Helix systems: An analytical framework for innovation policy and practice in the Knowledge Society. *Ind. High. Educ.* **2013**, *27*, 237–262. [CrossRef]

66. Nam, G.M.; Kim, D.G.; Choi, S.O. How Resources of Universities influence Industry Cooperation. *J. Open Innov. Technol. Market Complex.* **2019**, *5*, 9. [CrossRef]

67. Perkmann, M.; Tartari, V.; McKelvey, M.; Autio, E.; Broström, A.; D'Este, P.; Krabel, S. Academic engagement and commercialisation: A review of the literature on university–Industry relations. *Res. Policy* **2013**, *42*, 423–442. [CrossRef]

68. Pennacchio, A.B.L. University knowledge and firm innovation: Evidence from European countries. *J. Technol. Transf.* **2015**, *41*, 730–752.

69. Berbegal-mirabent, J.; Luís, J.; García, S.; Ribeiro-soriano, D.E. University–Industry partnerships for the provision of R & D services. *J. Bus. Res.* **2015**, *68*, 1407–1413.

70. Basit, S.; Medase, S. The diversity of knowledge sources and its impact on firm-level innovation: Evidence from Germany. *Eur. J. Innov. Manag.* **2019**, *22*, 681–714. [CrossRef]

71. Leydesdorff, L.; Woo, H.; Balazs, P. A routine for measuring synergy in university-industry-government relations: Mutual information as a triple-helix and quadruple-helix indicator. *Scientometrics* **2014**, *99*, 27–35. [CrossRef]

72. Thomson, K.; Lorenzini, F.; Markman, A.B.; Pogue, G.P.; French, R. Building an innovation coral reef. In *Open Innovation*; Oxford University Press: Oxford, UK, 2016; pp. 203–224.

73. Este, P.D.; Patel, P. University–Industry linkages in the UK: What are the factors underlying the variety of interactions with industry? *Res. Policy* **2007**, *36*, 1295–1313.

74. Schartinger, D.; Rammer, C.; Fischer, M.M.; Fröhlich, J. *Knowledge Interactions Between Universities and Industry in Austria: Sectoral Patterns and Determinants*; Springer: Berlin, Germany, 2002; Volume 31, pp. 303–328.

75. Etzkowitz, H. *Entrepreneurial Scientists and Entrepreneurial Universities in American Academic Science*; Springer: Berlin, Germany, 1983; Volume 21, pp. 198–233.

76. Clark, B. *Creating Entrepreneurial Universities: Organization Pathways of Transformation*; Guildford: Pergamon, Turkey, 1998; Available online: https://books.emeraldinsight.com/page/detail/creating-entrepreneurial-universities-burton-r-clarkCreating-Entrepreneurial-Universities/?k=9780080433547 (accessed on 1 October 2020).
77. Boliacino, F.; Pianta, M. The Pavitt Taxonomy, revisited: Patterns of innovation in manufacturing and services. *Econ. Politica* **2016**, *33*, 153–180. [CrossRef]
78. Costa, J.; Rodrigues, C. Why innovative firms do not rely on universities as innovation sources? *Glob. Bus. Econ. Res.* **2020**, *22*, 351–374. [CrossRef]
79. Bekkers, R.; Freitas, B. Analysing preferences for knowledge transfer channels between universities and industry: To what degree do sectors also matter? *Res. Policy* **2008**, *37*, 1837–1853. [CrossRef]
80. Santoro, M.D.; Chakrabarti, A.K. Firm size and technology centrality in industry–University interactions. *Res. Policy* **2002**, *31*, 1163–1180. [CrossRef]
81. Nutley, S.; Davies, H. Getting research into practice: Making a reality of evidence-based practice: Some lessons from the diffusion of innovations. *Publ. Money Manag.* **2010**, *20*, 954–962.
82. Chesbrough, H. *Open Business Models: How to Thrive in the New Innovation Landscape*; Harvard Business School Publishing: Cambridge, MA, USA, 2006.
83. Lozano, R. Collaboration as a pathway for sustainability. *Sustain. Dev.* **2007**, *15*, 370–381. [CrossRef]
84. Hossain, M. Open innovation: So far and a way forward. *World J. Sci. Technol. Sustain. Dev.* **2010**, *10*, 30–41. [CrossRef]
85. Wikhamn, B.; Wikhamn, W. Structuring of the Open Innovation Field. *J. Technol. Manag. Innov.* **2013**, *8*, 173–185.
86. Baldwin, C.Y.; Hienerth, C.; von Hippel, E. How user innovations become commercial products: A theoretical investigation and case study. *Res. Policy* **2006**, *35*, 1291–1313. [CrossRef]
87. Bruneel, J.; Este, P.D.; Salter, A. Investigating the factors that diminish the barriers to university–Industry collaboration. *Res. Policy* **2010**, *39*, 858–868. [CrossRef]
88. Etzkowitz, H.; Leydesdorff, L. The triple helix—University-industry-government relations: A laboratory for knowledge based economic development. *EASST Rev.* **1995**, *14*, 14–19.
89. Li, W.; Xu, J.; Zheng, M. Green governance: New perspective from open innovation. *Sustainability* **2018**, *10*, 3845. [CrossRef]
90. Rauter, R.; Globocnik, D.; Perl-Vorbach, E.; Baumgartner, J. Open innovation and its effects on economic and sustainability innovation performance. *J. Innov. Knowl.* **2018**, *4*, 226–233. [CrossRef]

MDPI

St. Alban-Anlage 66

4052 Basel

Switzerland

Tel. +41 61 683 77 34

Fax +41 61 302 89 18

www.mdpi.com

Sustainability Editorial Office

E-mail: sustainability@mdpi.com

www.mdpi.com/journal/sustainability